DATE DUE

APR 3 0 2003	
MAY 1 4 2003	
APR 2 4 2007	

DEMCO, INC. 38-2931

**Renner Learning Resource Center
Elgin Community College
Elgin, IL 60123**

JOB$ IN THE DRUG INDUSTRY

•••••••••••••••••

A Career Guide for Chemists

Richard Friary

Schering-Plough Research Institute
Kenilworth, New Jersey

ACADEMIC PRESS
A Harcourt Science and Technology Company

San Diego   San Francisco   New York   Boston   London   Sydney   Tokyo

This book is printed on acid-free paper.

Copyright © 2000 by ACADEMIC PRESS

All Rights Reserved.
No part of this publication may be reproduced or transmitted in any form or by any means, electronic or mechanical, including photocopy, recording, or any information storage and retrieval system, without permission in writing from the publisher.

Requests for permission to make copies of any part of the work should be mailed to: Permissions Department, Harcourt Inc., 6277 Sea Harbor Drive, Orlando, Florida 32887-6777

Academic Press
A Harcourt Science and Technology Company
525 B Street, Suite 1900, San Diego, California 92101-4495, USA
http://www.academicpress.com

Academic Press
Harcourt Place, 32 Jamestown Road, London NW1 7BY, UK
http://www.hbuk.co.uk/ap/

Library of Congress Catalog Card Number: 99-68354

International Standard Book Number: 0-12-267645-9

PRINTED IN THE UNITED STATES OF AMERICA
00 01 02 03 04 05 ML 9 8 7 6 5 4 3 2 1

CONTENTS

DISCLAIMER	xi
DEDICATION	xiii
ACKNOWLEDGMENTS	xv
FOREWORD	xvii
PREFACE	xix
ABOUT THE AUTHOR	xxiii

1. ENTICEMENTS: WHY ORGANIC CHEMISTS WORK IN THE PHARMACEUTICAL INDUSTRY — **1**

Introduction	1
Prerequisites	5
Primacy of the Pharmaceutical Industry in Drug Discovery and Development	*5*
The Importance of Chemical Synthesis	*7*
Future of the Pharmaceutical Industry	*9*
Altruism as the Unique Appeal of the Pharmaceutical Industry	12
Attractions of the Pharmaceutical and Chemical Industries	13
Benefits	*13*
Salaries	*28*
Sources and Supplements	38

2. ELEMENTS OF DRUG DISCOVERY AND DEVELOPMENT — **41**

Introduction	41
Profile of the Pharmaceutical Businesses	42
Time to Market	*42*
What Drug Discovery and Development Cost	*42*

Benefits of Patents	*42*
Likelihood of Success	*43*
Competition	*44*
Complexity	*45*
Kinds of Drugs, Companies, and Therapeutic Areas	46
Drug Categories	*46*
Nucleus of the Pharmaceutical Industry	*47*
Drug Discovery Firms and Start-Up Pharmaceutical Companies	*49*
Therapeutic Areas	*50*
Discovery and Development of Drugs	50
Stratagems	*50*
Tactics	*53*
Organization of Discovery and Development Chemical Research	66
Work Assignments	*66*
Transferability	*67*
The Project Team as an Organizing Principle	*67*
Function and Matrix Organization	*69*
Consolidation and Fragmentation in the Pharmaceutical Industry	70
Economic Forces	*70*
Outsourcing Science	*70*
Start-Up Companies	*71*
Sources and Supplements	71

3. JOBS IN THE DRUG INDUSTRY — 73

Introduction	73
Effects of Outsourcing on Employment Opportunities	*74*
Prerequisites and Experience	*76*
Over- and Underqualifications	*77*
Research Classifications	*77*
Discovery Research	78
Medicinal Chemistry	*78*
Natural Products	*78*
Structural Chemistry	*79*
Drug Metabolism	*80*
Radiochemistry	*81*
Service Groups	81
Analytical Chemistry	*81*

Compound Registration	*82*
Cheminformatics	*83*
Patent Coordinators	*84*
Patent Agents and Attorneys	*85*
Chemical Information	*85*
Synthetic Services	*86*
Profiling and Identification	*86*
Chromatography and Separations Science	*87*
Regulatory Affairs	*87*
Development Research	88
Natural Products	*89*
Bioorganic Catalysis	*89*
Safety	*90*
Temporary Jobs	*90*
Satellite Companies, Government Agencies, and Nonprofit Institutes	91
Service Firms	*92*
Scientific Advising	*94*
Chemical Development and Manufacturing	*94*
Fine Chemicals	*95*
Automated Synthesis	*95*
Software Development and Molecular Modeling	*95*
The Food and Drug Administration	*96*
The U.S. Patent and Trade Office	*96*
The Walter Reed Army Institute of Research	*97*
The National Institutes of Health	*97*
Miscellaneous Posts	*98*
Sources and Supplements	98

4. DISCOVERY AND DEVELOPMENTAL CHEMICAL RESEARCH: COMMON FEATURES — **101**

Introduction: What You Should Know about the Job You Seek	101
The Basics	103
Performance and Productivity	*104*
Aids to Productivity	*106*
Goal Setting	*107*
Budgeting	*108*
Internal Meetings	*109*

Communicating Results, Plans, Problems, and Persona	111
Reporting Results	*111*
Publishing	*112*
The Importance of Being Visible	*114*
Patenting Your Work	115
Patents Defined	*116*
Types of Patents	*118*
Patentability	*119*
Organizational Structure of Patents	*124*
Inventorship	*133*
Sources and Supplements	136

5. DISCOVERY RESEARCH: MEDICINAL CHEMISTRY — 139

Introduction: One Day in the Work of a Medicinal Chemist— A Fictional Account	139
Organization, or Who Chooses What	*149*
Getting Started	*149*
Discovery Research	151
Common Goals of Discovery Chemists	*151*
Lead Compounds	*152*
Making Target Compounds	*155*
Submitting Compounds for Biological Testing	*175*
Satisfactions and Successes	*177*
Metamorphosis: Synthetic Organic Chemist to Medicinal Chemist	178
Sources and Supplements	179

6. CHEMICAL DEVELOPMENT: CHALLENGE IN ORGANIC SYNTHESIS — 181

Introduction	181
Chemical Development	182
Information Resources	*184*
Scope	*184*
Timing	*185*
Purposes of Chemical Development	186
Bulk Supplies	*186*
A Manufacturing Synthesis	*187*
Organization	189
By Discipline	*189*
By Assignment	*189*

By Developmental Project Teams	191
Process Research and Development	192
Equipment	*195*
Changed Techniques and Conditions	*196*
Reforms	*198*
Elements of a Suitable Manufacturing Synthesis	199
Specifications	*199*
Safety	*200*
Soundness	*201*
Sourcing and Cost of Raw Materials	*203*
Costing a Synthesis	*204*
Efficiency, Brevity, and the Arithmetic Fiend	*205*
Throughput	*207*
Robustness	*208*
Regulation	*209*
Satisfactions and Successes	210
Sources and Supplements	210

7. QUALIFYING AND SEARCHING FOR JOBS IN THE DRUG INDUSTRY — 213

Introduction	213
What You Need to Know and Show	214
Common Prerequisites	*214*
Personal Qualities	*215*
Qualifications	*216*
Why Read about Visas?	221
Work Visas in the United States	*222*
Role of the American Chemical Society in Job Searches	228
Chemical & Engineering News (C&EN)	*228*
Career Services	*229*
Finding Job Openings	234
Print Advertisements for Chemists	*236*
Corporate Web Sites	*238*
Other Job Banks	*239*
Corporate Job Lines	*240*
Job Alerts by E-mail	*240*
Networking	*240*
Sources and Supplements	243

8. Evaluating Companies and Job Offers — 245
Introduction — 245
Criteria for Evaluating Companies — 248
Corporate Status and Future — *249*
Geography — *252*
Forms of Compensation Other Than Salary — *255*
Career Prospects — *258*
Benefits beyond Salary and Other Compensation — *263*
Sources and Supplements — 277

Appendix A — 279
Arthur C. Cope Award — 279
ACS Award for Creative Work in Synthetic Organic Chemistry — 279
Roger Adams Award in Organic Chemistry — 280
The *Ernest Guenther* Award in the Chemistry of Natural Products — 280
Tetrahedron Prize for Creativity in Organic Chemistry — 281
Herbert C. Brown Award for Creative Research in Synthetic Methods — 281
International Aspirin Prize for Solidarity through Chemistry — 281

Appendix B — 283
Introduction — 283
Kinds of Indexed Organizations — *284*
Contents of Each Entry — *285*
Excluded Information — *286*
". . . a tide . . . leads on to fortune" — *287*
Sources — *287*
Geographical Index of the North American Pharmaceutical Industry — 289
Canada — *289*
The United States of America — *289*
Name Index of the North American Pharmaceutical Industry — 297

Index — 349

DISCLAIMER

This book expresses its author's opinions and observations, not the views or policies of any company, except perhaps by coincidence. None of the writer's employers—past, present, or future—sponsored the work, nor did any of them endorse it.

The author has done his best to ensure the accuracy of the information presented in this book as of the publication date, drawing on sources believed reliable. However, the publisher and author take no responsibility for the validity or timeliness of all materials herein, and none for the consequences of using this information. Neither the publisher nor the author explicitly or implicitly promises that readers will find employment because of anything written or implied here.

This book is not intended, nor should you consider it, as legal advice applicable to your specific situation. Laws are frequently updated and are often subject to differing interpretations. You are solely responsible for your use of this book. The publisher and the author will not be responsible to you or anyone else because of any information contained in or left out of this book.

DEDICATION

To all my chemistry teachers and especially to
*Roderick Cyr, Wilmon B. Chipman, Thomas A. Spencer,
Richard W. Franck,* and *Robert Burns Woodward.*

ACKNOWLEDGMENTS

Few obligations are as pleasant to meet as my duty to thank the many people who so graciously contributed to this book. They include *Captain Kevin Pitzer, Ph.D.*, who described the medicinal chemistry practiced at the Walter Reed Army Institute of Research. From the National Institutes of Health, *Richard Drury* and *Dr. Kenneth Kirk* acquainted me with chemical research in their organizations. Respectively, *Frank E. Walworth* and *Dr. Mary Jordan* from the American Chemical Society kindly reviewed passages and generously contributed salary data from surveys. Novelist *Barry A. Nazarian* offered gentle criticism. So did *Dr. Thomas C. Nugent* of Catalytica, Inc., who read the whole manuscript and drew the distinction between process research and process development that appears here over his name.

Too many to be named yet too generous to be neglected, scores of personnel officers from as many companies responded to my questionnaire concerning summer jobs and internships for chemistry students. Two patent attorneys, *Drs. Thomas Hoffman* (Schering-Plough) and *Konstantinos Petrakis*, who are also organic chemists, troubled to read and correct Chapter 4. From a variety of pharmaceutical companies, other chemists improved chapters through their suggestions. These chemists include *Mathew Reese* (Pfizer), *Dr. Joseph Auerbach* (Merck & Co.); *Drs. David Provencal* and *Wayne Vaccaro* (Bristol-Myers Squibb); *Dr. William Metz* (Hoechst Marion Roussel); *Drs. Daniel Solomon, Martin Steinman,* and *Cheryl Alaimo* (Schering-Plough); and Drs. *Nick Carruthers* (Johnson and Johnson) and *Philip Decapite*. I thank them all.

My indebtedness also extends to *Drs. Peter Mauser* (Hoechst Marion Roussel) and *Robert W. Watkins* (Schering-Plough), respectively a physiologist and a pharmacologist, who assisted with Chapter 2. I am grateful to

Linda Klinger, Dr. David Packer, Kim Schettig, and *Mark Sherry,* all of Academic Press, for their efforts and encouragement in bringing this book to press.

Without the generosity of all these contributors, my endeavor would not have gone so far nor traveled so fast. Much of the credit for the published work therefore belongs to them.

FOREWORD

Friary's book is a much-needed treatise for those chemists, or for that matter any scientist, about to enter or contemplating entry into the pharmaceutical industry. Experience has shown that academe does not prepare us for the culture shock of moving into the multifarious environment of the pharmaceutical industry.

Drug discovery and development are a mix of disciplines and functions that excite and confuse the uninitiated, yet provide unparalleled opportunity for scientists who can master the tools at their disposal. At the beginning of a career in the industry, much of one's time is spent learning to chart a course through this complex milieu. And for those of us managing these organizations, our principal concern is to jump start our "tyros" so that they achieve the optimal blend of effectiveness and efficiency with the altruism that defines the motivation of pharmaceutical scientists. Friary's book provides the navigational instruments for the uninitiated. He defines the appeal and the expectations of the chemist's job and proceeds to an informed analysis of the discovery and development processes in enough detail to warrant scrutiny by medicinal and process chemists who are in position, but still learning their trade.

Friary's book is especially timely because chemistry is undergoing a renaissance during an age of industrialization of the scientific process. The industry has been enamored of the disciplines of molecular biology and genomics through the late 1980s and continuing into the 1990s. The fruits of those labors, i.e., new drug targets, and the consequence, i.e., screening of synthetic compound libraries, have provided more leads for chemical modification to create drugs than ever before. High-throughput biology has generated an unprecedented need for chemists, because they are the only scientists who can reduce this explosion of medical knowledge to practice.

Job$ in the Drug Industry removes the mysticism surrounding the pharmaceutical chemist's position. The book reveals how the flexibility of chemists trained in first principles of the physical world and the scientific process allows them to flourish in the pharmaceutical environment to become multifaceted drug hunters. Friary's book will be an important tool in the chemist's armamentarium, teaching chemists not only how to find the appropriate job, but also how their creations move to the status of an approved drug for preventing and controlling human disease.

William H. Koster
Senior Vice-President, Drug Discovery
Bristol-Myers Squibb Pharmaceutical Research Institute

PREFACE

> Those who cannot remember the past are condemned to fulfill it.
> —*George Santayana*

Plus ça change, plus c'est la même chose.—French proverb

Origins

This book originated in the summer of 1963, when I worked as a factotum in a textile mill. Having failed to find a temporary job doing chemical research, I strove to learn from an unsuccessful search for this preferred employment, hoping never to repeat mistakes. Dispatched early in the spring, my applications arrived too late and addressed too few research-based companies. That these failings sprang from unfamiliarity grew apparent later and provoked a resolution. If ever I could, I would help students find work, were they as much in need of assistance as I had been.

For more than 30 years I did not commit a word of this book to paper. In the meantime, that resolution sprang repeatedly to mind, usually in response to either of two spurs.

(1) As I published research, I often received job applications—and usually redirected them. Some job hunters, I learned, unwittingly apply to industrial researchers who have no positions to offer and little knowledge of any in the offing. So, résumés pop up in the laboratory as unexpectedly as burnt toast in the kitchen, where they can be as welcome as charcoal for breakfast. Such applications are better sent to directors familiar with openings and perhaps empowered to create or fill them. However, new graduates and even postdoctoral researchers frequently lack any network of helpful acquaintances, while anyone's efforts to create a network can go awry.

(2) On several occasions, I answered inquiries from students interested in summer or permanent jobs doing research in pharmaceutical firms. Some of my colleagues and I once participated in an informal series of informational interviews arranged by an undergraduate studying an applied science. Not a chemistry major, he wanted to work as an organic chemist in a large pharmaceutical company and wondered whether as a prerequisite he should earn a graduate degree in biochemistry or medicinal chemistry. Learn organic synthesis, we told him, giving him the names of chemists who teach our art in graduate schools. These students and job hunters, although they expressed interest in pharmaceutical industry careers, proved unacquainted with the hiring schedules, sizes, kinds, whereabouts, and doings of research-based drug companies, much as I had been. In the words of the proverb that opens this preface, the more things changed, the more they stayed the same.

Long-standing observations, as well as recent ones, suggested that newly graduated job seekers would benefit from guidance in finding employment. They also indicated that a need for a guidebook still exists, for none was in or out of print. Finally, a decision to write gained reinforcement from the Committee on Science, Engineering and Public Policy representing the National Academy of Sciences. In its 1995 report "Reshaping the Graduate Education of Scientists and Engineers," the Committee wanted students routinely to receive better information for making career choices than they now obtain.

Readership

The intended readership for this career guide comprises three groups of chemists, two kinds of student advisers, and a sixth, less easily classified group of people. Included in the first three groups are undergraduate students of chemistry; graduate students of organic, medicinal, or bioorganic chemistry; and postdoctoral chemical researchers whose dissertations concern the preceding three specialties. Chemists from these three groups are beginning or continuing their working lives, choosing between the pharmaceutical and other chemical industries or among industrial, governmental, or teaching careers.

Student advisers include university career development officers and professors of chemistry. The latter advisers are those who direct students' undergraduate or graduate chemical research or generally counsel undergraduate chemistry students.

The sixth group of readers includes personnel managers not trained as scientists but employed by the pharmaceutical or other research-based chemical industries. Examples are the agricultural, food, perfumery, animal health, and personal care industries, where it is also essential to understand and exploit molecular structure–function relations. Responsible for describing chemists' jobs and interviewing applicants, these recruiters take an interest in their work and qualifications. Other members of this diversified group are detailers, recruiters, physicians, biologists, biochemists, pharmacists, and physical, analytical, pharmaceutical, or manufacturing chemists. All these people can learn here what medicine discovery and development entail, what roles organic chemists play in these endeavors, and which companies make up the drug industry.

Scope, Purpose, and Means

Defining the book called for three restrictions. First, it would exclusively treat jobs in the North American pharmaceutical industry. This complex industry comprises not only fully integrated, long-established, and large drug houses, but also many other kinds of corporate and institutional employers. These features and the following two considerations were to lend adequate scope.

Second, the book would appeal primarily to the aforementioned three groups of chemists. Although the guide serves a broad readership of people employed in drug houses, its locus directed me to write about what I already professed to know or reasonably hoped to learn.

Finally, the book would deal with recent graduates' task of finding their first full-time jobs as chemists. It therefore says nothing about developing or changing a career or about retiring from one. Industry-experienced chemists need a guidebook less than newly graduated scientists, and career-change and –development books already exist.

Job$ solves the problem facing certain newly graduated chemists: how to find suitable jobs as researchers in the drug industry. This problem has topical as well as lasting aspects. That jobs are less visible because of the outsourcing that large companies practice is a topical difficulty. So are mergers, which bring cost savings at the expense of jobs. A timeless obstacle to securing employment speedily is lack of adequate familiarity with the many aspects of job hunting.

To accomplish its purpose, this book gathers in a single volume the fundamentals of getting a job as a medicinal or process chemist in the pharmaceutical industry. The only guidebook to do so, *Job$* shows why chemists join the industry and how drug discovery and development lead to chemists' jobs. It describes the kinds of work that chemists do and the goals their labors serve, and notes the satisfactions, salaries, pensions, and other benefits they can expect. It explains how to find and evaluate jobs and, in an appendix, shows where in Canada and the United States to look for them. A second appendix furnishes the locations, names, and contact information for more than 500 companies and other organizations that, among others, compose the North American pharmaceutical industry. Presented in an affordable and available format, this information will help prospective employees decide to join the drug industry.

Richard Friary

About the Author

Richard Friary, Ph.D., is a synthetic organic and medicinal chemist employed by the Schering-Plough Research Institute. This institute forms the discovery and development arm of a fully integrated multinational pharmaceutical company. The author's combined experience in this company and CIBA-Geigy (now part of Novartis) spans nearly 30 years.

As a senior principal scientist, the writer is an accomplished chemical researcher who discovered a safe and effective drug that relieves psoriasis and dermatitis. He was instrumental in steering the experimental medicine to clinical trials in human beings and in developing it afterward. Dr. Friary is among the few chemists ever to have made a drug that entered clinical studies in human beings, and among fewer still whose drug passed clinical trials. Eighteen patents and 31 articles name him as an inventor and an author.

Born in 1942, Richard Friary is a native of Biddeford, Maine, and a graduate of Colby (B.A., 1964) and Dartmouth Colleges (M.A., 1966) and of Fordham University (Ph.D., 1970). He is a veteran of the R. B. Woodward Research Institute in Basel, Switzerland, where he worked as a postdoctoral researcher from 1970 to 1973. There he learned medicinal chemistry by making cephalosporin C analogs as antibacterial agents, and organic synthesis through a total synthesis of prostaglandin $F_{2\alpha}$. Only a few chemists ever wrote as many as two articles with the finest organic chemist of all time, the late R. B. Woodward, and Friary is one of them.

He married the former Diane McKee of Berlin, New Hampshire, in 1968. Weather permitting, Friary sails on ice skates; a trade book he wrote on this winter sport was published recently. His other hobbies include listening to traditional jazz, splitting firewood, reading a newspaper while walking a dog, and writing about himself in the third person.

chapter

1

ENTICEMENTS

Why Organic Chemists Work in the Pharmaceutical Industry

> "Medicine is for the people the world over. . . . In this we share a special heritage of the ages [that] calls for the noblest effort of all of us to treat the sick and the infirm irrespective of boundaries and ideological differences."
> —Robert G. Denkewalter and Max Tischler

INTRODUCTION

Combining elements and compounds, organic chemists in the pharmaceutical industry create drugs that preserve, prolong, and improve lives. Many of them join the industry largely to do this work, and without their revolutionary labor of chemical synthesis our lives would be Hobbesian: nasty, brutish, and short. The modern occasion to discover and develop chemicals as drugs lacks any genuine parallel in the 10 millennia of human civilization. Only in the present century have science and medicine broadly realized Paracelsus' 400-year-old insight that chemicals cure disease.

Different chemicals, if developed as drugs, treat diverse diseases because chemical structure determines the properties responsible for biological activity and thereby for medical utility. Chemical and physical properties decide the nature, scope, and onset of the biological actions of drugs. They also govern the duration, intensity, mechanism, and reversibility of drug behavior. Embodied by compounds that are reaction products, these properties vary with the structural modifications achieved by synthesis. Some structural changes impart, improve, or abolish the usefulness of chemicals in treating one disease, while other alterations make chemicals unavailing or effective against another illness.

Announced in 1868 by the chemist A. *Crum-Brown* and the biologist T. R. *Fraser*, this principle—that chemical structure determines biological activity—underlies modern Western drug therapy. It fosters economic opportunities, so it inspired the present pharmaceutical industry to evolve from late-19th-century industrial-chemical companies. Implementation of the principle required chemists to synthesize organic compounds, an activity dating from 1828. In that year the immortal Friedrich Wöhler made urea from silver cyanide and ammonium chloride, demonstrating for the first time that organic synthesis was possible. The powers of synthesis expanded rapidly until, in 1888, a German dyestuff company introduced the first completely manmade drug, namely phenacetin. Created by chemists of Farbenfabriken vormals Friedrich Bayer & Company, it has analgesic and antipyretic activities like its successor, aspirin. Phenacetin, which is p-ethoxyacetanilide, provided relief in the multinational influenza epidemic of 1889.

With the advent of the synthetic sulfonamide antibacterial drugs about 1914, chemists everywhere began to decide the structures—and thus the properties—of matter that previously were accepted perforce. To do so, they deliberately created new materials, running scores of newly discovered chemical reactions. In a unique appeal, the industry still invites young chemists to relieve suffering tomorrow by making new drugs today. Culminating centuries of scientific progress, this endeavor enables humanity substantially to vanquish disease and restore health, powers hitherto left to indifferent fate, often with agonizing or appalling results.

In common with other companies in the chemical industry, drug houses make additional appeals to newly graduated medicinal and synthetic chemists. These attractions comprise three types, which are intellectual, educational, and financial. Any of them may exert an equal or greater pull than altruism. Different attractions appeal more or less to different individuals, who may spurn the allures described here even as they respond to others.

Synthetic and medicinal chemists within a pharmaceutical company enjoy an enduring intellectual preoccupation. It requires them to identify and make *lead compounds* and to advance them to clinically effective drugs. Such researchers therefore grapple urgently with intricate therapeutic problems to wrest practical, novel solutions from their inventions and experiments. Their efforts to discover and develop new chemicals as treatments for serious diseases are diverse in their origins, methods, and goals. Any project to find a drug for a given disease can last only 2–4 years. However, one such effort precedes another to discover a different treatment for the same condition or to find a new drug for another disease. Yet another short-lived undertaking soon follows the preceding one. Considered together, these efforts form a long series of short projects and intensely engage any inquisitive mind. Research directors choose no goals requiring career-long commitments, so intellectual stagnation need never attend a chemist's work.

Join a pharmaceutical company, and with its help you can climb the academic ladder. Such an employer customarily subsidizes the tuition costs of part-time, business-related studies leading to a master's or doctor's degree in chemistry. Sometimes it also underwrites the research needed to complete a doctoral dissertation. Then, it can furnish laboratory space and services, modern instruments and glassware, supplies of chemicals, or advice from experienced industrial researchers who are coworkers. With your doctorate conferred, you can seek an adjunct professor's post. It can allow you to teach part-time at a nearby university even while you pursue full-time research in industry. Employer-subsidized graduate degrees, which are not limited to chemistry but can include business or law, extend their recipient's horizon. They can facilitate advancement or career transitions.

Within a pharmaceutical company, mid-career transitions to research administration, drug metabolism, molecular modeling, research management, regulatory affairs, and patent agenting or liaison are certainly feasible. Occasionally a chemist finds work in administration, as distinct from management. For example, vice-presidents for research or development in some companies employ organic chemists as administrative assistants. Departments of drug metabolism hire B.S. and M.S. chemists. For doctoral researchers, an opportunity to move from research into management usually follows several years' work. It rewards technical or scientific accomplishments rather than managerial training or predilection, so it does not precede an apprenticeship. This and other moves ordinarily require exposure to, knowledge of, and visibility in pharmaceutical operations beyond chemical discovery research or development. These three prerequisites attend the progress of a drug that enters clinical trials. Such an entry represents an extraordinary achievement since so many compounds are called to human studies but so few are chosen for them. Although a chance to change careers indirectly rewards such a success, on-the-job scientific achievement is not the sole qualification for research management or work other than industrial chemical research. Therefore, the chance rarely greets an arriving chemist, despite a prestigious Ph.D. or a renowned thesis adviser. High stakes forbid heavy wagers on dark horses.

> *Promising graduates establish careers rather than escape routes, and those contemplating employment in discovery or development research resolve to master and exploit synthetic chemistry. They appreciate that a researcher's work demands the productivity that comes from experiments. A*

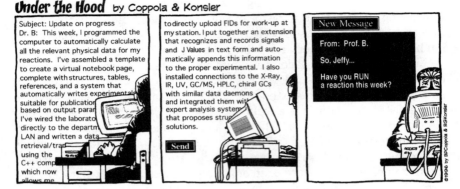

Figure 1-1 Computer tweaking

> drug house thrives on massive samples and large numbers of novel compounds more than it flourishes with weighty volumes of new knowledge.

It is not an intelligence agency. Gathering and analyzing data, therefore, and tweaking computers and writing reports are not exclusively a synthetic chemist's work, although these hands-off activities are needed (Figure 1-1). But they are subordinate to the chemical reactions that occur in fume hoods to create new forms of matter with specified biological activity. Such results reward an industrial chemical researcher, just as robberies enriched the notorious American thief *Willie Sutton*. Asked why he stole from banks, the astute felon reportedly said, "That's where the money is."

The financial attractions of an industrial chemist's work embrace benefits, retirement packages, and substantial salaries. In 1998, median salaries throughout the whole chemical industry ranged from $51,200 for B.S. chemists through $65,000 for M.S. chemists to $85,000 for Ph.D. chemists.

Chemical researchers hunting jobs within this industry enjoy an advantage over their counterparts seeking employment in academia and government. Drug houses hire more medicinal and synthetic chemists at all degree levels than other employers do. Their growing work forces employ one-fifth to one-quarter of all industrial chemists. These scientists number nearly twice as many as those working for any other group of employers in the chemical industry. Therefore, the chance that a recently graduated chemist will find employment is higher with a pharmaceutical company than elsewhere. Reasons to seek work there appear in the next three sections.

PREREQUISITES

Primacy of the Pharmaceutical Industry in Drug Discovery and Development

Who Sponsors Research?

Pharmaceutical companies provide nearly all safe, effective drugs throughout the industrialized world. Their role in discovering, developing, introducing, and supplying such remedies predominates, although the plaudits are sometimes misdirected to government or academic laboratories. For example, scores of drug firms throughout the world introduced 229 new drugs from 1983 to 1993. These drugs, known as new chemical entities (NCEs), represent nearly 93% of all new medicine launches in that period, which came to 247. Such a high percentage is typical. As a result, most chemists who contribute directly to drug discovery and development do so within the private sector of the pharmaceutical industry. Alternatives certainly exist to make contributions elsewhere, but their number is comparatively small. One observer, *B. Testa*, Professor and Head of Medicinal Chemistry at the University of Lausanne in Switzerland, writes, "Pharmaceutical research in communist countries has been all but sterile, and major governmental programs (for example, the search for antimalarials) have been disappointing [elsewhere]."

Only about 18 NCEs arose from all other sources during 1983–1993. They include one drug from each of six state-owned enterprises in mainland China and eastern Europe, where no private companies operated during most of the same time. Government and academic institutions in the rest of the world introduced the remaining 12 new drugs over 11 years. These institutions comprise the public sector of the pharmaceutical industry because, directly or indirectly, they draw much of their financial support from taxes. Generally, in the West and particularly in North America, their record for drug launches reflects a minor but laudable role.

According to *David Cavalla*, author and industry observer,

> ... *invention of new chemical entities [that is, drugs] by universities is extremely rare.* While the seed corn of drug research is often found in universities, these places do not have the persistence or commercial resources to carry this through to a commercial product. (Cavalla, 1997, italics added)

Cavalla does note three exceptions, namely the drugs diphenhydramine, atracurium, and norgestrol, all of which university researchers discovered.

Drug houses play the leading role now. In 1994, for example, the 25 best-selling drugs came exclusively from 15 pharmaceutical firms. They brought their sponsors over $31 billion in 1 year, although their income represents only 12% of total drug sales in the same period.

At any one time, the number of drugs being developed crudely measures the contributions made by various participating institutions. In 1995, the worldwide total equaled 3169. Private pharmaceutical laboratories supplied over 96% of this number. The National Institutes of Health, which includes several U.S. government-sponsored laboratories, contributed 3%. All participating university laboratories added no more than 0.09%. This distribution is telling but crude for two reasons. First, it neglects the relative importance of any of the drugs being developed. Second, few developed drugs can be launched. At present, about 1 in 5000 newly made compounds merits some preclinical development. But only 1 in 10,000 is worthy enough to progress through pre- and postclinical development to the marketplace. The estimated cost and time of this research equal $600,000,000 and 10–12 years.

What Research Costs

The foregoing ratios, cost, and time explain the minor role played by university and government research in drug discovery and development. Even large, wealthy universities cannot coordinate and complete the countless needed tasks without compromising academic freedom. They cannot afford the requisite sums. For example, in 1992, several of the largest pharmaceutical companies spent more than $1 billion on research and development alone: Roche, $1.21, Bristol-Myers Squibb, $1.08, and Glaxo, $1.05 billion. By 1996, Glaxo and Roche respectively allotted $1.89 and $1.66 billion to research and development. Such amounts exceed the reach of the best endowed universities or the gross national products of some countries. Academic scientific research therefore occurs on a smaller scale and seeks fundamental knowledge rather than practical applications. Often the crucial scientific understanding gained there steers and spurs drug discovery and development. But they usually take place elsewhere, often decades later and sometimes in unforeseen ways.

The sums required to discover, develop, and launch drugs fall within the grasp of governments. With few exceptions, however, they are not invested in drug research. Democratic governments lack the political will needed to choose long-lasting, commercial, and worthy research projects. This deficiency means they cannot easily allocate huge financial resources among the chosen few enterprises. The short tenure of elected officials combined with their desires for reelection makes them mercurial rather than resolute, except in war making and defense spending. Of course, governments invest vast amounts in defense research, but such projects may have commercial value partly because their sponsors are also their consumers.

Industrial Diversity

The international pharmaceutical industry is multifaceted, as Chapter 2 shows. It contains private drug firms, government and university laboratories, custom-synthesis houses, regulatory agencies, and other essential institutions. This variety permits the most successful job hunters to make choices about their careers. Whether they seek employment in one kind of institution or another can therefore depend partly on their own desires. Yet, for the greatest chance of directly participating in successful drug discovery and development, a starting-out chemist must work in a pharmaceutical company. It's the best game in town.

The Importance of Chemical Synthesis

An invitation to join a pharmaceutical company, which is not given lightly, represents an opportunity to join a complex, cooperative enterprise. Within it, synthetic and medicinal chemists are not optional but indispensable, and their role has been likened to a mother's: *"The chemist gives birth to the drug* [italics added], but the [physician] supports its first steps (attributed to *Fourneau* by *A. Maurois*)." Indeed, the labor of birthing drugs can earn not only a chemist's salary but a clinician's respect.

Source of Drugs

Writing in the *New England Journal of Medicine, Irvine H. Page, M.D.,* paid tribute with "Antihypertensive Drugs: Our Debt to Industrial Chemists." Until the 1940s, he pointed out, few physicians and patients thought high blood pressure deserved treatment. But the next decade inaugurated a golden age of antihypertensive therapy. For the first time useful blood-pressure-lowering drugs were invented on paper, created in fume hoods from other chemicals, and tested with animals. Not merely were these drugs numerous, but they also acted by different physiological mechanisms. Different modes of action were advantageous partly because they allowed physicians to help patients with one drug who did not respond to another. Culminating with the introduction of the drug captopril in the mid-1970s, this era dramatically reversed the prospects of hypertensive people from morbid to hopeful. "Within about 15 years, a panoply of highly effective drugs had been made available wholly through the skills of chemists in the pharmaceutical industry. It is important," *Page* concluded, "to realize that *drug innovation begins with investigators who work in the chemical laboratories . . .*" [italics added].

Chemists' Work

These researchers alter the structures of marketed or experimental drugs, using chemical reactions to create new compounds with improved sets of physical and chemical properties. Alternatively, they combine simpler molecules to create complex new drugs. The discovery researchers strive in both cases to identify and change the chemical structures that determine biological activities. They labor to impart therapeutic pharmacological effects, to increase established activities, or to reduce undesirable side effects. The work shows promise if it meets stringent criteria. Structurally novel, patentable chemicals must exert one or more of the desired biological effects at the predetermined potency level. Developmental chemists must make them in sufficient quantities with high purities and known contaminant profiles and on strict schedules. The synthetic routes they ultimately adopt to manufacture tens or hundreds of kilos of drugs must be conservative, unlike the liberal approaches tolerated in discovery research. Developed syntheses cannot create high operating risks or costs, nor entail large amounts of wastes. Discovery and development chemists' labors triumph when their drug—which satisfies many other criteria—enters the international pharmacopoeia.

Chemists' Accomplishments

Molecular modification, as it is known, represents some of the work that industrial organic chemists do. It attracted the notice of the cardiologist and Professor of Clinical Medicine *Michael J. Halberstam.* He wrote, "New molecules and drugs derived by chemical modification of active drugs have for 200 years been the source of virtually all important pharmacotherapeutic advances. While most molecular modification does not result in clinically dramatic improvement," *Halberstam* went on, *"most clinically significant improvement results from molecular modification"* [italics added].

This work, necessary as it is to drug discovery and development, does not suffice to bring commercial and therapeutic success, as two industry executives note. *Bert Spilker,* President of Orphan Medical, and *Pedro Cuatrecasas,* former President of Parke-Davis, note that a single scientist rarely if ever discovers, develops, and launches a new drug. "[Nevertheless,] though the breakthroughs come by a series of small steps, *the original idea of a chemist to pursue a specific type of chemical or to find a chemical with specific properties may represent great insight and creativity"* [italics added].

Chemical synthesis remains as important to society as it became during World War II, when the modern pharmaceutical industry originated in its efforts to develop penicillins. Ever since, organic chemists literally have made the drugs that relieve our illnesses. Further examples include the antiinflammatory cortisone, developed in the 1940s to treat arthritis, and the protease inhibitors indinavir, ritonavir, and saquyinavir, introduced in 1995 to suppress the human immunodeficiency virus (HIV).

Future of the Pharmaceutical Industry

"Organic and medicinal chemists can take heart in the knowledge that the traditional skills and chemical knowledge, which led to [the antiulcer drug] cimetidine [in 1972] ... still have a part to play today," says *Dr. Roger Brimblecombe,* Chairman of the Vanguard Medical Group, a contract research organization. "Now, as then, a chemical lead is necessary," he went on, "to [begin] developing a useful new medicine." The pharmaceutical industry therefore needs research chemists, and so offers them a promising future. Regardless of recurrent cost-cutting pressures, its research intensiveness makes its outlook bright. Indeed, the willingness of pharmaceutical executives to invest corporate earnings in research reflects their eagerness to employ researchers. In 1992, for example, one company, Eli Lilly of Indianapolis, devoted as much as 15% of annual income to research and development.

Research Investment

As a whole, the pharmaceutical industry spent an average of 11.5% of its 1992 income on research. This expenditure was greater than any other that U.S. industry made for the same purpose. Next came the semiconductor and computer industries, which allocated 9.4 and 8.8% of sales proceeds to research. The all-industry average equaled a mere 3.7%. In 1996, each of 10 drug houses spent $1 billion or more on research and development, amounts that represented 13–21% of sales income. The industry as a whole plans to spend $18.9 billion for these purposes during 1997. This multibillion-dollar amount, according to the Pharmaceutical Research and Manufacturers Association, exceeds what about half of the state governments spend each year. It is nearly 12% higher than the 1996 investment. Marked by vast expenditures on innovative research, the recent past of the pharmaceutical industry justifies optimism, despite the turmoil created by consolidation.

Job Availability

The industry sustains work force growth. Since 1990, its employees increased in number by 10%, which represents an exception to a trend in other sectors of the chemical industry. Although the number of workers in the plastics, synthetic fiber, and rubber industries fell by 25% since 1980, the pharmaceutical industry work force rose by 30%. An annual growth rate of 2–3% typifies pharmaceutical companies.

The U.S. Department of Labor foresees plentiful jobs for chemists in the pharmaceutical industry. It attributes the need for chemists partly to sustained competition among firms striving to bring innovative drugs to an aging population. Organic chemists, among others, will enjoy the best job prospects.

The chemical industry as a whole offers reasonable prospects for full-time, long-term employment of newly hired chemists. Those researchers working in the pharmaceutical industry can expect to share these prospects. From 1975 to 1995, the unemployment rate among chemists seeking work in the United States ranged from a low of 0.8% to a high of 2.6%. Although the high occurred as recently as 1994, it was comparable to the rate among all professional workers. And, it was less than the 5.5% rate for the whole national work force. The rate fell slightly to 2.5% in 1995. Although this rate is low, any chemist today—like many other types of employees—risks a layoff and a hunt for another job. However, an employable chemical researcher can expect to remain within the chemical industry if she so desires. It is stable, productive, and innovative.

Job Stability

The international pharmaceutical industry is stable because it makes established, needed products that are effective and safe. It depends upon sales of the many small molecules that compose most marketed drugs. Not in 2 centuries, for example, has there been any substitute for the low-molecular-weight chemical morphine for relieving the pain that accompanies grave illness. In 1994, sales of this drug and scores of other standbys brought $258 billion to the pharmaceutical industry. Its products meet scant external competition from health foods or homeopathic remedies. Competing drugs developed within pharmaceutical or biopharmaceutical companies—for example, large, genetically engineered molecules—remain too few to displace traditional medicines from pharmacy counters or hospital shelves. For several reasons, therefore, the social importance of small-molecule drugs remains pervasive. They are not threatened by the commercial obsolescence that, within a few decades, befell audio recordings on polyvinyl chloride disks. Indeed, in 10 of the largest U.S. drug houses, average sales growth in 1995 doubled the previous year's value. It equaled 19% most recently. Small-molecule drugs also offer promise for the future:

> *New chemical entities ... remain the mainstay of most pharmaceutical research, because, by and large, organic molecules with molecular weights below 500 are those most likely to possess the capacity to provide blockbuster drugs of the next century.* (Cavalla, 1997; italics added)

On the other hand, *"Only a few companies are profitable consistently"* [Cavalla, 1997; italics added]. Most of the 30 major biotechnology companies regularly suffer financial losses measured in millions of dollars. In 1995, for instance, 23 of these firms reported deficits ranging from 1.4 to 48.4 million dollars. The first three quarters of 1996 were no exception: 70% of the 30 companies again incurred losses.

> *A job hunter therefore must judge the chance that any nascent biopharmaceutical company will survive its infancy.*

Innovation

Loss and profit aside, innovation in drug discovery remains high in the United States. The number of new medicines reaching the market from 1975 to 1994 equaled 1061, and U.S. pharmaceutical companies produced a large share of them. These firms launched 26% of all new drugs in the years 1975–1979, and their share rose to 31% in 1990–1994. Only Japan's share, also 31%, was comparable recently. Pharmaceutical companies in six other countries—the United Kingdom, Switzerland, France, Germany, Italy, and Belgium—accounted for 32% of new drugs launched in 1990–1994.

Although Japan's output of new drugs reached 90 in 1990–1994, most of these medicines (80) attained local sales only. By contrast, new medicines from the U.S. drug houses provided 35% of sales in the global marketplace, making this country the world's leader in these years. Global sales to seven major markets play an important role in determining corporate performance, as does the speed with which new drugs diffuse abroad. About seven-eighths of all global drugs reach seven national markets in less than 5 years.

In 1995, output figures resembled the values for 1990–1994. The year saw 39 new medicines launched, with United States and Japanese pharmaceutical companies respectively contributing 13 and 11 of them. Eight other countries introduced the remaining 15 drugs: 4 from the United Kingdom; 3 from Switzerland; 2 each from Germany and France; and 1 each from Canada, Italy, India, and Spain. Measuring innovation, new mechanisms of biological action characterized 6 of the drugs launched by U.S. companies. This important kind of novelty leads to fast diffusion of new drugs to the global marketplace. It also prepares for the valuable discovery of structurally related successor drugs with incremental therapeutic advantages over the pioneering medicine.

Job Suitability

Consider again 1994's best-selling medicines: all but one best-seller possesses a molecular weight of less than 1000. The cyclic peptide cyclosporin A, weighing in at 1201, is the only exception. The others are small and relatively simple nonpeptide molecules rather than vast and complex proteins. With an average molecular weight of 395, their diminutive size and few degrees of structural freedom are important to those who practice the art of chemical synthesis. Because of their training, what many newly graduated organic chemists do best is to make small molecules, not large ones. Regardless of whether they enter development or discovery research, pharmaceutical companies hire organic chemists initially to work at what they do best. This match represents one of the immediate appeals of employment as a researcher in the chemical industry—and its other attractions come next.

Altruism as the Unique Appeal of the Pharmaceutical Industry

Chemicals cure disease, a profound truth that explains the past and predicts the future of the international pharmaceutical industry. It arose to exploit opportunities to ameliorate disease with naturally occurring, semisynthetic, and synthetic chemicals. Once familiar but now too often forgotten or neglected, the history of penicillins provides arresting examples of benevolent chemicals. Although the tale of the enchanted β-lactam ring within penicillins and related antibiotics begins nearly 75 years ago, it offers instructive and still fresh evidence of human indebtedness. Many of us now alive owe not only our health but our existence to penicillins, as our unborn children will do. In the future, the industry will extend its efforts to attack a wider front of illnesses with drugs from an ever-expanding pharmacopoeia.

Organic chemists employed in the pharmaceutical industry create health-giving, pain-relieving, and lifesaving drugs for people threatened or injured by disease or trauma. Their efforts represent the unique appeal of lifelong research in a drug firm, namely altruism. It is one creature's selfless efforts to help other members of the same species and a defining feature of human civilization. International in scope, its timeless allure transcends allegiances to any particular science.

What a young scientist, *Dr. Daniela Salvemini* (1996), said of her choice to work for the Searle drug company speaks for many newly recruited and veteran chemical researchers:

> *... working as a pharmacologist in a suitable pharmaceutical [company] will give me the opportunity ... to fulfill my dream. I want my research*

> *to . . . contribute to the development of a new drug, and there is no better place for me to do so than in the pharmaceutical industry.* (italics added)

Participation in such an endeavor offers chemists an opportunity to aid others in a way that is analogous to that enjoyed by their medical counterparts, physicians and nurses. Distinct ramifications however, attend the different contributions made by the two groups of workers. Throughout their working lives, medical practitioners can hope to aid thousands of patients directly. Jointly discovered or developed by chemists, biologists, and pharmaceutical scientists, a successful drug can help sufferers numbering in the hundreds of thousands. It can benefit patients during and after the researchers' lifetimes. To make a successful medicine can give life to its recipients and immortality to its discoverers. These consequences are appealing features of a career open only to a few people. So, an enthusiastic applicant for a chemist's job punctuated a recent interview with "I want to make a drug!"

ATTRACTIONS OF THE PHARMACEUTICAL AND CHEMICAL INDUSTRIES

Benefits

Young chemists preparing to seek their first positions familiarize themselves with statutory and fringe benefits granted to employees. Because statutory benefits are common but few while fringe benefits are numerous and varied, this discussion concentrates on the latter type (Table 1-1). Healthy young people may be unlikely to need medical insurance early in their careers, but this and other valuable fringe benefits help define good jobs. Knowledge of them assists new graduates in deciding whether to work in industry, government, or academia. In part, it directs their searches for employment within the pharmaceutical industry and equips them to recognize the better of two offers.

Indeed, deferred benefits rather than starting salaries are sometimes all that distinguish one job offer from another. The competitive starting salaries proposed by large pharmaceutical houses to equally qualified candidates can be similar and largely nonnegotiable. Outstanding salaries, if they were widely known, would vitiate the intricate calculations that inform compensation structures throughout the industry. They would set unwelcome precedents. Consequently such salaries generally lie outside the control of anyone empowered to offer a predetermined position to a particular candidate at a certain time. Despite exceptions, compensation specialists fix

Table 1-1 The Twenty Most Common Benefits[a,b]

Unpaid personal leave
Employee help
Coverage for routine gynecological examinations
Spending account for dependent care
Coverage for healthy-baby care
Personal days off work
Vacation day transfer
Help with closing costs
Wellness program
Relocation days off work
Gradual return to work
Banking nearby
Surveys of employee attitudes
Birth control coverage
Adult dependent health coverage
Housing referral and resource
Early planning for retirement
Health benefits for part-time employees
Flexible working hours
Tuition reimbursement for nonbusiness courses
Health care for dependents of part-time workers
Part-time employees' prorated benefits
Child care referral and resource

[a]In ascending order of frequency.
[b]Sources: Towers Perrin via Gillian Flynn in *Personnel Journal*, October, 1995, p. 77.

within narrow limits the starting salaries associated with specific positions before candidates can learn of the openings.

Another reason to learn and judge the benefits associated with job offers is too important to neglect although it is negative. The crucial, initial choice of one employer over another may be irreversible. In mid-life, as a result, changing employers without changing professional discipline is arduous if it is feasible at all. Chemical research and development experience is undervalued like other work experience. Here is what an independent pharmaceutical industry recruiter, *Greg Clark* (1995), says:

> *Avoid staying on the scientific ladder. Career change for those who have been on the scientific ladder for years can be difficult . . . move into a management position, or . . . stay very current on developments in your field of expertise.* (italics added)

The newspapers, magazines, and job banks carry few advertisements for industrial chemists with 5 years or more of experience. Chemists' salaries rise fastest early in their careers, so veterans rapidly become more expensive and, in one respect, less credible than neophytes. As in any industry, old-timers evoke disbelief who offer to accept newcomers' salaries for the same work. The tarnish of age soon coats the luster of youth.

Fringe benefits vary from company to company in the same period and from time to time within one company. Different companies within the chemical industries, however, offer their employees certain common advantages. At any particular time, a variety of factors determine the number and nature of benefits conferred within any one company. Important factors include corporate size, profitability, competitiveness, and prospects; but they are not limited to these variables. As a part of an employee's total compensation, however, even variable benefits are important. They deserve careful comparison when a successful applicant weighs two offers because the job accepted may last a working lifetime.

In considering benefits of any kind, job hunters understand that private companies are oligarchic rather than democratic in organization, action, and attitude. And the chemical and pharmaceutical industries offer scant exceptions to this observation. Even as corporations confer some benefits on every one of their employees, they award many privileges only to some workers. One company, for example, may provide free, on-site parking for all, yet reserve nearby parking spots for only a few. For the favored recipients, such an advantage can motivate or inspire loyalty. It saves time and rewards success.

So numerous, extensive, and comprehensive are the genuine benefits of full-time employment within the chemical industry that they demand separate expositions. They fall into five categories: educational, financial, social, temporal, and miscellaneous. Temporal benefits pertain to working times like business hours and yearly vacations. This account of job benefits inevitably reflects personal experience and perspectives, so it strives not for universal, timeless accuracy but for representativeness.

Educational Benefits

Educational opportunities abound within pharmaceutical companies. An employee can complete a coherent, formal education, take courses without academic credit, or learn chemistry informally from a variety of other sources.

Tuition reimbursement. Drug houses, as do other large employers, commonly subsidize the costs of formal education for employees who satisfactorily complete the work. Tuition subsidies help recruit and retain valuable workers. Passing grades suffice to earn reimbursement, but employees initially lay out the money needed for tuition. Consequently, some savings are needed at the outset.

New employees dependent not on savings but on salary need to understand how the IRS treats tuition subsidies. Some subsidies can be taxable at the federal level and subject to Social Security taxes and withholding. Under legislation passed in 1978, the U.S. Congress periodically reexamines the taxability of tuition subsidies. It sometimes rescinds taxes retroactively, leaving to taxpayers the burden of recovering withheld funds. Depending on what they expect Congress to do, some employers withhold money from salaries to pay the tuition tax, while others do not. Eager students would therefore be wise to ascertain their employer's practice, lest they suffer paycheck stroke.

Extended horizons. Through part-time study and research, college graduates who are chemists can earn master's or doctor's degrees, usually but not exclusively, sought in chemistry. Accepting tuition subsidies does not entail restrictions to certain scientific specialties, so analytical chemists, for example, study their own subject or organic chemistry. Biologists, if they acknowledge that chemistry is the only worthwhile human endeavor, may become chemists.

Some students take advanced degrees in business or law, which prepare them to make careers outside science. Others find work that uses all their qualifications. Both kinds of career transition can occur within one company if it is large or growing. But student-employees must be alert to the possibility that to work immediately in their new fields they may have to find employment in a different corporation or location. No company that initially hires a chemist thereby guarantees later employment to the same person as a lawyer. A new attorney wishing to continue his employment in law but not in science must hunt work like any other job applicant.

Who needs a Ph.D.? The reward for spare-time study leading to an advanced degree is a higher salary. As the following section shows, salaries of chemists with doctor's or master's degrees exceed those of baccalaureate chemists. Formed by the American Chemical Society (ACS), a Committee on Professional Training indirectly explains this situation:

> *Why graduate school? Modern chemistry is a vast field for which an undergraduate degree provides only a framework on which to build. Graduate training is a virtual necessity if one is to become truly proficient with the subject and possess a good working knowledge of chemistry.* (italics added)

Given a choice, therefore, many student-employees seek a doctorate degree for the greater challenges, rewards, and opportunities that it brings. And what journalist *Spencer Klaus* (1968) noted 30 years ago remains apt:

> *... a scientist without a PhD is like a brother in a Cistercian monastery. Generally, he labors in the fields while others sing in the choir.* (italics added)

This situation persists today. Advising newcomers, a B.S. chemist with nearly 50 years of service to a drug house recently said, "Get a Ph.D.!"

In industry, the totaled financial rewards of a Ph.D. chemist surpass the lifetime earnings of a B.S. scientist. Using 1996 figures, the former can expect to earn nearly 54% more in his or her career: $2,985,000 versus $1,935,000. This percentage arises from an analysis that makes four assumptions. (1) Both chemists continuously work 43 years, from the ages of 22 to 65. (2) The B.S. chemist earns a steady $45,000 yearly, which represents a median, pharmaceutical industry salary. (3) The Ph.D. chemist makes $15,000 annually during 4 years of graduate studies and then works for $75,000 each year for the next 39. (4) Income taxes, bonuses, pay raises, and savings from salaries are all neglected for simplicity. The Ph.D. chemist's employer effectively pays him $262,000 for each year spent in graduate school, where research is fun.

Rising to a higher salary level, however, may require not only a graduate degree but a matching promotion. For various reasons, receipt of the latter can lag behind award of the former. To understand how this situation originates, consider the case of a B.S. scientist who earns a Ph.D. while he remains employed as an assistant chemist. His new diploma qualifies him for a senior post. Let us suppose that no opening exists in his present company when he receives his degree. Unless attrition opens a suitable position there, one has to be created and justified in terms of the needed work and resources. Convincing reasons should show that employing an assistant chemist is no longer necessary. Absent a strong argument, a replacement must be found after interviewing candidates and mustering resources again. Meeting all these requirements is certainly feasible, but voraciously consumes time and sufficiently explains delay.

> *Cast your net broadly to snare the senior position for which a Ph.D. qualifies you, if you earn this degree part-time. Restricting your search to your present employer sometimes entails a long and irksome wait.*

During this time, your known faults can weigh more heavily than your newly acquired but lesser known virtues. At worst, however, another drug house views you only as another fresh Ph.D. chemist; at best, it sees you only as an experienced, determined, and newly qualified doctoral researcher.

Graduate school. Employed chemists who embark on graduate study attend night school. Most of them, even while they go to graduate school, continue the corporate research that earns their livelihoods. They dedicate much of their spare time—but few, if any, of their paid working hours—to course work and dissertation research. Like other employees, they keep the usual working hours, but devote weekends and vacation days to their graduate studies. Many student-employees do their dissertation research in their employers' laboratories and with their consent, which the students obtain beforehand.

Full-time graduate study during the day requires a resignation or leave of absence, which is sometimes available to a few highly valued employees. However, a guarantee to protect a job is rare, and an unpaid leave without one is undesirable. Employers believe that an employee's work is crucial, and use the importance of the work to create the job initially. So to grant a leave of absence entails contradiction, risks the attendant embarrassment, and requires a perhaps impolitic campaign to secure the needed agreements. A leave usually creates a need for a successor to do the departing employee's work. When a replacement emerges from flat-beaten bushes, then the labor of filling his position will effectively have been done twice. Personnel recruiters do not necessarily welcome such a task. For all these reasons, a leave of absence resembles a chest-high bank of wild, flowering rosebushes—delightful to view but distressful to enter.

With the thorniness of long leaves understood, prospective students must appreciate another point concerning short-term residency requirements. Universities that award graduate degrees in science normally require a residency term, which sometimes can last only a few months. To fulfill residency requirements, graduate students make prior arrangements with their employers and schools. During this time, they pursue their research on campus and beside faculty members, postdoctoral researchers, and full-time graduate students. Doing so is advantageous to the student-employee but arranging it with an employer demands delicacy and planning.

Combining full-time employment with part-time study creates other requirements. A long-lasting commitment and unflagging resources of energy and enthusiasm are among them. Employees contemplating graduate studies need stamina sufficient for two jobs; they cannot neglect either their employment or their studies. Part-time study unavoidably increases the time required to gain the degree sought. For a Ph.D., it may extend the term beyond 5 years. Such a stretch—perhaps trifling from the standpoint of a humanities graduate student—weighs heavily on a chemist's scale. A married colleague, for example, earned a doctoral degree in night school. Begun so long before his degree was conferred, his valiant 15-year course of study and research drew comment during the celebration marking his 30th year of service to the company. His dissertation research was said to have begun before Mendeleyev invented the Periodic Table of the Elements.

Part-time students make welcome additions to academic research groups. Needing no financial support from teaching or research assistantships, they do not draw on university resources or research grants. They arrive paying their own tuition and earning their own livelihoods, unlike full-time students, who are dependent on assistantships. Regardless of any effusive welcome, however, student-employees beginning dissertation research make thoughtful choices of topics and mentors. They select closed, definite problems and tolerant faculty members experienced in training part-time students. A part-time student wisely seeks a place in the laboratory of an established and preferably tenured professor. Such a professor's career prospects are not contingent on his student's graduate research. Otherwise, the pace of part-time research can foster conflict.

What to study. The drug industry employs more synthetic chemists than any other kind, regardless if they work in chemical discovery or development groups. Competition for places is therefore keen, so objective qualifications like research success help recruiters hire chemists. Publications and particularly patents express success in chemistry. Desirable experience includes relevant summer internships and on-campus undergraduate research. It especially improves B.S. graduates' prospects for employment in the pharmaceutical industry.

Learn synthesis if you want to work or advance in the pharmaceutical industry and demonstrate your proficiency respectively with journal publications or patents.

Many B.S. chemists increase their skills and extend their experience in graduate schools that confer M.S. degrees upon them. They do research, write theses, and take courses. This work, most of it in organic chemistry, makes them better candidates for jobs in the pharmaceutical industry. But the choice of graduate school is critical. By definition, a good one helps a student escape the undeserved, long-lasting stigma of a master's degree. Schools sometimes award this degree to scientists who fail or withdraw from Ph.D. programs. Such a sheepskin becomes indelibly branded as the drop-out's consolation prize. Perhaps the best school to attend for a master's degree is one that does *not* award doctor's degrees and that *does* require a master's thesis describing original and successful research that is published.

In any case, no B.S. chemist should believe the enduring myth that misinforms many American undergraduates in the humanities and sciences. A master's degree is not *prerequisite to a doctor's degree.*

Obtaining the former can needlessly and expensively delay getting the latter. It prolongs the time you earn a student's pittance and postpones the moment you start working at a doctor's salary. Go directly into a Ph.D. program, if that's the degree you want.

> *To increase your industrial marketability as a recent Ph.D. recipient, attend a distinguished graduate school of chemistry within a famous university. Your dissertation should concern practical organic synthesis directed by a professor eminently successful in that art.*

It can only help if your thesis results advance his or her research or if your postdoctoral synthetic work does. All the better for you if this chemist won the Nobel Prize, as did *G. Olah, E. J. Corey, D. J. Cram,* and *H. C. Brown*, or received another award (Appendix A). To track prize winners, look at the free annual issues of *American Chemical Society Awards*. Each edition names the following year's recipients.

Expertise in organic synthesis obviously qualifies a researcher to work in a chemical development group. But, surprisingly, the same expertise is also prerequisite for work within drug discovery, even though training in medicinal chemistry might be thought preferable. For decades, however, synthetic chemists have found favor over medicinal ones, and hiring practices are unlikely to change. Here is what a recent survey of attitudes and practices revealed about the desirability of employing synthetic versus medicinal chemists:

> *... the pharmaceutical employer generally seeks out the brightest and most dynamic synthetic organic chemists, most often with a Ph.D. and postdoctoral experience in the laboratory of well known and highly regarded professors of organic chemistry.* (Ganellin, Mitscher, and Topliss, *1995;* italics added)

Pharmaceutical companies surveyed in the United States, Europe, and Japan were asked what qualifications, expertise, and knowledge they sought among organic chemists (*Ganellin, Mitscher, and Topliss,* 1995). Over 80% of the respondents wanted advanced knowledge in organic synthesis, while others sought expertise in synthesis. More than half of them looked for Ph.D. chemists who had done synthesis as postdoctoral researchers. In reply to another question, companies stated a strong preference (>90%) for hiring organic chemists additionally educated as medicinal chemists. Most of them rejected the other suggested answer, which was to employ specialists in medicinal chemistry with acquired expertise in synthesis. Although medicinal chemistry and organic synthesis are both necessary within discovery research, employees learn only the former subject on the job.

Realism will help you plan a graduate education leading to a Ph.D. while you are working in a pharmaceutical company. A sound plan takes five points into account. (1) The graduate school that you attend must lie reasonably close to your home or workplace. (2) It must offer the evening courses that you need. (3) The professor who serves as your thesis adviser must be active and perhaps young enough to take on new graduate students. (4) Before accepting an offer to attend a particular school, you will want to ascertain the professor's plans. Is he soon to retire completely or easing into retirement by restricting his coworkers to postdoctorate researchers? Has he accepted or sought an offer to teach at another university? If so, you may find yourself working for a different faculty member in the same graduate school. Alternatively, you may have to move to another school to remain in the same research group. (5) Your chosen adviser's renown will have attracted graduate students to his school, while the federal funds that support research in organic synthesis are diminishing at other universities, a situation that repels determined students. So, you should expect to compete vigorously for a place in a laboratory devoted to synthesis. Be aware that you may face a difficult or impossible task: winning or avoiding the lottery by which popular departments of chemistry assign graduate students to sought-after faculty mentors. Otherwise, you will draw an adviser whom you do not desire.

In-house courses. Without pursuing a graduate degree, any chemist can continue his education within a pharmaceutical company. Employers sponsor isolated, tuition-free courses which take place during working hours in company lecture halls. Such courses fill gaps in academic training and present new techniques and topics from emerging areas of research. They excite and satisfy professional interest, while the knowledge gained soon finds use on the job. They are not normally semester-long courses but last a week or less. Completing such a course may require prolonged independent study of a photocopied text distributed by the lecturer. A student's diligence earns no academic credit in these circumstances, so the acquisition of knowledge and the prospect of applying it must suffice.

Corporate employers encourage expert employees to choose the subjects and to lecture, and they also hire teachers extramurally. For example, nuclear magnetic resonance spectroscopists, chromatographers, or mass spectrometrists present useful new techniques that they make available to fellow employees. Invited consultants and chemistry professors teach their specialties, so industrial research chemists can enjoy learning from internationally recognized practitioners. Often the academics draw from treasured stocks of superb examples, giving polished lectures that evoke earned applause.

Chemical development, heterocyclic chemistry, and surveys of the pharmaceutical business are popular topics for in-house courses. All marketed drugs undergo chemical development, and many of them are heterocyclic compounds; so such courses have relevance on the job. Nonetheless, chemists' formal schooling often neglects both topics, which are extensive, intricate, and practically applicable in discovery and developmental chemical research.

Evening courses. Industrial organic chemists avail themselves of evening courses offered by local schools, where they enjoy subsidized tuition. Because they work for pharmaceutical companies, many of these chemists densely populate northern New Jersey. Consequently, even a local high school there taught introductory pharmacology in its adult education curriculum. A Ph.D. biologist, whose full-time employer was likewise a pharmaceutical company, served as the part-time instructor. His well-presented, highly regarded offering was, alas, little understood in some quarters. The town sewerage authority once directed a local sewer contractor to pass the course. It was hoped that a knowledge of pharmacology would help the baffled builder tell putrefaction from purity.

Other offerings relevant to chemists appear in the catalogs of nearby universities. Science offerings accompany courses from other areas: negotiating, public speaking, effective writing, using English as a second language, and reducing foreign accents. In New Jersey, where much of the pharmaceutical industry concentrates, nearby schools include the Stevens Institute of Technology in Hoboken, Rutgers University on the Newark and New Brunswick campuses, and Drew University in Madison.

Some courses are rigorous, lasting one or two semesters and requiring periodic examinations. They can offer academic credit, often by arrangement. Open to students not enrolled in any graduate program, such courses ordinarily take place in urban areas. Cities and their surroundings contain large populations of interested chemists as well as numerous universities. Large populations provide sufficient numbers of students to justify the course offerings. The presence of several universities makes it likely that one of them presents evening courses.

Employment in the pharmaceutical industry continues an organic chemist's education by other means than classes. Weekly, biweekly, or monthly in-house seminars attract renowned academics who give hour-long chemistry lectures to discovery and development researchers. On one occasion an award-winning chemist's address bore the title "How to Win a Nobel Prize." Attending other regular seminars, given by biologists or physicians, also characterizes a career spent in a large pharmaceutical company. Opportunities for discussion arise at the ends of the lectures when the speakers answer questions from the audience. Some listeners privately consult the speakers during their visits.

Chemical societies. Drug firms typically pay the costs of belonging to a scientific association. Examples are the American Chemical Society, the American Association of Pharmaceutical Scientists, and the Royal Society of Chemistry. Membership reduces the expense of individual subscriptions to the chemical journals that these societies publish. Society members read these journals to learn technological developments and rival firms' doings. Partly as a result, pharmaceutical companies bear the costs of their employees' personal copies.

Books. If you like a wealth of chemistry texts, you'll like working in this rich industry. Large pharmaceutical companies furnish relevant science books to their research workers. They make personal copies available, retaining ownership or discounting costs. Discounts at some times or in some firms amount to 100%, while the book topics range widely through chemistry.

Libraries. The library of a pharmaceutical company represents an educational benefit as well an indispensable tool. The attractions of a corporate library are not limited to its physical holdings, but also embrace the knowledge of its staff and the scope of its services. Large pharmaceutical companies maintain good collections that combine many valuable features of university science and public libraries. They keep as many as 10,000 books that include chemistry texts, monographs, and series. Even popular corporate histories like *Barry Werth's* 1994 *The Billion Dollar Molecule* appear on the shelves.

Periodical collections run to bound or microfilmed issues of hundreds of scientific journals from all disciplines relevant to the pharmaceutical business. Bound journals like *Chemical Abstracts* need physical storage space that aspires to *Coleridge's* "caverns measureless to man," so modern libraries are changing to microfilmed and electronic versions. Further items from specialized electronic databases include United States and European patents, all stored on compact disks. Although microfilmed journals discourage the browsing that bound ones inspire, electronic databases like "Pharmascope" permit complex computer searches. Easy to carry out, such searches represent an indispensable advantage of a modern library, which puts software-laden personal computers in the hands of its patrons.

A library loans books to readers, of course, but a good corporate library provides many more services than this. For example, it keeps long hours. It puts its catalog on line, which allows chemists in remote offices to search for a book using networked microcomputers. Such networks let researchers place electronic requests for photocopied journal articles. A good corporate library might furnish daily stock market information, *Books in Print* as a compact disk, facsimile copies of articles ordered from another library, and searches of the chemical literature executed with "Chemical Abstracts Service On Line." Staff librarians with degrees in library science or chemistry do much of this work at chemists' requests, often finding pearls in clamshell heaps.

Scientific meetings. Pharmaceutical companies here and abroad encourage their employees to attend chemists' meetings, which help them keep their knowledge and skills current. Valued for another reason, attendance allows chemists a chance to develop and maintain a network of professional acquaintances. Lecturing and presenting posters, which are also encouraged by employers, bring recognition to individuals and corporations, and they aid recruiting efforts. They place an emerging drug before the attention of the medicinal chemical and biological communities. Familiarity with such a drug, especially in academic circles, can stimulate external studies that accelerate its internal development. Trip reports distributed by returning participants offer valuable insights into competing companies' efforts and accomplishments. They summarize exploitable new chemical developments from these and other sources. Such knowledge can inspire scientists to take new research directions or dissuade them from following bare-trodden paths. Chemists' attendance at meetings, therefore, serves both their own and their employers' interests. Industrial employers consequently reimburse their researchers for the costs of going to meetings. They cover the income-deductible expenses of registration, travel, lodging, food, entertainment of guests during business discussions, and lost time at work.

Conferences of chemists take place locally, regionally, nationally, and internationally. Scientists' associations organize the gatherings, and a few of the sponsors include the American, Canadian, and Mexican Chemical Societies. Other sponsors are the International Society for Heterocyclic Chemistry, the European Chemical Society, the Royal Society of Chemistry, the International Union of Pure and Applied Chemistry, and the American Association of Pharmaceutical Scientists. Each year, divisions of the American Chemical Society organize national organic and medicinal chemistry symposia, which supplement regional and other national conventions. Private organizations as well as university departments of chemistry arrange other conferences. The Wesleyan University Department of Chemistry, for example, sponsors the annual *Leermaker* Symposium. Some medicinal chemists attend conferences of biologists; for example, those gatherings sponsored by the Federation of American Societies for Experimental Biology.

Chemists' off-site meetings typically last 1 day to 1 week. Day-long conferences often treat a single subject; for example, recent progress in solid-state organic synthesis. Week-long meetings like the many Gordon Conferences offer more varied fare to a few hundred chemists, with exemplary conferences dedicated to medicinal chemistry, stereochemistry, and chemical development.

Attended by 10,000 chemists or more, the greatest topical variety characterizes the colorful, frenzied national congresses of the American Chemical Society. Lectures start Sunday morning. They last 15 to 40 minutes and relentlessly succeed one another from early morning until late afternoon. Later in the week, thronged poster sessions, mixers, and black-tie dinners diversify the schedule. A rich commercial exposition demonstrates intricately wrought chemical glassware, slick analytical instruments, and pastel molecular models on computer screens. Competing to snatch passersby from the aisles, exhibitors dart from their booths as grasping lobsters spring from stony lairs, only to bestow souvenirs in plastic shopping bags. Earnest instructors, networking editors, eager students, aspiring authors, and resolute bibliophiles crowd the booksellers' stalls to heft and sample the year's offerings from chemical publishers.

For various reasons, large meetings like these take place in the host country's biggest cities. Chicago, Los Angeles, New York, Orlando, New Orleans, Denver, and other cities vie to become the sites for such annual national congresses. At longer intervals, Mexico City plays host to joint meetings of the American, Canadian, and Mexican Chemical Societies, as do Toronto and Québec. The organization sponsoring a large conference arranges tours as well as day-long employment, technical, and scientific workshops.

Rarefied settings here and abroad accommodate smaller numbers of conferees at equally well-established annual meetings. Schedules call for poster sessions and lectures in the mornings and evenings, which frees afternoons for recreations and discussions. Favored spots range from Spartan boarding schools near the White Mountains through resorts in the Rocky Mountains to grand hotels atop Bürgenstock, an alp overlooking Lake Lucerne in Switzerland, where arriving conference-goers cross the lake by ferry boat and ascend to their lodgings by cable car. Uplifting congresses like these and others are so intellectually nourishing that they stimulate annual migrations resembling the mass movements of Alaskan caribou to their summer feeding grounds in the Arctic. Beginning each June, nations and continents exchange their academic- and industrial-scientist populations: Canadians venture South; Mexicans journey North; Europeans travel West; and North, South, and Central Americans fly East.

Financial Benefits

Second in importance only to salary are the financial benefits of employment in the pharmaceutical industry, especially within the largest firms. Detailed in Chapter 8, these benefits include insurance, pensions, 401k accounts, profit-sharing arrangements, and company stock option and stock purchase plans. Other gains from industrial employment are not only monetary but social.

Social Benefits

Industrial employers furnish many benefits that improve daily life. These benefits include subsidized car and van pooling, child and elder care, adoption consultation and referral, retirement planning, and health care. Health-related benefits comprise physical examinations and educational programs. Periodic, on-site examinations detect diseases like cancer, measure systemically circulating cholesterol, and record blood pressure, to cite only three examples. Educational programs that include informative leaflets and lunchtime seminars teach the care of allergies, the effects of diet, and the value of cardiovascular exercise. By prohibiting smoking indoors and offering free smoking-cessation courses, employers discourage a dangerous habit.

Temporal Benefits

Among the advantages of employment in the pharmaceutical industry can be a 5-day, 40-hour week; flexible working hours; compressed work weeks; long holiday closings; tenure-related vacations; and vacation-day carryovers to the next year. Not every firm, however, offers all of these features, and some give few. Nevertheless, enough pharmaceutical companies do provide many of these benefits to make all of them noteworthy.

Flexible scheduling. If an employer offers flexible working hours (*flextime* or *flexitime*), then it requires its employees to work during certain invariant periods of each 8-hour working day. Otherwise it allows them to begin and end each day's work as they please, with little or no supervision of trusted employees. Core hours, for example, might span the time from 10 AM to 4 PM, except for, say, a 30-minute lunch. With $5\frac{1}{2}$ hours of any 8-hour working day accounted for, one employee might begin work at 7:30 AM and leave at 4 PM. Another, however, might prefer to depart at 6:30 PM, which would require a starting time of 10 AM, and such a late start time would be permissible. Flexible hours like these allow any employee to vary his working schedule from day to day, if he works 8 hours and meets the core-hour requirement. Such schedules make it easy to run the personal errands that daily life demands, so flexible working times are valuable. Arrivals and departures that are late or early let automobile commuters escape much rush-hour traffic. They shorten travel times wherever motor vehicles congest roads.

Holidays. Pharmaceutical companies ordinarily expect their employees to work only 5 days per week and normally observe most of the individual national holidays. However, some reduce the number of these observances and instead suspend work during the winter holidays. They close before December 25 and reopen on January 2, giving their employees a stretch of free time not counted as vacation.

Some companies offer their employees compressed working weeks. In a span of 2 weeks amounting to 80 hours but only 9 days, an employee works eight 9-hour days and one 8-hour day. He then has Friday off and makes 18 trips instead of 20. This schedule saves two trips or 10%. It reduces motor vehicle use and abates air pollution.

Vacations. Two- to 3-week-long vacations are standard for new employees, although their annual holidays are shorter than European scientists' vacations. (In Europe, a 6-week-long stretch is not uncommon.) In the United States, vacation lengths depend upon the highest academic degree attained, the number of years worked, and the employer. Newly hired researchers with doctorates can receive, from the beginning of their employment, vacations 50% longer than the annual holidays granted to their counterparts with master's and bachelor's degrees. Paid vacation lengths rise in steps: 2 to 3 weeks during the first 15 years of service, 3 to 4 weeks in the next 5 or 10 years, and so on. Job seekers should therefore consult personnel departments to learn the prevailing vacation policies—as well as sabbatical eligibility requirements.

Miscellaneous Benefits

At least one pharmaceutical company announces and grants sabbatical leaves. On its Web site, Genentech, Inc., a biopharmaceutical firm that employs organic chemists in South San Francisco, assures that sabbatical leaves supplement paid vacations and holidays.

Other extraordinary benefits from California companies include the assistance Genelabs Technologies gives its employees. It helps them select, buy, and set-up home computers. COR Therapeutics in South San Francisco offers public transportation assistance, while Gilead Sciences employees enjoy subsidized gourmet dining in its Tides Café. "Set in tranquil Japanese garden surroundings," the Karl Strauss Brewery benefits workers of Molecular Simulations and other companies (www.msi.com).

More prosaic benefits include subsidized safety shoes and spectacles. Because of their steel toe caps and strong lenses, respectively, they can be necessities rather than benefits of employment. Nevertheless, they prevent injuries off the job, so they offer take-home value. Many firms also provide gymnasiums on-site or nearby. Perhaps no miscellaneous benefit is valuable enough to make it decisive in refusing or accepting employment. But these benefits nonetheless demonstrate a pharmaceutical company's regard for its employees, so they partly define the prevailing corporate culture.

Making the move. When you accept a job, many companies provide relocation assistance. It covers the costs of moving household goods from your old home to your new one. It can include the lodging expenses incurred while you look for a new home. In some cases, employers help in the sale of your previous home, occasionally buying it. If you are moving from an inexpensive to an expensive area, your new employer may subsidize the rent you pay in a new apartment. Employers offer to pay the difference between the low rent for your previous home and the high rent for your new one. The subsidy comes with time and money limits but can amount to thousands of dollars: you should ask about it lest no one offer it. Items covered by relocation assistance differ from one company to another, so it is wise to discuss this coverage with a personnel representative.

The company store. Additional benefits with readily calculated dollar values come from the stores operated by large pharmaceutical companies. The companies make their own products available at substantial discounts to employees. Not only prescription drugs but over-the-counter remedies, cosmetics, medical devices, and eye-care products are stocked. Other establishments that serve employees occupy company premises and include hair salons and credit unions. With nearly all the services of commercial banks, credit unions offer competitive interest rates for automobile loans, mortgages, checking and saving accounts, and certificates of deposit.

Communications. Modern telecommunications, business machines, and computer equipment make it easier than ever before for chemists to participate in the activities of chemical societies. Chemical researchers compose slides, posters, articles, and reviews; serve as journal referees, honorary editors, and directors; and help organize meetings of academic and other industrial chemists. Today, industrial chemists type and illustrate using microcomputers, word processors, structure-drawing applications, and optical scanners. They do so most easily if they possess the requisite keyboard skills, especially 10-finger typing.

Personal computers and modern communications equipment inexorably spread through pharmaceutical companies during the past 20 years. In providing such equipment, these employers indirectly but effectively sponsor much of the day-to-day business of chemical and other scientific societies. Sponsorship benefits all their members, especially those industrial chemists who publish their research.

Salaries

"How much will I make?" is the burning question answered here. This section surveys the salaries paid to veterans and newcomers in the chemical and pharmaceutical industries. It presents the factors that determine a veteran chemist's pay and that go beyond his performance (Table 1-2). Such factors offer guidance to new graduates deciding to seek work in the pharmaceutical

Table 1-2 Selected Factors Determining Veteran Chemists' Salaries

Industrial, governmental, or academic nature of employment
Highest academic degree attained
Division of the chemical industry
Sex
Geography
Company size
Years of service
Starting pay
Individual performance on the job

industry or to look elsewhere. The section also sets forth the current median salaries paid in the pharmaceutical industry to newcomers at any degree level. Medians represent the values attained by at least one-half of the chemists responding to an annual survey carried out by the ACS. Knowledge of the figures helps set the lower limit acceptable to a job hunter. To estimate the upper limit that a job hunter might expect, the section discusses publicized salaries recently offered to certain nonimmigrant chemists.

Upper and lower salary limits, which a job seeker must set herself, have an enduring importance. They enable a candidate to compare and contrast offers from two firms, aid in negotiations, and cultivate realistic expectations.

To discuss the factors influencing salary ideally demands the concreteness and timeliness lent by the latest available wage figures. However, unavoidable publishing delays mean that a book cannot compete with the Internet or weekly newsmagazines in presenting the latest available values. As a result, the following discussions include data drawn from the now-dated 1998 and 1996 ACS salary surveys. More recent data appears in weeklies like *Chemical and Engineering News*.

Employer Type

Different kinds of employers place varying values on the services of chemists with diverse academic degrees and accordingly compensate them more or less. Historically and currently (1998), the chemical industry pays the most for chemists at any degree level (Table 1-3). The smallest salaries customarily accrue to chemists employed in academia, with the wages of chemists working in government falling in the middle.

Academic Degree

Within any of the traditional three employment categories—industrial, governmental, and academic—median annual starting salaries increase in the order B.S. (or B.A.) < M.S. (or M.A.) < Ph.D. (Table 1-3). Within the chemical

Table 1-3 Median 1998 Starting Salaries of Inexperienced Full-Time Chemists (in $1000s)[a]

Degree	Industry	Government	Academia
B.S.	32.0	28.0	25.0
M.S.	40.2	36.8	30.4
Ph.D.	60.0	53.0	35.5

[a]Source: M. Heylin, *C&EN*, March 1, 1999, p. 17.

industry, doctoral chemists' wages in 1998 were 88% more than baccalaureate chemists' salaries. The spread was somewhat greater in government service (89%) but least in academia (42%).

Industrial Division

The chemical industry traditionally forms many divisions (Figure 1-2), each paying chemists according to the highest degree attained. Doctoral chem-

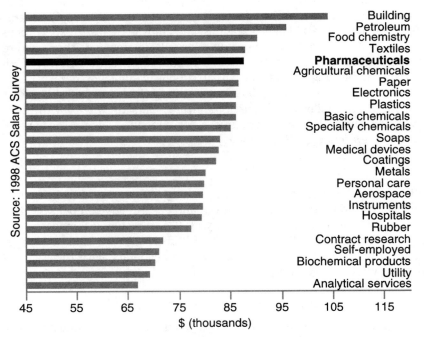

Figure 1-2 Median 1998 salaries for experienced Ph.D. chemists employed full-time

ists' pay in 1998 made pharmaceutical companies fifth in salaries to building, petroleum, food chemistry, and textile firms. But M.S. and B.S. chemists working in pharmaceuticals fared worse. Their salaries ranked 17th in 1998 (Figures 1-3 and 1-4).

Sex

"A surprising result from the [1998] survey is the salary edge that male chemistry graduates with full-time jobs apparently have over their female colleagues at all degree levels..." (*Heylin, 1999*). Male Ph.D. chemists throughout the chemical industries earned median salaries 5% higher than their female counterparts. The corresponding differences for M.S. and B.S. chemists respectively were +9% and +5%. In the pharmaceutical industry, the 1998 survey showed no difference between the salaries of male and female B.S. chemists, but substantial differences at the M.S. and Ph.D. levels (Table 1-4). These differences, respectively, equaled +7.5% (M.S.) and +9% (Ph.D.), with the men paid more than the women.

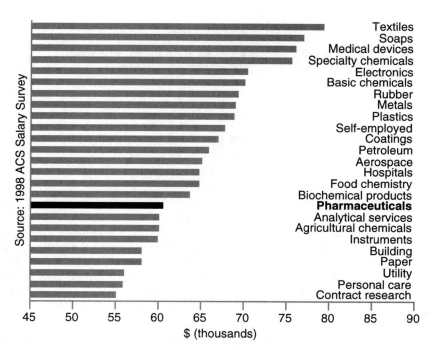

Figure 1-3 Median 1998 salaries for experienced M.S. chemists employed full-time

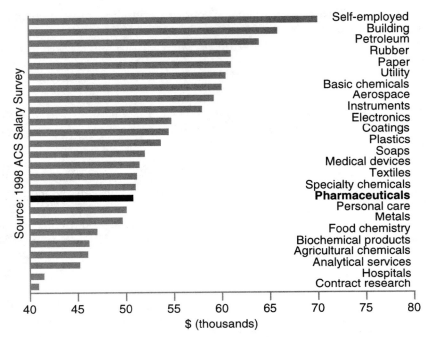

Figure 1-4 Median 1998 salaries for experienced B.S. chemists employed full-time

Geography

Where you work partly determines how much you are paid, which is illustrated by data from the 1996 salary survey. In that year, industrial chemists employed in Delaware, New Jersey, New York, and Pennsylvania received greater median salaries than their counterparts elsewhere in the United States. They did so regardless of which academic degree they held, so that chemists in the foregoing Mid-Atlantic States who held doctor's, master's, and bachelor's degrees generally were the best paid throughout the country. Employers in the West Northcentral States paid their B.S. chemists less than these workers would have received anywhere else in the country. Iowa, Kansas, Minnesota, Missouri, Nebraska, North Dakota, and South Dakota compose the West Northcentral region. Chemists with master's degrees earned the smallest salaries in the Mountain and Pacific States. The Mountain States comprise Arizona, Colorado, Idaho, Montana, Nevada, New Mexico, Utah, and Wyoming; while the Pacific States include California, Oregon, and Washington. Doctoral chemists earned the least in Alabama, Kentucky, Tennessee, and Georgia. Across the country, median salary

Table 1-4 Median 1998 Starting Salaries of Full-Time Chemists in the Pharmaceutical Industry (in $1000s)[a,b]

Degree	Women	Men
B.S.	32.0	32.0
M.S.	40.0	47.0
Ph.D.	56.8	62.0

[a]Source: 1998 ACS Salary Survey.
[b]Rounded in the first decimal place.

ranges in 1996 equaled $72,200–61,400 (Ph.D.); $55,900–48,400 (M.S.); and $46,500–40,000 (B.S.). With the differences between the highest and lowest salaries in a category reaching $10,800–6,500, job hunters at every degree level must give these geographic wage-spreads consideration. The great cost of living in the Mid-Atlantic States effectively reduces the higher salaries paid there. It is as high in New Jersey, for example, as it is in Hawaii and Alaska: these three states suffer the greatest living costs in the United States.

Company Size

The largest employers pay the highest median salaries throughout the chemical industry. In 1996 B.S. chemists' salaries in companies with fewer than 500 workers reached $40,000. Companies with 25,000 or more workers paid B.S. chemists median salaries of $50,200. For M.S. and Ph.D. chemists the corresponding ranges in the smallest and largest companies equaled $49,000–60,000 and $70,000–80,000. Consequently, any chemist who accepts employment in a small company may wish to seek compensation in benefits for the money forgone in salary. The reduced costs of living in some parts of the United States can also be compensation.

Years of Service and Inflation

In the chemical industry, median salaries paid to full-time chemists ordinarily increase with time on the job. In 1996 wages rose 5, 4.4, and 4.7% for B.S., M.S., and Ph.D. chemists. Within any degree category, the youngest members received the greatest median increases, measured in percentage terms. For example, B.S. chemists aged 20 to 29 received increases of 6.6%. Their 60- to 69-year-old baccalaureate coworkers' raises reached only 4.2%. Doctoral scientists' increases equaled 3.4 and 7.0% for those ages 60–69 and 20–29, respectively. Younger chemists' labors do not necessarily excel older scientists' work, so their larger raises do not invariably reflect a link between greater pay and better performance. Higher percentage raises may be affordable to employers because their basis is lower salaries paid to fewer workers.

Anyway, in and before 1996, increases in chemists' median annual salaries, and especially young chemists' impressive raises, meant little on the long term. Expressed in constant dollars, chemists' wages have failed significantly to rise or fall. When dollar deflation is accounted for, industrial chemists' salaries remained in 1996 what they were in 1985.

Salary calculator. Another source of salary information for industrial scientists working in the United States is an online calculator posted by *R&D Magazine*. To reach this calculator, visit www.rdmag.com. The calculator requires your age, scientific discipline, years of experience, and highest academic degree. It also needs the number of employees in the kind of company that interests you and the state where you would like to work. When you enter this information, the calculator displays an annual salary based on a survey of magazine subscribers. This figure may exceed what a newcomer can expect to earn, for the respondents are "seasoned pros."

Starting Pay

Any scientist seeking work in the North American pharmaceutical industry should do so in the spirit that predominates there, which is not altruism but capitalism. A chemist's labor is as much a commodity as pork bellies, so he should trade it as dearly as a trader sells futures. He has at least three reasons to sell high.

1. The surging economy demands chemists, while fewer students are willing to become experimentalists.
2. A chemist's marketability declines steeply with time, much like the value of a used car or a computer. The descent begins when he receives her highest academic degree and reaches a nadir within 5 to 10 years after commencement ceremonies. Few companies advertise for chemists whose experience spans more than 5 years. Falling marketability soon erodes an initially strong bargaining position, which a successful negotiator must exploit promptly, like a sailor who departs with the tide thrusting and the wind blowing.
3. Today's starting salary influences tomorrow's wages. Barring extraordinary personal success or corporate generosity, annual percentage raises in pay take present salary as their basis. Therefore, wages 30 years hence depend partly on the salary accepted initially, as does retirement income.

Going offers. Periodicals make estimates available each year. And the Federal Immigration Act of 1990 obliges employers to display on their premises semipublicly certain exact, position-specific salaries. (The following section entitled *"Nonimmigrant Wages"* discusses these salaries.)

> *In hunting a job or negotiating an offer an applicant wisely ascertains the prevailing wage for the position sought. This information prepares him to weigh and negotiate one offer and to compare two. An assertive, self-interested job hunter probably risks no harm to his prospects by seeking a wage exceeding what he can expect to receive. Without current salary data, a too-trusting applicant may accept a wage less than the going one.*

ACS salary survey. Foremost among the sources of salary information is the yearly survey carried out by the ACS Office of Career Services. Each year this office polls thousands of society members, many of whom complete and mail an extensive questionnaire. A social science methodologist analyzes the results and compiles them into annual reports typically entitled *Starting Salaries: 1996* and *Salaries: 1996*. These reports are detailed and broad and are most informative when hiring is vigorous, especially of inexperienced new graduates.

Starting Salaries: 1996 reports starting wages by divisions of the chemical industry like plastics and pharmaceuticals. It also presents them by the highest academic degree granted to employed chemists who completed the questionnaire. *Salaries: 1996* analyzes chemists' employment rates and other data. Staff writers at ACS excerpt both reports for periodical publications like *Chemical and Engineering News*, *Today's Chemist at Work*, and even the British journal *Chemistry and Industry*. Chemical libraries offer all these periodicals to readers. The survey typically harvests more information than the reports can hold, so the ACS Office of Career Services makes the overflow available to inquirers.

During 1998, inexperienced chemists taking employment in the pharmaceutical industry accepted the following median starting salaries: B.S., $32,000; M.S., $43,000; and Ph.D., $61,600. The mean starting salary and standard deviation for chemists with doctoral degrees equaled $62,375 ± 8106. The sample size, however, was small. Only 54 new Ph.D. holders responded to the ACS survey. Of all 1998 Ph.D. winners who responded, 35% of them found work in the pharmaceutical industry. The percentages of newly graduated inexperienced chemists with bachelor's and master's degrees who went into the drug industry are 32 and 49%, respectively.

Because relevant surveys of chemists' wages are rare, *Starting Salaries: 1996* or later issues is indispensable, so it deserves several comments.

1. Priced at about $30, it lies within reach of student budgets, while the $150 *Salaries: 1996* may fall beyond them.
2. However, its information can become obsolete within a year.
3. The survey does not specify the company proposing a given salary, so a job hunter cannot compare what one firm offers to what another does. (For the same reason, however, *Starting Salaries: 1996* protects the anonymity of respondents to the annual survey.)
4. Certain job categories may lack salary data in some years because too few newcomers who found employment completed the survey questionnaire.
5. The survey does not ascertain the effects of certain factors on chemists' income specifically in the pharmaceutical industry. These factors comprise geography, company size, annual raises, inflation, and bonuses.
6. Published salary information can be difficult to interpret. For example, many veterans employed full-time earn advanced degrees in certain years. Their experience brings them higher salaries than newcomers receive. To include their wages tends to raise the median figure beyond what an inexperienced beginner can expect. Also, does a given starting wage reflect what is paid to a Ph.D. chemist with no postdoctoral experience? Postdoctoral training ordinarily brings a higher salary, and including many such salaries would increase the median wage.
7. Although any year's published survey may lack data sought by job hunters, often this information has been gathered and can be analyzed as needed. The ACS Department of Career Services welcomes society members' inquiries about unpublished material collected for the annual surveys.

Nonimmigrant wages. Offers made to selected nonimmigrants entail valuable postings of little-known and perhaps hard-to-get starting salary information. In passing the 1990 Immigration Act, Congress required employers openly to disclose offers made to holders of H-1B visas. Called a labor condition application (LCA), each disclosure pertains only to a single individual and a certain job, whereas the aspects covered include the job title, employment term and place, and exact starting salary. Personnel departments post notices of these jobs on certain corporate bulletin boards, thereby meeting legal requirements. A single notice of an individual job posted for 10 days suffices. Any visitor or employee can read the notices, despite their restricted distribution. Their revelations jump from chemist to chemist as forest fires leap from treetop to treetop, partly because newcomers' offers can equal or exceed veterans' salaries.

Department of Labor regulations 507.760 and 655.760, which can be found at www.dol.gov, require employers to make documents supporting an LCA publicly available. These documents reveal the methods used to fix the actual wage and to determine the prevailing wage in the geographic area. The actual wage represents the salary received by an employee whose training, skills, and experience resemble the offer recipient's qualifications. The prevailing wage may represent an average. Under the terms of the Immigration Act, employers must pay H-1B visa holders the higher of the actual and prevailing wages. Labor condition applications therefore reveal the salaries paid chemical researchers by a particular drug company or nearby firms in the pharmaceutical industry.

Job hunters who do not visit a particular corporate site can perhaps rely on networking to elicit LCA salary figures from acquaintances employed when and where the posted notices appear. Alternatively, a job hunter can contact the Department of Labor (200 Constitution Avenue, NW, Room N-4456, Washington, DC). It makes a national list of labor condition applications available for public examination.

> *Salary information disseminated through H-1B notices benefits both native and foreign job hunters. It helps set a reasonable upper limit to any job hunter's salary expectation and can prevent a citizen from accepting too low an offer.*

In 1996–1997, for example, a large pharmaceutical company offered one Ph.D. organic chemist a $67,000 staring salary for entry-level employment in discovery research; another received an offer of $66,000. Other immigrant researchers with doctoral degrees received offers of $64,000 and $61,000. (The top offer made late in 1998 to an immigrant Ph.D. chemist rose to $71,000.) During 1996–1997, a drug firm offered $45,000 to an inexperienced chemist with a master's degree or to a chemist with a bachelor's degree and 2 to 4 years' experience. A second such chemist received an offer of $46,500, while a third chemist's offer was $48,000.

Viewed through the magnifying lens of compound returns on investment, salary differences of $6000 and $3000 loom large. The future values of these amounts equal hundreds of thousands of dollars after 30 years. If the highest paid Ph.D. and B.S. chemists pay income taxes of 30%, they could save the remaining $4200 and 2100 annually. A 7% average yearly return from stock would bring them inflation-adjusted capital gains of $396,735 and $198,368 (*Siegel*). Consequently, it is foolhardy to dismiss a $3000 salary difference as only a few percentage points above the lowest wage and wise to consider the future value of money.

SOURCES AND SUPPLEMENTS

American Chemical Society; "Current Trends in Chemical Technology, Business, and Employment," American Chemical Society, Washington, DC, 1994.
American Chemical Society Awards, Bulletin 7, 1999 ed., Washington, DC.
Anonymous; "How much are you worth?" *R&D Magazine,* Sept., 1998, S-17.
Brimblecombe, R.; "25 years of pharmaceutical R & D," *Scrip Magazine,* April, 1997, 32–33.
Busse, W. D., Ganellin, C. R., and Mitscher, L. A.; "Vocational training for medicinal chemists: Views from industry," *European Journal of Medicinal Chemistry* **31**, 747–760 (1996).
Cavalla, D.; "Modern Strategy for Preclinical Pharmaceutical R & D: Towards the Virtual Research Company," Wiley, Chichester, 1997.
Clark, G.; "What does the future hold for careers in the pharmaceutical industry?" *Inflammation Research Association Newsletter* **4**(2) June, 1995, 1–10.
Committee on Professional Training; "Planning for Graduate Work in Chemistry," American Chemical Society, Washington, DC, 1990.
Crum-Brown, A., and Fraser, T. R.; *Transcripts of the Royal Society of Edinburgh* **25**(151), 693 (1868).
Denkewalter, R. G., and Tishler, M.; "Drug research—whence and whiter," *Progress in Drug Research* **10**, 11–13 (1966).
Ganellin, C. R., Mitscher, L. A., and Topliss, J. G.; "Educating medicinal chemists," Chapter 33 in *Annual Reports in Medicinal Chemistry* **30**, 329–338 (1995).
Halberstam, M. J.; "Too many drugs?" *Forum on Medicine* **2**, 170–176 (1979a).
Heaton, A. (ed.); "The Chemical Industry," 2nd ed., Chapman & Hall, London, 1994.
Heylin, M.; "Chemists' employment situation continues to worsen, salaries weak," *Chemical and Engineering News,* July 29, 10-16 (1996).
Heylin, M.; "More Jobs, Better Pay for 1998 Chemistry Grads," *Chemical and Engineering News,* March 1, 14-17 (1999).
Klaus, S.; "The New Brahmins," Wm. Morrow & Co., New York, 1968.
Liebenau, J.; "Medical Science & Medical Industry: The Formation of the American Pharmaceutical Industry," Johns Hopkins University Press, 1987.
Mahoney, T.; "Merchants of Life: An Account of the American Pharmaceutical Industry," Ayer (date not supplied, ISBN 0-8369-2608-0; LCCN: 77-167381; reprint).
Maurois, A.; "The Life of Sir Alexander Fleming," Penguin Books, Harmondsworth, Middlesex, UK, 1963, p. 158.

Owens, F., Uhler, R., and Marasco, C.; "Careers for Chemists," American Chemical Society, Washington, DC, 1997.

Page, I. H.; "Antihypertensive drugs: our debt to industrial chemists," *New England Journal of Medicine* **304**, 615–618 (1981).

Salvemini, D.; "Preclinical industrial research," *Trends in Pharmaceutical Science* **17**, 58 (1996).

Sheehan, J. C.; "The Enchanted Ring: The Untold Story of Penicillin," The MIT Press, Cambridge, 1982; reviewed by F. P. Doyle and J. H. C. Nayler in "Differing perspectives on penicillin," *Chemistry and Industry*, 4 April, 275–276 (1983).

Siegel, J. S.; "Stocks for the Long Run," 2nd ed., McGraw-Hill, New York, 1998.

Spilker, B., and Cuatrecasas, P.; "Inside the Drug Industry," Prous Science, Barcelona, 1990.

Testa, B.; "Missions and finality of drug research: a personal view," *Pharmaceutical News* **3**(6), 10–12 (1996).

Thayer, A. M.; "Drug, biopharmaceutical company sales rise in second quarter and first half," *Chemical and Engineering News*, August 19, 27 (1996).

Weatherall, M.; "In Search of a Cure: A History of Pharmaceutical Discovery," Oxford University Press, Oxford, 1991.

Wiggins, S. N.; "The Pharmaceutical Industry," S. Pejovich and H. Dethloff (eds.), Private Enterprise Research Center, 1985.

Wilson, D.; "In Search of Penicillin," Knopf, New York, 1976.

chapter

2

ELEMENTS OF DRUG DISCOVERY AND DEVELOPMENT

"... the discovery of truly new agents is totally unpredictable and highly risky."—Bert Spilker and Pedro Cuatrecasas

"... in pharmaceutical research ... it is axiomatic that 'you cannot guarantee success, you can only attempt to minimize the risk of failure.'"

—Barry Cox

INTRODUCTION

Efforts to create drugs originate a variety of jobs for organic chemists, many of whom work in fully integrated pharmaceutical companies (FIPCOs). Their employers engage in six basic operations: discovering, developing, registering, manufacturing, marketing, and monitoring drugs. The first four of these efforts employ nearly all chemists who work in FIPCOs. Only two tasks—discovery and development—occupy most entry-level scientists. Seeking employment as a chemist in a pharmaceutical company therefore calls for learning about drug discovery and development as well as the other pharmaceutical businesses, which this chapter surveys selectively. It opens with some distinctive features of the drug industry and closes with a forecast. In between, the chapter summarizes the strategy and tactics of drug discovery and development and sketches one organizational structure in which pharmaceutical industry chemists commonly work. Appreciation of this work, and acceptance of its rewards and risks, underlie a happy and productive career.

Profile of the Pharmaceutical Businesses

Time to Market

To bring a new medicine to the marketplace currently takes 10–12 years. By contrast, an industrial chemist devising a new adhesive can expect to see the fruits of this work 3 months after beginning it. The longer interval arises partly through the variability of pharmacological and clinical measurements made in animal and human studies of potential medicines. It stems from the many necessary biochemical, animal, and human trials from which other kinds of manufacturers are exempt. The FIPCO Pfizer, for example, sponsored 85 independent clinical studies of its quinolone antibiotic trovafloxacin. The longer interval for medicines versus other goods also springs from the difficulty of discovering drugs, notwithstanding efforts to design them completely. The requirement that drugs, manufacturing plants, and laboratory and manufacturing practices secure approval from government agencies also contributes to the long project lifetimes. The agencies include the Food and Drug Administration (FDA) and its counterparts abroad, the U.S. Environmental Protection Agency (EPA), and the Occupational Safety and Hazard Administration (OSHA).

What Drug Discovery and Development Cost

Long project lifetimes partly explain the high cost of the pharmaceutical enterprise (Figure 2-1). Including lost opportunity costs, the expense of soup-to-nuts drug discovery and development equaled $54 million in 1976. It rose steeply to $231 million in 1990 and, in 1999, held at some $600 million (in 1997 dollars). One new drug alone costs this amount, and clinical development accounts for about three-fifths of the total. For a variety of reasons, the total cost will increase unless the number of medicines approved annually rises. From the international high, which reached 60 in the 1960s, the number of new chemical entities per year now remains steady at about 46.

Benefits of Patents

Long project lifetimes and high costs explain why the pharmaceutical industry depends on patents to protect its drugs. Patents forbid competitors to use the originators' inventions. They give the greatest possible assurance

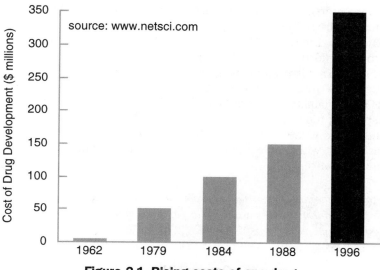

Figure 2-1 Rising costs of one drug

that costs can be recovered and profits secured, thereby stimulating medicine research. Patents benefit the public by disclosing valuable inventions that might otherwise remain trade secrets. They compensate pharmaceutical companies for risking research projects to discover new drugs, given the likelihood that competitors' medicines will soon supplant their own. In the 1980s, 5 to 7 years elapsed before a competing drug appeared. In the early 1990s, pursuing medicines arrived in 3 years.

Likelihood of Success

Alack, if launching a drug measures success, then failure concludes most efforts to create one. Only one new compound in about 5000 reaches clinical trials, while fewer than 1 in 10,000 advances to marketing. Making 10,000 new compounds, however, traditionally requires more than the career-long efforts of a single chemist employed in discovery research. It demands the output of an estimated four to seven chemists each working 30 years in conventional organic synthesis. The most productive four researchers would each have to make over 80 target compounds yearly for one to reach the marketplace. If these scientists were equally lucky or astute in choosing compounds to synthesize, each chemist would have about a 25% chance of making a drug in his working lifetime. (Newer, faster methods of synthesis like combinatorial chemistry may circumvent the historical throughput barriers to new drug discovery.)

Efforts to discover or develop drugs sometimes halt abruptly, and reasons to suspend them abound. Examples include ineffective biological assays; lack of *in vitro* or *in vivo* activity or of selectivity; impotency; and toxicity of the drug or its metabolites in animal or human trials. Others are lack of bioavailability, oral activity, or patentable novelty; change of managerial mind; and too small a market or too low a return on investment. These or other faults vitiate most efforts to create innovative medicines. On the chemist's part, withstanding failure takes resilience, while awaiting success demands patience.

> *For any chemist pursuing a career in the pharmaceutical industry, renewable enthusiasm is a desirable personal quality, especially in discovery research.*

Competition

Eternal competition among FIPCOs is the price of profits, and it occurs in various arenas. Different companies compete to sell drugs that serve the same therapeutic area, like cardiovascular medicine. Within respiratory medicine, for example, FIPCOs also strive to discover similarly acting drugs for the same disease. Beginning in the early 1980s, for example, many companies throughout Europe and North America sought an antiasthma drug that antagonized the effects of leukotriene D_4. The Zeneca company (now AstraZeneca) won this race by launching its zafirlukast in 1996. The next year, Merck & Co. (Whitehouse Station) introduced the leukotriene antagonist montelukast.

Pharmaceutical companies compete not only within a given therapeutic area to find a medicine for a specific disease. They also contend with one another in seeking drugs that act by novel pharmacological mechanisms. From 1993 to 1997, for example, 13 firms secured 82 patents for drugs to treat asthma by four new mechanisms of biological action. Moreover, competitors vie to discover new drugs acting by mechanisms known for decades. For example, medicines that agonize certain β-receptors treat asthma and are the subject of three companies' six patents granted from 1993 to 1997.

The established FIPCOs face competition not only from one another but also from start-up pharmaceutical and biopharmaceutical companies, which number about 1300 in the United States. Competition surfaces from other quarters, too. Companies like DuPont and Monsanto, which manufactured chemicals in bulk, recently entered the fray, respectively forming the DuPont Pharmaceutical Company and Monsanto Life Sciences. The opportunities to profit attracted Bristol-Myers (now Bristol-Myers Squibb) and Procter & Gamble, hitherto consumer products companies.

Complexity

Integrated pharmaceutical companies require the minimum biomass of 600–800 employees, judiciously distributed among the many job functions. In output, for example, the chemists creating new compounds must match the biologists carrying out the assays. The chemists' work otherwise overwhelms the biologists' capacity. Alternatively, the latter's demand for samples tries the former's productivity.

Imbalances in staff, though they hinder progress, can also offer paradoxical business opportunities. The flow of new drugs from discovery to development research sometimes rises or falls. An uneven output taxes or idles developmental researchers like process chemists and formulations scientists. In preference to laying employees off, some pharmaceutical companies—for example, Abbott and Boehringer Ingelheim—sell their own services to other firms with which they might have competed rather than cooperated. Other companies maintain a small permanent headcount during fallow periods. When demand increases, they give temporary employment to, for instance, chemists working as process researchers. Some firms adopt both tactics of contracting their services and temporarily hiring scientists.

Perhaps the greatest sign of complexity in the pharmaceutical business issues from the obligatory regulatory filings. Late in 1997, for example, the Pfizer company received FDA approval for the previously mentioned quinolone antibiotic trovafloxacin. To apply for this approval, the company presented the results of clinical trials taking place in 27 countries and requiring 13,500 patients.

Other signs lie in the variety of disciplines needed (Table 2-1) and in the armies of people employed within the largest FIPCOs (Figure 2-2). For example, to bring Merck's dihydroavermectin B into human therapy, where it prevents river blindness, required 17 kinds of scientists: biologists, biochemists, biochemical and chemical engineers, biostatisticians, chemists, clinical researchers, computer scientists and engineers, entomologists, microbiologists, parasitologists, pharmacists, pharmacologists, regulatory affairs specialists, toxicologists, and veterinarians. In the largest FIPCOs, the numbers of employees of all kinds range from as few as 25,000 (Schering-Plough) to as many as 118,000 (Hoechst AG, the parent of Hoechst Marion Roussel). Half of these selected firms employ 41,000–58,000 people, including but not limited to researchers. These researchers strive to make different types of medicines that serve various therapeutic areas.

Table 2-1 Scientific and Selected Professional Staff Employed in Pharmaceutical Companies

Discovery	Development	Discovery or Development
Medicinal chemists	Biochemical engineers	Analytical chemists
Biochemists	Bioorganic chemists	Radiochemists
Natural products chemists	Chemical engineers	Drug metabolism experts
Physical chemists	Manufacturing chemists	Computer scientists
Synthetic chemists	Pharmaceutical chemists	Information scientists
Theoretical chemists	Physical organic chemists	Patent attorneys, agents, and coordinators
Dermatologists	Synthetic chemists	
Entomologists		
Histologists	Clinical physicians	
Immunologists	Epidemiologists	
Microbiologists	Pharmacists	
Molecular biologists	Pharmacoeconomists	
Mycologists	Pharmacokineticists	
Neuroscientists	Statisticians	
Nutritionists		
Parasitologists		
Pathologists		
Pharmacologists		
Physiologists		
Registrars		
Taxonomists		
Toxicologists		
Tumor biologists		
Veterinarians		
Virologists		

KINDS OF DRUGS, COMPANIES, AND THERAPEUTIC AREAS

Drug Categories

Medicines fall into three broad functional categories: diagnostic, prophylactic, and therapeutic. Diagnostic products, as their name suggests, detect or

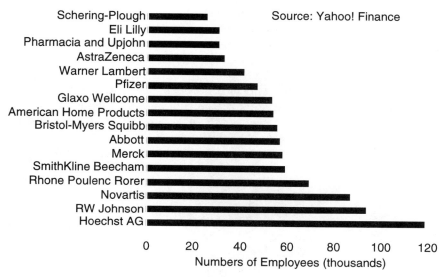

Figure 2-2 Employees of selected FIPCOS

probe diseases, while prophylactic agents prevent them. Many prophylactic agents are biological products like vaccines, although some are small molecules like ascorbic acid, which is vitamin C. Vaccines rank among the triumphs of modern medical science, and one of them recently eradicated the scourge of smallpox from the world. Thanks to vaccines, the incidences of polio and measles have also fallen. Therapeutic agents treat diseases and are usually small molecules, but not always. Interferons, which represent a class of large protein molecules, exemplify therapeutic drugs that are biological in natural origin and biotechnological in manufacturing source. They treat hepatitis C, a viral disease, and certain cancers.

Nucleus of the Pharmaceutical Industry

To manufacture drugs in these three classes, special companies exist throughout the international pharmaceutical industry. Diagnostic companies make their products with biotechnology and organic chemistry. Companies dedicated to producing biological drugs use the methods of biotechnology, which yield vaccines, peptides, proteins, and nucleotides, among other products. Fully integrated pharmaceutical companies arose in the late 19th century to produce small-molecule therapeutic agents using organic chemistry. Today, some FIPCOs employ both biotechnology and organic chemistry to manufacture diagnostic, biological, and therapeutic agents. Other companies, begun exclusively as biotechnology firms, evolved into biopharmaceutical companies. They use not only biotechnological methods but also organochemical ones to manufacture or seek drugs. Despite fluid boundaries among the three kinds of companies—diagnostic, biopharmaceutical/biotechnological, and FIPCO—these firms compose the nucleus of the North American pharmaceutical industry.

Fully integrated pharmaceutical companies, however, are not alone in constituting the drug industry. Across the continent, this industry includes hundreds of other companies. Many of them employ organic chemists but do not engage in all the businesses of a fully integrated drug company. Yet other firms within the industry do not require such chemists at all. Job hunters therefore need to learn which kinds of other drug houses do hire scientists, which ones require their skills as researchers, and which do not.

Clinical Research Organizations

Some drug companies carry out no preclinical chemical or biological research at all, nor any process chemistry or production. These companies are clinical research organizations (CROs), which contract to take experimental drugs through phases I–III of human studies. They need no discovery, development, or manufacturing chemists, but deserve mention, if only as an alert. Some of their names contain the words *pharmaceutical company* or *pharmaceuticals,* which can mislead a job hunter unfamiliar with this subdivision of the drug industry. Some CROs, however, do study the metabolism of the drugs they are developing, so they do employ chemists.

Generic Drug Companies

The industry includes generic drug houses that register, manufacture, formulate, package, and sell medicines. In general, these firms do not engage in the kind of discovery research that leads to new chemical entities. Their business opportunities arise when the patents covering FIPCOs' valuable drugs expire. Patent expirations can cast these drugs into the public domain. Any company may then make and sell them as long as doing so meets regulatory agencies' requirements. Since these drugs originate elsewhere, generic companies do not hire organic chemists to discover the medicines. Instead, the companies need small numbers of chemical operators and manufacturing chemists to supervise large-scale preparations.

Generic pharmaceutical companies sometimes do employ organic chemists as process chemists. To appreciate this situation, imagine that a FIPCO's compound patent expires, depriving a valuable medicine of further composition-of-matter protection. Also suppose that a separate process patent still protects the originator's commercial synthesis. In these circumstances, the generic firm wishing to sell this medicine is free to do so if its production synthesis circumvents all applicable process patents. But, if the generic firm does not license production rights, it must devise, develop, and seek regulatory approval of its own manufacturing process. This work constitutes chemical development, so doing it requires process chemists. It fosters employment opportunities.

Drug Discovery Firms and Start-Up Pharmaceutical Companies

Other pharmaceutical companies exist not to market drugs without discovering lead compounds, but, conversely, work to find lead compounds without selling finished drugs. Drug discovery firms are newcomers to the industry, created partly by technological innovation. Many specialize in fast, high-output organic synthesis. Their methods comprise solid-state and solution-phase combinatorial chemistry or semiautomated parallel synthesis. Their business plans call for them to prepare new organic compounds and to sell them or license rights to them. Such compounds attract established FIPCOs, agricultural chemical companies, or animal health firms, all of which test the substances for biological activities. Discovery companies thrive on such agreements and do not all strive to evolve to complete pharmaceutical firms. They make a quickly growing and widely spreading market for organic chemists' skills.

Yet other small companies—usually start-ups—do carry out both biological and chemical discovery research. They limit their efforts, perhaps to a few related diseases within one therapeutic area in which the founders— who are often academics or erstwhile researchers from large drug houses— have expertise. Examples from Massachusetts include Phytera, which seeks innovative drugs in plants, and MycoPharmaceuticals, which genetically engineers microbes to make small-molecule drugs. British Columbia is home to AnorMED, which seeks to treat life-threatening diseases with small-molecule metal-binding drugs. Microcide Pharmaceuticals in California concentrates on finding medicines to treat drug-resistant bacterial infections.

Many start-up firms plan to grow into large, fully integrated pharmaceutical companies, but initially employ six organic chemists or fewer. Such small firms have no need for chemical development until they select their first drug candidate for clinical trials. When this crucial choice looms, the company may assign responsibility for chemical development to a discovery researcher or it may seek a process chemist extramurally. In either case, the effort creates jobs for organic chemists. As is shown in Chapter 3, many specialized firms contract to carry out process research and development and to make bulk supplies of intermediate and final products.

Therapeutic Areas

Western medicine recognizes nearly 300 categories of therapeutic agents, while the largest markets comprise only 8 kinds. Arranged in order of descending market size, the major kinds of therapeutic drugs include central nervous system (CNS), gastrointestinal (GI), anti-infective, respiratory, cardiovascular (CV), metabolic, vitamin, and dermatological. The CNS drugs relieve pain, nausea, depression, or anxiety, to cite only a few examples; GI drugs serve patients with irritable bowel syndrome or ulcers. Anti-infective drugs treat bacterial, fungal, and viral infections as well as oncological diseases. Patients with asthma, allergies, cystic fibrosis, and emphysema find relief in respiratory drugs; while CV agents ameliorate circulatory and heart diseases like high blood pressure, heart disease, and arrhythmia. Diabetic people, for example, suffer a metabolic disease and benefit from insulin therapy. Vitamins B_{12}, D, and C respectively cure folate anemia, rickets, and scurvy, which are deficiency diseases as diabetes is. Cyclosporin A, a cyclic peptide, recently entered the dermatological market as a treatment for psoriasis, a disfiguring skin disease.

DISCOVERY AND DEVELOPMENT OF DRUGS

This section introduces the fundamentals of drug discovery and development, summarizing the roles of certain scientists who work within FIPCOs. It surveys preclinical discovery and chemical development, where many chemists find work. It also introduces other development efforts that occur when a clinical candidate has been chosen or when the choice nears. These efforts include drug metabolism, radiochemistry, and radioimmunology, and they also employ organic and medicinal chemists.

To look for an innovative drug is not a decision that any pharmaceutical company takes hastily. A good choice can ultimately improve or save patients' lives and influence scores or hundreds of careers for decades. Other stakes are billions of dollars in sales and hundreds of millions in expenses (Figure 2-1). Successful projects therefore meet four strategic preconditions.

Stratagems

Medical Needs

(1) A suitable therapeutic area affords opportunities to find drugs that serve existing markets. An area lacks promise if this condition goes unfulfilled. It is

the more inviting if there is no preexisting therapy for a single disease, if one drug relieves several diseases, and if the markets for each disease are large. Some medicines, known as orphan drugs, are developed for small markets.

To seek a drug within a certain therapeutic arena represents a high-level managerial decision. Such a decision costs money, deploys personnel, occupies equipment, and requires a corporate infrastructure to set timetables and priorities. It draws on contributions from clinical medicine, sales, marketing, pharmacoeconomics, biology, and chemistry. The decision may constitute an initiative in a therapeutic area new to the company and thereby call for a knowledge of the competition. How many other companies are working in the area? When did they begin, what direction did they take, and what success have they had? The decision may capitalize on a newly feasible approach to treating a known or new disease. For example, consider virology, a subdivision of the anti-infective therapeutic area discussed earlier. Inhibiting an enzyme partly responsible for reproduction of the immunodeficiency virus HIV represents a novel approach to therapy of a viral disease.

Therapeutic Approaches

(2) The decision makers specify the disease(s) or condition(s) that the desired drug is to treat before focused biological and chemical research can begin. (3) It must also be possible for them to recommend a particular approach to treatment. To do so demands an understanding of the underlying biology and pathology associated with the disease. Ordinarily, much of the prerequisite knowledge already exists. It may lie within corporate archives, the open scientific and medical literature, or consultants' and employed researchers' experience. FIPCOs also acquire know-how through licensing agreements, partnerships, and other strategic alliances. Commonplace approaches on the molecular level that concerns organic chemists involve enzymes or receptors. Small-molecule drugs can inhibit or stimulate the former and agonize or antagonize the latter. If the enzyme or receptor plays a role in the disease, one of these biochemical events can ultimately produce a therapeutic effect.

Approaches to therapy differ from one disease to another, and they also change as knowledge about a certain disease accumulates. Years ago, any company entering the gastrointestinal area might have sought a drug specifically to treat stomach ulcers. Several firms did so, looking for antagonists of the histamine-2 (H_2) receptor to reduce the excess acid secretion that exacerbates stomach ulcers. They sought to impair receptor functioning by blocking the receptor with the antagonist, much as a blank key occupies a lock without moving the tumblers. The efforts of biological and chemical discovery researchers culminated first in the small-molecule drug cimetidine and later in its successor ranitidine. They won commercial and therapeutic success, respectively, for Smith Kline (later SmithKline Beecham) and Glaxo (now GlaxoWellcome). Another approach, however, now complements theirs. Remarkably, treatment with a cocktail of three previously known antibiotics—that are *not* H_2 antagonists—also cures stomach ulcers. The three antibiotics are amoxicillin, clarithromycin, and lansoprazole. This subsequent finding, from which neither SmithKline nor Glaxo could benefit, shows that infection with the stomach bacterium *Helicobacter pylori* also causes ulceration. The infection responds to treatment with the antibiotics.

Therapeutic Advantage

(4) A new medicine that treats the same disease as an established drug must offer a therapeutic advantage to benefit patients, win the attention of prescribing physicians, and capture a market share. Finding a medicine superior to an extant drug can represent a valuable opportunity. However, the superiority of the new drug for treating the same disease should reside largely in its therapeutic advantage(s) over the old one. A different mechanism of action can impart such an advantage, perhaps relieving a greater fraction of the patients, prolonging the benefits of therapy, or averting side effects. Affordable development costs and patentable structural novelty are of paramount importance in a successor. But, without the promise of therapeutic advantage, they would not justify or inspire a search for a new drug.

Contemporary efforts to discover medicines exploit modern understanding of the mechanisms of drug actions as well as the biochemical and genetic foundations of diseases. Minute sample quantities suffice for biochemists to test almost astronomical numbers of mixed compounds in a variety of automated enzyme- and receptor-based assays *in vitro*. More often then ever, animal models of human diseases are available or devisable and useful in selecting orally active compounds from those active *in vitro*. Modern methods and knowledge make it feasible to search for drugs offering therapeutic advantage rather than mere clinical effect. They now address the causes of diseases, which represents an advance over earlier efforts merely to treat symptoms.

Research Goals

The aims of a typical discovery project call for finding one small molecule as a clinical candidate and another as a so-called back-up drug. Small molecules are desirable because they frequently offer the good oral activity that leads to high bioavailability. A back-up drug offers a fallback position in case the original compound fails in the obligatory and expensive long-term toxicology studies or in clinical trials. In a successful project, the two compounds represent different structural classes and are solids possessing patentable structural novelty. They show potency, oral activity, selectivity, and suitable duration of action. These features appear in decisive live-animal models of one or more human diseases, all within a certain therapeutic area. The two compounds also display potency in a mechanism-based, *in vitro* assay indicative of the desired clinical effect. Found within about 4 years from the start of the project, they demonstrate safety in short-term live-animal toxicological assays before long-term toxicology studies begin *in vivo*.

Tactics

Roles of Biological Discovery Groups

In a global war against disease, therapeutic areas are the theaters of conflict, specific illnesses represent the assault objectives, and particular therapeutic approaches correspond to the military tactics. Then the researchers are the foot soldiers, who wage the battle wielding scientific knowledge. Biologists lead the attack.

A biological discovery group devises a working hypothesis with three elements. (1) A substance potent in the primary, mechanism-based assay produces the expected effects in secondary, functional assays that quantitate physiological events. (2) It predictably exerts therapeutic effects in decisive live-animal models of the human disease and shows appropriate selectivity. Finally, (3) this substance or one derived from it brings significant, measurable relief to human patients in controlled trials. The biologists assigned to the project experiment to confirm the first two elements of their hypothesis, using a standard compound or a new one. Such a substance, if it were highly potent in the animal models and patentably novel in structure, might become the lead compound and the first clinical candidate. It would serve only as the lead compound without patentability.

Finding proof of concept can be a difficult feat requiring months of labor by many researchers. It sometimes requires a fresh start with another compound or with other techniques applied to the original one. For example, if the would-be lead compound lacks oral activity, perhaps because of poor absorption, then a change in the mode of administration merits a trial. Intravenous or intramuscular dosing may bring decisive *in vivo* activity, so proving the concept. However, oral activity is so desirable, and clinical trials so costly, that a compound lacking it might not advance to human testing.

Biological discovery groups find and import suitable assays from the literature of biology, devise new tests as needed, and carry them out. Their role includes validating the assays established intramurally. They accomplish this task by studying standard compounds possessing the activities of interest, especially in mechanism-based primary assays.

With adequate quantities of enough active compounds from a certain structural class, the work of composing a short list of potent substances can begin. To establish potency the project biologists carry out rising-dose experiments with several or many active compounds. These experiments succeed when incremental increases in dose produce proportional rises in activity over a wide dose range. The biologists measure the potency of each sufficiently active compound in the mechanism-based primary test and perhaps in secondary functional assays. They carry out live-animal experiments modeling various aspects of the human disease of interest. Scores or even hundreds of compounds are called to these trials but few are chosen to progress. Most compounds tested in a given primary assay are biologically inactive, as a rule, and most compounds active in that assay are impotent in secondary tests.

Highly potent compounds advance to a battery of selectivity assays. Any clinical candidate must not only demonstrate effectiveness in the desired therapeutic area but lack side effects in other areas. A cardiovascular drug, for example, which patients may take for decades, should not show unwanted central nervous system, gastrointestinal, or respiratory activities. Comprehensive selectivity profiles *in vitro* and *in vivo* allow pharmacologists to compare and contrast potent compounds and thereby to choose among possible clinical candidates. Profiling reduces the number of candidates still further.

The remaining compounds must afford several that are suitably long acting. A marketed drug with a long-lived effect induces greater compliance to the dosing regimen than a shorter-acting one. Adherence to such a regimen can be crucial to successful therapy. Patients prescribed β-lactam antibiotics, for example, who cease taking their drug when their symptoms disappear, often suffer reinfection. Failing to complete the course of therapy, they unknowingly defeat its purpose.

To ascertain duration of drug action in live-animal trials, pharmacologists measure the therapeutic effects of fixed doses at changing intervals. In separate experiments, they increase and decrease the dose, making measurements at similar time periods. They repeat these experiments until they can estimate the minimum dose needed to reach the desired therapeutic effect after a certain time. A knowledge of this dose helps establish the margin of safety. Suppose that, to maximize the therapeutic effect 12 hours after dosing, the minimum drug dose is 30 milligrams per kilogram of animal body weight. Then the dose at which any side effects appear might have to be 10 times higher.

Contributions from Radiochemistry and Drug Metabolism

Radiochemists and drug metabolism experts work at the interface between discovery and development research. The former scientists are basically organic chemists who synthesize labeled samples of the clinical candidate(s) for the latter. For biological researchers, they also prepare radiolabeled samples of receptor ligands that possess high affinity and specific activity. The traditional labels are the radioactive isotopes carbon-14, tritium, and iodine-125, often because they combine sensitivity of detection and ease of synthesis. Carbon-14, for example, can be incorporated within a benzene ring and distributed randomly there. As long as important metabolites retain this aromatic ring, exact knowledge of the isotope's molecular whereabouts is unnecessary. However, environmental pressures are leading drug developers to label experimental medicines with stable isotopes like carbon-13 rather than radioactive ones. This practice reduces the burden of radioactive waste disposal.

Experts in drug metabolism carry out time-course studies with radioactive samples. They give a variety of doses and, at intervals, measure plasma radioactivity rather than therapeutic effect. Sensitive instruments allow for quantitative measurements; for example, of the time and concentration at which radioactivity reaches a maximum. To lay a basis for selecting the animal species used for long-term toxicology studies, these pilot probes of duration may employ several mammalian species. Different species metabolize a single drug dissimilarly, so a suitable animal species handles it as the human one does.

Both the toxicological and human species may convert the medicine to several metabolites with varying structures. Drug metabolism experts—who are sometimes trained as physical organic, bioorganic, or medicinal chemists—elucidate the structures of these compounds, isolate them, and ascertain metabolic profiles. They establish the number and relative amounts of metabolites excreted, for example, in urine, and measure the urine radioactivity to account for all that the original dose contained. Their profiling work in a given project begins with several animal species and culminates with our own. In general, a knowledge of metabolic fate—absorption, distribution, storage, metabolic reactions, and excretion—helps demonstrate the effectiveness of drugs. Contract research organizations exist to gather this knowledge, which many FIPCOs also seek intramurally.

The past saw a sharp distinction drawn between discovery and development research in a pharmaceutical company. For example, discovery researchers working on a given project once brought one compound—usually the most potent one from a series—as the clinical candidate to development scientists, who endorsed or spurned the choice. Today, discovery researchers bring several potent compounds to drug metabolism and toxicology studies before committing the success or failure of their project to any one substance. They choose a clinical candidate not only for its pharmacological and pharmacokinetic profiles, but also for its physicochemical properties, which govern cell penetration and oral absorption. Important properties include crystal form, particle size, water solubility, partition coefficient, acidity or basicity, molecular weight, and the ability to accept or donate hydrogen bonds.

Progress and Success in Clinical Development

Certain milestones mark progress in the journey from an experimental drug to a marketed medicine. They include Investigational New Drug (IND) status, Phases I to III of clinical testing, New Drug Approval, and Phase IV of human trials and monitoring. To study an experimental drug given to humans, a pharmaceutical company in the United States needs the approval of the Food and Drug Administration of the federal government. The company seeks approval to treat a particular disease by submitting a written application, presenting evidence of the safety and effectiveness of the drug. It discloses what is known of the mechanism of drug action and of any adverse effects. Animal experiments, often in several species, furnish this evidence. The company also discloses its plans for human trials as well as the structure and synthesis of the drug. It automatically receives approval if the agency offers no objections within 30 days, and the compound then attains Investigational New Drug status. This standing permits human clinical trials to begin.

Phase I trials see the experimental drug given to a small number of healthy volunteers, often men. Typically 20 subjects take part in each trial, receiving single or multiple doses of the drug formulated variously; the number of volunteers can reach 100. Patients—for example, cancer sufferers—participate when it would be unethical to restrict the studies to healthy volunteers. The object of these trials frequently is to ascertain safety; rarely is it to determine effectiveness. The tests probe the range of doses that subjects tolerate and include pharmacokinetic measurements of absorption, distribution, metabolism, and excretion. Phase I lasts about a year, and drug sponsors annually report results from these and later trials to the FDA. A drug demonstrating safety progresses to the next stage of clinical development.

As many as 300 patients, or as few as 100, participate in Phase II trials, which take some 2 years. They establish the effectiveness of the drug in relieving the disease of interest, determine the margin of safety, and measure the therapeutic range of doses. To profile the dose–response relationship, the trials include rising-dose regimens. Organizers may conduct all or some the trials strictly, giving placebos to some of the patients and to others a standard drug that serves as a positive control. Because one drug may show effectiveness in different ways, the trials may use varying patient populations; for example, juvenile or adult. Use of varying drug formulations establishes which one is best for a given indication.

Effectiveness and safety in Phase II trials, and safety in Phase I, are necessary but not sufficient for regulatory approval. They do allow the project to advance, and the decision to proceed with development often follows a successful, rigorous Phase II trial.

Large-scale, long-lasting trials characterize Phase III of clinical development, which continues for about 3 years. The numbers of patients range from 200 to 4000, with some studies continuing for years. Sponsors restrict them to the patient population to be treated, ceasing to look for groups that would benefit most, which were sought in the previous phase. They decide and use the final drug formulation given the chosen patient population for the selected indication. The objective of these Phase III studies remains a showing of safety and effectiveness; and, if taken, leads to regulatory approval. Some Phase III trials may continue or begin during deliberations by the FDA.

Sponsors of drugs successful in Phase III approach the FDA seeking New Drug Approval (NDA). Their application contains all pharmacological and toxicological data from preclinical animal studies and all the results from clinical trials that bear on safety and efficacy. Applicants provide profusely detailed manufacturing and formulating specifications. For indications, contraindications, dosing, and side effects, the NDA application proposes the information that the package label displays.

An FDA review of the documents takes several years and lasted 2–3 years in or before 1991. Approval is needed before sales may begin. However, before a pharmaceutical company secures approval, 6 to 11 years will have elapsed since clinical development began.

Phase IV trials and surveillance begin after marketing starts. The trials continue much of the work of earlier studies that evaluated safety and efficacy. They also search for drug–drug interactions and side effects not previously detected. Because certain adverse effects afflict only tiny fractions of the treated population, some Phase IV trials necessarily involve many thousands of patients. Other studies compare and contrast the new drug to any existing ones for the same disease, search for new indications or patient populations, or look for information to assist in marketing.

Chemical, pharmaceutical, and clinical development make up the future of an effective, long-acting compound promising safety and offering high plasma levels and nontoxic metabolites. Chemical discovery efforts leading to such a potential medicine are contemporaneous with much of the biological discovery research, but begin later and last longer.

Roles of Chemical Discovery Groups

Organic chemists beginning work in a given project represent a second assault wave. They discharge two principal duties, one creative and one logistical. (1) Chemists conceive patentably novel examples of structurally related compounds, predicting they will exert in humans the desired biological activity. Such acts of conception suffice to make these scientists the only inventors named by the chemical compound patents that protect marketed drugs. (2) In a successful project the inventors or other chemists go on to create these compounds physically and to make a subset of potent ones with the preferred biological profile. The number and quantity of compounds prepared for testing are crucial. Hundreds or thousands of them lack the activity, potency, or selectivity desired and merit no further consideration. The chemists obviously must make more than this number, and doing so occupies most of their working time. Their other scientific duties can include resynthesis of potent new compounds, preparation of standard ones, and synthesis of metabolites and degradation products. Figure 2-3 summarizes this and other work of chemical and biological discovery groups, presenting the sequence of events that compose a successful discovery project.

Making or buying supplies. With potent compounds, the biologists' material demands soon escalate, and it often falls to the assigned discovery chemists to satisfy them. Screening with primary, *in vitro* assays normally requires only minute weights of each compound; for example, 1 or 2 milligrams or less. Confirmatory retesting of active compounds needs a few milligrams more of sample. The material requirements of *in vitro* assays understandably increase with the number of tests carried out—for example, to determine selectivity—but they are rarely burdensome to the responsible chemists. Often the first run-through of a short synthesis provides enough material for extensive biological screening in different therapeutic areas. *In vitro* assays demand smaller sample weights than do *in vivo* tests, a welcome advantage.

In the latter tests, however, oral screening doses sometimes range as high as 100 milligrams per kilogram of animal body weight. Such doses help toxicologists ascertain safety margins and help pharmacologists find proof of concept. If the disease model also uses massive animals or multiple doses, then resynthesis often becomes necessary. So important is resynthesis that Monsanto and Wyeth-Ayerst devote several chemists' efforts to this work. Similarly, Pfizer operates a synthetic services group charged with characterizing biologically active compounds and preparing them on a suitable scale. Making the required quantity of the drug can demand considerable labor, as the accompanying example portrays.

Resynthesis: How Much and How Soon?

Suppose that a pharmacologist proposes to give a 100-milligram-per-kilogram dose to each of three 25-kilogram animals. If little or none of the drug remains in stock, the duty of supplying more falls to one of the chemists, who embarks on resynthesis. In planning this work, the chemist considers how much the pharmacologist needs and how soon.

project lifetime:
- therapeutic hypothesis validated in live-animal model of human disease of interest
- lead identified by screening file compounds with a mechanism-based in-vitro assay
- lead compound proved active and selective in a battery of functional assays
- synthesis of lead-compound analogs begun
- high in-vitro potency found associated with a novel structural class
- composition-of-matter patents drafted and sought
- in-vivo biological profiling of potent compounds started
- re-synthesis of potent compounds commenced
- diastereomers chemically separated and biologically tested
- promising racemates resolved or their enantiomers independently prepared
- stereospecific biological activities determined with separate enantiomers
- U.S. and other composition-of-matter patents issued
- single-enantiomer clinical candidates proposed, and their pharmacological and pharmacokinetic profiles compared and contrasted
- clinical candidate selected, and Investigational New Drug status sought
- biological and chemical discovery efforts shifted to optimizing a structurally different lead compound as a back-up

Figure 2-3 Stages and idealized timing of preclinical biological and chemical discovery

The experiment requires 7.5 grams of drug, which requires a substantial synthetic effort if the arithmetic demon besets the work, as usually it does. An unoptimized six-stage synthesis might proceed in 50% yield per step. Making 7.5 grams of the drug demands 500 grams of the starting material, assuming equal molecular weights for these two compounds and five intermediate ones. Ninety-eight percent of this mass becomes waste, yet often there is no need to improve the synthesis. Because the outcome of biological studies with the final product is unpredictable, the inefficiency of a synthesis hardly justifies improvements during preclinical development. Preventing losses en route is less important than delivering material to the front. So, there is always time pressure to provide the required amount of experimental drug.

No prudent chemist would begin the labor with as little as 500 grams. Catastrophe might reduce the product weight to less than the amount needed. This embarrassing shortfall would delay the biologist's work, spotlight the chemist's, and require the latter to begin again. To carry out the resynthesis twice wastes time. A thoughtful chemist increases the scale, perhaps doubling it to 1 kilogram of starting material, despite the need to employ dilute solutions in large flasks. Because low throughput typifies the reactions run by discovery chemists, solution-phase synthesis entails concentrations as low as 10 milliliters per gram of starting material. Half empty to permit refluxing, the flask used in the first reaction envisioned here occupies 20 liters. Maneuvering such flasks and dealing with the contents can be imperative in meeting the demands of preclinical biological development.

Chemical work on a multigram scale offers advantages. For the biologists, it offers unlooked-for opportunities to reproduce single-dose experiments and to carry out rising-dose trials, should they prove desirable. It prevents crippling material losses, so permitting the chemist to complete the work.

To warrant synthesis, the potent compound invoked here need not be slated for human trials. It might only offer biologists a chance to confirm their working hypothesis. Alternatively, the compound might serve as a standard routinely useful in calibrating their assays. The job of procuring and replenishing it goes to chemical discovery groups, which may have to make multigram quantities of the substance. Before beginning any synthesis, however, they seek other means of obtaining the needed supply.

The assigned chemist ascertains if fine-chemical vendors sell the compound, inquiring about the quantity stocked and the schedule of future shipments. Certain vendors, which employ organic chemists, make a business of supplying standard compounds and emergent drugs to pharmacologists. Academic laboratories of organic chemistry can occasionally provide a standard compound, perhaps as a synthetic sample of a natural product that plays a pathophysiological role in a human disease.

Also the chemist looks into a custom synthesis done extramurally. Many chemical companies bid to serve the pharmaceutical and other chemical industries in this way. They offer to prepare a fixed quantity of a given compound, deliver the product on an agreed date, and keep the matter confidential, all for a predetermined fee.

The chemist tasked with procuring the standard compound asks about getting it made in-house. Depending on the division of responsibilities made within a particular pharmaceutical company, the preparation might be carried out by a kilo laboratory serving discovery researchers. A developmental chemical researcher might also prepare the compound. Finally, the discovery chemist responsible for the project may do so.

Identifying and synthesizing metabolites. An organic chemist employed in a drug metabolism laboratory, or seconded to one, identifies metabolite(s) of experimental medicines advancing to clinical trials. Identification is a prerequisite to synthesizing the metabolite and testing it for the desired biological activity or for an undesirable toxicity. To appreciate these practices requires an understanding of the importance of drug metabolites.

A metabolite rather than the parent compound may be largely responsible for the desired therapeutic effect. Far-reaching economic consequences can attend this happenstance if the active metabolite escapes patenting by the company that devises the parent drug. In such circumstances, a competitor that secures a valuable patent protecting the metabolite can threaten to seize the market.

Furthermore, a metabolite can be more toxic than the parent drug. If so, it is important to test and identify the metabolite. A knowledge of its structure allows discovery chemists to try making a safer drug. Molecular modification can prevent the biochemical reaction that changes the parent substance into a toxic metabolite. Before any metabolites can be tested and identified, however, they must be isolated. Testing a metabolite for toxicity or the desired biological activity may require a larger quantity of the substance than clinical trials or animal experiments can provide. So, it falls to synthetic organic chemists, perhaps with expertise in biotransformations, to prepare the substance.

Making substrates for radioimmunoassays. Organic chemists aid other development researchers. For example, it may be desirable to devise a radioimmunoassay to detect in various biological fluids the levels of a drug undergoing clinical development. This requires the developer to raise antibodies to a modified protein embodying a nearly complete fragment of the drug structure. The substrate chosen must offer a functional group that can bond to the unmodified protein selected. Since proteins contain primary amino groups, a carboxylic acid moiety in the fragment can usefully link the two structural units through an amide bond. A carboxyl group, however, may not already exist among the other compounds of an active series originally prepared for biological testing. If not, then chemists synthesize a suitably substituted new compound structurally analogous to the drug being developed.

Identifying degradants. Although pharmaceutical scientists bear primary responsibility for establishing the stability of drugs progressing to clinical trials, organic chemists also contribute to this aspect of drug development. They help elucidate the structures of *degradants*, which are impurities possibly arising during storage or transport of the active substance. These compounds can result from chemical reactions that the drug undergoes in ambient conditions: hydrolysis, photolysis, thermolysis, or aerial oxidation. A successful medicine must withstand heat, oxygen, water, and light for an adequately long shelf-life, and its packaging must offer protection from these chemical influences. A pharmaceutical scientist therefore tests stability by stressing the drug and then analyzing for the presence of degradants. Typically, he irradiates the drug with light from fluorescent lamps. These lamps also radiate some ultraviolet light, which can cause photochemical reactions of the drug.

A reactive drug leads to a known or novel photoproduct. Structure elucidation ordinarily begins with a comparison of high-performance liquid chromatograms. With any luck, a chromatogram of the photoproduct matches one of a known compound belonging to the series of analogs that includes the drug. Such a match can quickly lead to a satisfactory structural assignment when a sample of such a known substance is available for comparison. Sometimes, however, biological testing exhausts the supplies of likely degradants, so synthetic organic chemists replenish them. If no chromatographic comparison yields a match, then the pharmaceutical scientist studying the drug stability must provide enough of the degradant for the more extensive characterization that leads to structure elucidation.

Identification and synthesis of degradation products and metabolites are some of the other services offered by certain contract research organizations. Both start-up and integrated pharmaceutical companies avail themselves of such services. Some of these companies as well as other FIPCOs also carry on this work themselves.

Roles of Chemical Development Groups

Integrated pharmaceutical companies carry on chemical research in development as well as in discovery. The two groups, both employing organic chemists, have separate staffs and management and sometimes work in different locations to accomplish distinct goals. From its practitioners, successful chemical development projects demand scrutiny of chemical reactions and knowledge of mechanistic organic chemistry. They need talent and experience to devise problem-solving experiments and to execute them. Chemical development efforts represent applied science; they are market driven, perhaps even more than their counterparts in discovery research.

Only those syntheses merit development that offer patentable novelty as chemical processes and that furnish an important product. At the outset of development, and at its conclusion, the structure and properties of the final product are already known. It possesses the desired therapeutic effect in an animal model of the human disease and the potential to serve an existing market. Development efforts are *not* devoted to improving the biological profile of a clinical candidate by finding a successor or a back-up drug.

Finding a manufacturing synthesis. Chemical development projects have dual aims. First, if clinical trials succeed and the Food and Drug Administration approves launching, then development chemists must be ready with a tested manufacturing process. To devise and demonstrate such a process can represent a formidable task, if tens or thousands of tons of drug are needed as the so-called bulk active ingredient. Because of its large scale, a good manufacturing process draws on the talents of organic chemists, chemical engineers, and pharmacoeconomists, to name only a few kinds of experts.

Evaluating hazards. When a drug begins chemical and clinical development, the final choice of a manufacturing synthesis may lie far ahead. Nevertheless, the initial demand for large amounts of an experimental drug raises safety concerns that are satisfied early in chemical development. Phase I and II trials can easily require many kilos of the drug, which cannot be made to schedule on the small scale that satisfied the requirements of preclinical studies. Consequently, the task of making multikilo batches falls to the pilot plant rather than to a research or kilo-laboratory. Work in this plant can entail a 100-fold increase in scale from laboratory research. Scale-up profoundly affects heat transfer, and exotherms offer a likely source of runaway reactions. Such reactions sometimes endanger employees' lives or health, outrage the community, damage the plant, or delay progress. Development chemists therefore evaluate the hazards of all the reactions that compose the synthesis adopted or invented. With accelerated rate calorimetry, for example, they look for dangerously exothermic reactions; they also search for explosion and reagent hazards.

Making clinical supplies. The second aim of chemical development is to provide clinical supplies of the drug. It is a short-term goal more urgent than the first. Appreciating this urgency calls for understanding the discovery efforts to date in a given project and the human clinical trials to come. When a project crosses the interface to chemical development, biological and chemical discovery efforts have already cost 3 to 6 million dollars. Moreover, the company faces an outlay of several hundred million dollars before launching the medicine. Nonetheless, no one knows what concerns everyone: *can the drug candidate safely exert the desired therapeutic effect in humans?* No trial with humans has begun, however, so no answer is available. Nor will anyone learn whether the medicine will bring any therapeutic effect until the *second* clinical trial ends (see the accompanying example). "Safety first!" is the rule with which pharmaceutical companies and regulatory agencies work. The first clinical trial always probes safety but only occasionally does it test effectiveness. Often the second clinical trial is placebo controlled and double blinded. If so, then no one learns until the code is broken whether the drug is superior to the placebo, a result that can take months or a year or more to arrive.

Material Demands of a Clinical Dose–Response Study

Suppose that 60 adult male patients participate in a rising-dose trial of an experimental drug, calling for three equally populated groups. The members of each group weigh an average of 90 kg and receive the drug twice daily for 30 days. One group receives a 1-milligram-per-kilogram (mpk) dose per person, another takes 10 mpk, while the third group is given 100 mpk. Multiplication reveals that the three groups require 0.108, 1.08, and 10.8 kilograms each to complete the month-long trial, which demands a total of nearly 12 kilograms.

The chemists responsible for providing these 12 kilos consider using the unoptimized synthesis devised by discovery researchers. If the process chemists do operate it, they do so because time constraints prevent them from improving or circumventing it. It entails six chemical steps, we again suppose, each one proceeding in an average yield of 50%.

How much starting material do the chemical developers need? Bedeviled by a scant overall yield of 1.56%, the undeveloped synthesis requires 768 kilograms of the starting material to furnish 12 kilograms of the final drug product. The first step alone demands 7680 liters of solvent, if the discovery chemists' low-throughput procedure remains unchanged. A research or even a kilo laboratory containing 22-l glass flasks cannot practically accomplish chemical work on such a scale. The job calls for a pilot plant equipped with a 10,000-liter glass-lined steel reactor. Further reflection confirms what an awesome, somber enterprise this chemical development is. Seven thousand seven hundred liters of a hydrocarbon solvent contains more than enough energy to drive a compact automobile across the North American continent—*20 times.*

For the initial human trials to begin as soon as clinical physicians enroll the needed volunteers, development chemists provide the experimental drug quickly. The initial amount the chemists make allows the clinicians to demonstrate its safety. Later the process chemists make larger amounts ready for other trials, which establish that the drug brings a therapeutic effect. Additional studies help clinicians fix the dose and dosing schedule and learn the therapeutic dose range. A rising-dose study probes this range. (The accompanying example shows how much drug can be needed for a hypothetical but realistic trial.) Altogether, clinical studies require surprisingly large quantities, especially when a drug enters Phase II testing, which escalates material demands. At this stage of development, for example, Pharmacia and Upjohn chemists made *tons* of their reverse-transcriptase inhibitor atevirdine, which treats autoimmune diseases (AIDS).

The outset of development sees demands for high-purity clinical supplies that are available only through the research synthesis. These demands may conflict with the need to find a manufacturing synthesis. A suitable manufacturing synthesis ordinarily represents the result of time-consuming process research, not a prerequisite that is met before chemical development begins. Consequently, developers must strike a balance between making urgently needed supplies by the laboratory synthesis and devising a suitable new route. Often the laboratory synthesis, effective as it was, cannot meet one or more requirements of a manufacturing process. Furthermore, to compare and contrast two or more manufacturing processes before making the final choice of route can be necessary or desirable.

Adopting drug specifications. With the final synthesis not yet decided, developed, or even devised, the quality of drug used in clinical trials is crucial. It can determine the specifications that the drug must later meet as it emerges from a manufacturing plant. Regulatory agencies link permission to market a drug partly to the nontoxicity of the manufacturing impurities it inevitably contains. Although clinical trials show that the drug and its trace contaminants are safe, the particular synthesis used to prepare clinical supplies determines the contaminants. In a sense, they are tied to it. This method, however, may not represent the final manufacturing synthesis, which is the ultimate arbiter of the drug impurities.

A drug synthesis, like any other, leaves trace impurities in the final product. When humans are to take the medicine in clinical trials, however, each impurity must be individually identified. This task of elucidating contaminant structures and measuring their concentrations falls to development chemists. Isolating trace impurities calls for elution chromatography of mother liquors, sometimes on impressive scales, while measuring concentrations requires analytical chemists to develop liquid chromatography assays. Development chemists repeat their profiling work whenever the synthesis changes significantly, especially in the last step. If several syntheses are developed to afford the best choice of the final manufacturing process, then the chemists investigate the impurities that come from each synthesis and accumulate in the final product.

When human trials begin, the contaminants of a clinically safe drug total no more than 1–2%, and their number is always few. In general, the number and nature of impurities in an experimental drug often change as development advances. It may be preferable to allow a greater content and number of nontoxic impurities during early development. This tolerance allows chemical developers to tighten drug standards later, until the total impurity content declines 10- to 20-fold, reaching 0.1% or less. Such a drop becomes feasible as process chemists refine the manufacturing synthesis that they finally adopt.

ORGANIZATION OF DISCOVERY AND DEVELOPMENT CHEMICAL RESEARCH

Company goals pervade corporate research, which unites creativity with practicality. Pharmaceutical companies do not lack for worthy projects, but entertain so many that they deliberately assign priorities. Consequently, they hire entry-level chemical researchers to begin or continue already-vetted projects. They do not initially expect newly hired researchers to help devise new undertakings. And for good reason, because even postdoctoral expertise in synthetic organic or medicinal chemistry avails little in judging whether proposed new ventures meet criteria important to pharmaceutical companies. Such standards concern the therapeutic area to enter and the economics of entering it; the specific disease to treat, prevent, or diagnose; and the particular biochemical approach to follow. These far-reaching goals dictate strict organization of discovery and development research, which leads to closely focused efforts. Newcomers to the industry can consequently expect to do the work for which their academic degrees qualify them. However, as they gain relevant knowledge and experience, they earn the chances to propose research and to work independently and creatively.

> *Join the pharmaceutical industry if you want to participate in structured chemical research. But take a job knowing how the organization of drug discovery and development affects chemists' careers. An industrial position liberates senior researchers from the fund raising they would undertake in academe but entails a sacrifice of freedom uncommon in university science.*

Work Assignments

Staffing needs of existing projects govern personnel assignments. Entry-level process chemists, for example, do not choose to work on making clinical supplies of a cardiovascular drug instead of a dermatological medicine. Nor is the choice theirs to help develop a manufacturing process for a clinically safe and effective gastrointestinal compound. They instead do jobs assigned according to corporate needs and chemists' versatility and aided by their own experience.

Newcomers accepting assignments in discovery research similarly exert little or no influence over the therapeutic fields they are sent to till. This situation arises because many or most therapeutic areas require no specialized chemical knowledge of, say, macrocyclic compounds, but are well served by chemists' broadly applicable skills. Chemists' abilities to make, purify, and identify organic compounds assist either discovery or development groups. They lead to structurally diverse new compounds for the many therapeutic areas where new molecules are sought as drugs. These skills are also prerequisite to work in chemical development, where they help prepare clinical supplies and improve existing syntheses or devise new ones.

Transferability

Within a pharmaceutical company, a certain versatility distinguishes organic chemists from biologists. The biologists' skills confine them to therapeutic areas determined by their academic or on-the-job training. An immunologist, for example, trained to devise cell-based assays of proteins, cannot easily do a cardiovascular pharmacologist's work. The latter scientist, in contrast, measures drug effects in live-animal models of cardiovascular disease but lacks the former's training.

> *In contrast, employed chemists enjoy a certain career mobility that their colleagues miss. They can move from one therapeutic area to another or transfer between development and discovery research. Chemists' career shifts can occur within one company or between companies, if they take place early enough in a working lifetime and meet other criteria. They can also include transitions from a pharmaceutical company to a law firm or government agency.*

The Project Team as an Organizing Principle

The introduction of a new drug to medical practice culminates a vast project lasting thousands of human-years and drawing on hundreds of employees, contractors, and patients. Their contributions, like those of any other workforce with a common purpose, benefits from organization, which the project team provides. Among researchers in a large FIPCO, (1) this commonplace approach can involve both discovery and development scientists, as well as other professionals, in a single interdisciplinary team. Alternatively, (2) it can structure the efforts of either discovery or development personnel. In this case too, a project team includes not merely chemists but also personnel from other sciences and from other disciplines like patent law.

Consider the first alternative. The members of a project team comprehending both discovery and development researchers range from biologists, chemists, and attorneys; proceed through drug metabolism experts, toxicologists, pharmaceutical scientists, and clinicians; and include chemical engineers and regulatory affairs specialists. One lawyer, or one group of them, composes all needed applications for composition-of-matter, process, and formulations patents. A single group of chemists begins the work by making new compounds for biological testing, proceeds with resynthesis and scale-up, and ends by demonstrating a manufacturing process. Neither the attorneys nor the chemists have any permanent assignment to any one therapeutic area, but switch as needed from one to another. A chemist who attends the meetings and reads the reports of such a team gains exposure to all facets of chemical discovery and development research and to other aspects of pharmaceutical operations. Team responsibilities may begin with the choice of a therapeutic area or approach, continue through an Investigational New Drug filing, and conclude with New Drug Approval. So broad and important are the duties of such a team that it includes one or more directors and a vice-president.

In the second alternative, project team members comprise discovery or development personnel, but not both. The staff of a discovery team remains interdisciplinary but represents fewer scientific and professional occupations. Their team work begins similarly but largely ends with the selection of a drug for clinical trials, when a development team takes over. Discovery team members receive lasting assignments to therapeutic areas, while development team personnel do not. This organizational scheme leads to the greater specialization that characterizes medicinal and process chemists.

The project-team approach features a famous success. In 1972, a team led by organic chemist *Robert Ganellin* and pharmacologist *James Black* discovered the antiulcer, H_2-antagonist cimetidine at Smith Kline and French (now SmithKline Beecham). Their scientific and commercial success culminated 15 years of research. During this time, some 20 biologists and 50 chemists tested and made 700 compounds, the first 200 of them in 4 years. (By today's combinatorial chemistry standards, a set of 700 new compounds is minute.) Two of these compounds failed one or more clinical studies before cimetidine—the *second* back-up drug—passed human trials. The American Chemical Society and the Royal Society of Chemistry (UK) designated this achievement an International Historic Chemical Landmark. "The event," said *Peter J. Machin,* SmithKline vice-president for medicinal chemistry, speaking at the 1997 ceremony, "provides an opportunity to sing the praises of chemistry."

Function and Matrix Organization

Two other organizational principles characterize chemical discovery and development research in a large FIPCO. One of them is the functional organization depicted in Figure 2-4. The other principle is a matrix organization imposed over project teams; it is detailed in Chapter 6.

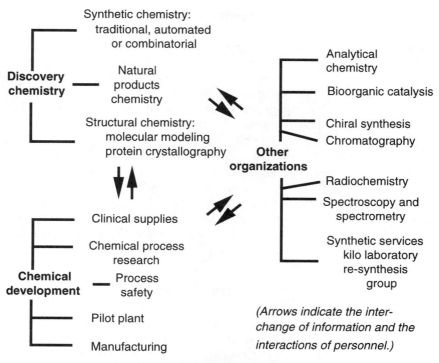

Figure 2-4 Functional organization of chemical discovery and development research within a FIPCO

Consolidation and Fragmentation in the Pharmaceutical Industry

If recent history influences the near future, industry observers may expect two trends to continue. Some large companies will merge with one another, and they and other established firms will outsource to procure chemical discovery and development services. Mergers, temporary alliances, licensing arrangements, and contract research originate in the economics of the pharmaceutical businesses. This origin fosters outsourcing to specialized companies, leads to temporary employment, and indirectly creates start-up firms. Start-ups and outsourcing, which is defined below, are important to new graduates seeking work because they influence the numbers and locations of available jobs. Their effects on employment opportunities are detailed in Chapter 3.

Economic Forces

Consolidation by mergers and fragmentation by outsourcing and starting up respond to forces increasing the costs of research and decreasing the returns on investment. In FIPCOs, these forces include widespread and stringent regulations to assure drug safety and environmental protection; numerous and challenging clinical trials to demonstrate drug efficacy; and enlarged outlays to meet contemporary requirements for laboratory safety and to hire and pay clinicians. Other factors comprise price concessions to reimbursers; concentration in the exacting therapeutic areas that offer lucrative opportunities; and vigorous competition from generic drugs, but shortening of useful patent lifetime by prolonged clinical development. Competition and shortening close the window of greatest profitability. Financial pressures find relief in mergers and restructurings, which bring economies of scale and opportunities for synergism to the new company, but layoffs to some employees.

Outsourcing Science

Pharmaceutical companies need not be self-sufficient in all their operations all the time, so they obtain certain goods and services from external suppliers, in a practice known as outsourcing. These goods and services do not merely represent, say, office supplies and corporate advertising; remarkably, they also include scientific research. External suppliers furnish to some FIPCOs the results of chemical and biological drug discovery efforts as well as chemical, pharmaceutical, and clinical development work. Practiced for reasons of opportunity, time, and money, outsourcing lets companies take avenues otherwise closed by a lack of in-house personnel, equipment, or skill. It allows them to exploit new technology without having to finance its development, thereby offering a valuable time advantage. Quickly and widely embraced, new technology like combinatorial chemistry promises more new medicines faster than older methods provide. Large FIPCOs seek more clinical candidates each year because so few new chemical entities reach the market. Procuring more drug candidates entails more risks, which can be shared among small companies through the outsourcing that large firms practice. Smaller companies offer lower costs because they incur fewer and lesser overhead expenses, including salaries.

Start-Up Companies

The lesser overhead costs of smaller companies partly explain the blizzard of pharmaceutical start-ups evident across the continent. These costs make it possible for them to seek drugs that would bring $25–100 million in annual sales, too little to appeal to the largest drug houses. Larger firms' higher fixed expenses would unacceptably reduce expected profits. This situation gives the advantage to smaller companies, which a better selling drug would not do. In that case, a small start-up might find itself competing with the vast sales force and advertising budget of a large company.

SOURCES AND SUPPLEMENTS

Anonymous; "The process of drug development: A course," http://www.awod.com/netsci/Courseware/Drugs/Intro/top.html.

Brown, P.; "A difficult year ahead," *Scrip Magazine,* January, 72–74 (1997).

Campbell, P. (ed.); "Intelligent Drug Design," Supplement to *Nature* **384**, 1–26 (1996).

Cavalla, D.; "Modern Strategy for Preclinical Pharmaceutical R & D: Towards the Virtual Research Company," Wiley, Chichester, 1997.

Cox, B.; "Strategies for drug discovery: Structuring serendipity," *The Pharmaceutical Journal* **243**, 329–337 (1989).

de Stevens, G.; "Serendipity and structured research in drug discovery," *Progress in Drug Research* **30**, 189–203 (1986).

Fusfeld, H. I.; "Industry's Future: Changing Patterns of Industrial Research," American Chemical Society, Washington, DC, 1994.

Hamnier, C. E. (ed.); "Drug Development," 2nd ed., CRC Press, Boca Raton, 1990.

Malick, J. B., and Williams, M. (eds.); "Drug Discovery and Development," The Humana Press, Clifton, NJ, 1987.

Machin, P. J., quoted in Freemantle, M.; "Milestone in drug discovery," *Chemical and Engineering News,* January 5, 31–32 (1998).

Spilker, B., and Cuatrecasas, P.; "Inside the Drug Industry," Prous Science, Barcelona, 1990.

Spilker, B.; "Multinational Pharmaceutical Companies: Principles and Practices," 2nd ed., Raven, New York, 1994.

Spilker, B.; "Phases of clinical trials and phases of drug development," *Drug News & Perspectives* **9**(10), 601-606 (1996).

Wierenga, D. E., and Eaton, C. R.; "The drug development and approval process," http://www.allp.com/drug_dev.htm.

chapter

3

JOBS IN THE DRUG INDUSTRY

> *"It is important to the future of this company and to the pharmaceutical industry in general that institutions of higher learning continue to produce the well trained and high-quality synthetic organic chemists needed to ensure a steady stream of new and innovative medicines."*
> —*Dr. Peter S. Ringrose,* President of the Bristol-Myers Squibb Pharmaceutical Research Institute

INTRODUCTION

This chapter summarizes organic chemists' jobs in the drug industry. Most of these scientists prepare, isolate, purify, characterize, and identify organic compounds. In a fully integrated pharmaceutical company (FIPCO), their skills serve several internal organizations, including radiochemistry or kilolaboratories and process or medicinal chemistry research groups. A FIPCO, because it discovers, develops, registers, and manufactures drugs, offers the greatest variety of jobs to new chemistry graduates. Baccalaureate chemists, whose laboratory training represents a basic vocational qualification, enjoy the widest spectrum of choices.

Other companies, although they engage in fewer than the six businesses of a FIPCO, also hire organic chemists. So do government agencies, patent law firms, and nonprofit research institutions. Familiarity with types of employers and the work they do can help a chemist seeking an entry-level job. It reveals whether the job sought is one done in a particular firm and can lend direction to the search. This appreciation enables an applicant to seek work deliberately, make choices intelligently, and impress interviewers favorably. Foreknowledge, which no one need flaunt, demonstrates willingness to prepare, which is crucial in doing chemical research.

> *Indeed, an applicant who mentions during an interview the variety of jobs that organic chemists do in the pharmaceutical industry distinguishes herself.*

Effects of Outsourcing on Employment Opportunities

To maximize the chance of finding employment, a determined job seeker learns the kinds of institutions that employ organic chemists within the medical industry. Such organizations, if they are not FIPCOs, can be hard to identify partly because of outsourcing.

Outsourcing effectively transfers some organic chemists' jobs from the largest pharmaceutical companies to small start-up firms or service companies within the industry. For example, to exploit developments like combinatorial chemistry a FIPCO would, until recently, have hired chemists to apply the method to drug discovery. Now, however, it faces other choices. It employs a service company specializing in combinatorial chemistry *and* hires chemists to exploit this technical development. Alternatively, it hires few chemists if any for this purpose, but engages or buys the specialized company to provide both the technology and the compounds it yields. Although the small numbers of large FIPCOs continue to employ hundreds of chemists each, large numbers of smaller firms give work to far fewer scientists. Outsourcing tends to reduce the opportunities for employment in any one large company, leading the job seeker to approach more employers.

To identify companies employing and hiring chemists is one of the tasks faced by a new graduate seeking work. The largest FIPCOs remain easy to identify, partly because their products reach consumers as over-the-counter or prescription drugs labeled with their makers' names (Table 3-1). Individual firms advertise on television and radio and in magazines and newspapers. Any patient, pharmacist, physician, watcher, or reader can therefore name many firms employing chemists.

Other employers are not as obvious. Today, large FIPCOs contract some drug discovery and development research to start-up and other service firms. These companies therefore employ a certain number of chemists who in years past would have worked for easily identifiable, large FIPCOs. Now, however, vacancies arise in small, lesser known start-up and service firms. Newcomers to the industry, the former have no products to advertise. The

Table 3-1 1998 Rankings of Top-Ten Pharmaceutical Companies[a,b]

Company	Sales[b,c]	Sales Growth[d]	Profitability	R & D Investment[e]
Merck & Co.	1	4	8	5
Novartis	6	2	10	2
Pfizer	7	9	3	4
AstraZeneca[f]	4		6, 5	
Bristol-Myers Squibb	5	7	7	
Glaxo Wellcome	3		1	1
Johnson & Johnson	9		2	8
Lilly	10	8		6
Roche		6		3
Schering-Plough		3	4	
SmithKline Beecham			9	9
Abbott		10		
American Home Products	8			
Aventis	2			
Bayer		5		
Hoechst				7
Pharmacia & Upjohn				10
Warner-Lambert		1		

[a]The number of rankings for each company determines its location in the varying gray bands. For example, the first three companies earned four rankings each. To take another instance, Roche, Schering-Plough and SmithKline Beecham won two rankings each and so are grouped in the same gray band as one another.
[b]Latest complete data; source: *Scrip Magazine,* January, 1999, pp. 39–40. Each listed company ranks among the top 10 in one or more of the four tabulated categories.
[c]*Sales* are referred to in U.S. dollars.
[d]Represents the percentage increase of pharmaceutical-sector sales since 1997.
[e]Research and development; *investment* is referred to in U.S. dollars.
[f]Before announcing a merger in 1998, Astra ranked 6th in profitability while Zeneca came 5th.

latter sell their services only to pharmaceutical companies, not directly or indirectly to consumers. This situation brings them little name recognition.

Difficulties aside, recent business changes make outsourcing frequent and start-up firms widespread, especially in California and Massachusetts. Outsourcers and start-ups are here to stay and deserve consideration as potential employers because they bring opportunities for permanent positions.

Prerequisites and Experience

Chemists' posts in the pharmaceutical industry usually require a baccalaureate degree, and some of them demand a graduate degree or its equivalent in work experience. Generally, the lesser is an applicant's academic degree, then the wider is the variety of entry-level jobs on offer. Inexperienced baccalaureate chemists enjoy the largest number of initial choices within a FIPCO because they lack specialized skills directing them to specific careers. From the outset, however, these chemists earn lower salaries than those with experience or with master's and doctor's degrees. The latter's graduate education adds value to the training that B.S. scientists receive.

> *Nearly all the posts open to inexperienced new graduates demand experimental work for which applicants develop and voice enthusiasm during interviews.*

Within limits, the greatest consideration is the applicant's highest degree; the smallest factor is the work experience that employers expect. For example, entry-level offers often go to researchers holding Ph.D. degrees but lacking industrial experience or postdoctoral training. The advanced training received by Ph.D. and M.S. scientists substitutes for time spent on the job.

Chemists with bachelor's degrees lacking research experience but seeking employment can meet delay in finding work. To speed the search for work calls for baccalaureate chemists judiciously to select the jobs they seek and the employers they approach. Not all employers are willing or able to hire and train aspiring researchers. Large FIPCOs do hire B.S. chemists lacking research experience, partly because they employ some group leaders who agree to accept the tasks and risks of mentorship. Those B.S. chemists who have undergraduate research experience or worked as summer interns in a pharmaceutical or another chemical company stand the best chance of winning interviews.

Companies wanting experienced candidates typically require more years at work from bachelor's degree chemists than from master's degree holders: 2 to 4 years, for example, versus 0 to 2, respectively. Incidentally, this practice explains some of the shorthand found in advertisements, where the abbreviation "B.S./M.S." signals opportunity for holders of either degree, not indecision regarding qualifications.

Employers similarly prefer hiring M.S. chemists who wrote dissertations to employing M.S. scientists who presented none. A suitable thesis demonstrates an achievement made possible by experimental techniques that are essential to many drug house operations. Certainly it may review parts of the chemical literature, yet a valued dissertation does not confine itself to literary efforts. It is experimentalists' training that makes chemists employable because a need for their skills pervades discovery and development efforts from onset to outcome.

Technological developments create new requirements for many positions, so qualifications change with time, becoming more stringent. Certain work once done by B.S. chemists now requires advanced degrees in subjects other than chemistry as well as experience or special training. Years ago, a baccalaureate degree sufficed to qualify a chemist for an industrial research librarian's position. Such an entry-level job may now call for a master's degree in library science, but still require a B.S. in chemistry. It may also necessitate familiarity with the special computer programs and databases that the pharmaceutical industry offers its researchers, patent attorneys, and clinicians. Examples include STN, Derwent, and MEDLINE.

Over- and Underqualifications

Graduate degrees and research experience can represent overqualifications for some positions. For example, few FIPCOs (if any) advertise to hire Ph.D. chemists as new-compound registrars (see text below). They seek B.S. or M.S. chemists for these posts and train them intramurally.

Doctoral chemists are rarely sought or employed for the work that B.S. and M.S. chemists do, even though their qualifications are more than ample. Employers fear that dissatisfaction will eventuate or that abuse or culpability will result. To conceal a doctor's degree and present instead a genuine master's or bachelor's degree misrepresents one's attainments. Such deceit establishes grounds for dismissal.

For other positions, even a Ph.D. and postdoctoral training cannot substitute for industrial research experience. A chemist needs experience to find work in a patent or regulatory affairs department. Patent coordinators (see text below), who are not lawyers but chemists, learn the basics of their future liaison positions as researchers whose inventions are patented. Their experience in assisting attorneys qualifies them to devote all their efforts to such work.

Research Classifications

Most organic chemists' entry-level jobs entail chemical discovery or development research. Other positions fall into a service-group category, which covers jobs in chemical development within one company, chemical discovery in a second firm, and in both development and discovery within a third drug house. Different firms organize the same work in dissimilar ways, so a degree of arbitrariness attends this or any allocation of chemists' jobs to discovery or development categories. What is more important than classification is knowing what jobs chemists do and what the work entails; which jobs are entry-level positions; what experience and academic degrees are prerequisite; and which positions offer permanent or temporary work. Although temporary posts are discussed in the section of this chapter devoted to chemical development, such jobs are also found in discovery research in medicinal chemistry.

DISCOVERY RESEARCH

Medicinal Chemistry

Discovery research is one of the two largest sources of entry-level jobs for synthetic and medicinal chemists, most of whom work in medicinal chemistry. Hired at all degree levels, often as inexperienced new graduates, they make new compounds for biological evaluation and do so by traditional or new methods. These latter methods, which some chemists are hired to invent, develop, and implement, include combinatorial techniques and solid-state and automated synthesis. Responsibilities for automation can include devising, modifying, and maintaining equipment used for carrying out reactions, and adapting known organic reactions to solid-state conditions. In some cases they involve programming. At present, groups employing combinatorial chemists are growing in number and size, although no slackening of demand for conventionally trained researchers is evident. Often the former chemists work to identify a few lead compounds among hundreds or thousands of compounds produced combinatorially. The latter scientists optimize lead structures by making analogous compounds with superior properties, as Chapter 5 describes.

Natural Products

Originated more than a century ago, the quest for new drugs still leads chemists to examine plant sources of organic compounds. Found in willow bark, salicylic acid exemplifies an early success as the historical precursor of aspirin. In the first half of the 20th century, microbial fermentations brought forth a cornucopia of valuable new drugs including penicillin and tetracycline antibiotics. Synthesis transformed naturally occurring raw materials into the first steroidal anti-inflammatory drugs by the 1960s. Semisynthetic steroids still serve humankind as valuable medicines, for example, as therapy for asthmatics. The past 25 years offered marine organisms as the origin of novel, bioactive substances in abundance. More recent times saw a renaissance in natural products research in the pharmaceutical industry. For example, Merck negotiated an agreement with the government of Costa Rica, allowing the firm to prospect for drugs in that country's tropical rain forest.

Some pharmaceutical companies employ entire departments of B.S., M.S., and Ph.D. organic chemists to screen natural products for novel, pharmacologically active compounds. These natural products may originate in plants, fermentation broths, or animal matter. Scientists make the extractions and prepare the extracts for screening. Working in these departments, they also isolate, purify, characterize, and identify the compounds responsible for the activity. Biological assays direct the work, so that inactive substances squander no efforts or resources.

Recent advances in automation raised the efficiency of biological assays, leading to high-throughput natural-product screens. They afford more frequent findings of biological activity *in vitro* (*hits*) than were realized hitherto. Such developments in biology and biochemistry lead to increasing specialization. Consequently, some natural products chemists devote most of their efforts to isolating active compounds and elucidating chemical structures.

To work in a natural products department, however, may still require some chemists to practice organic synthesis. It has always been advantageous to modify the structures of naturally occurring compounds. Structural changes can increase the potency of semisynthetic derivatives or impart suitable physical properties like water solubility or membrane permeability.

Structural Chemistry

Discovery research offers entry-level posts to computational chemists and protein crystallographers, who usually hold Ph.D. degrees. Some of these jobs go to bachelor's and master's degree holders with 4 and 2 years' experience, respectively. Computational chemists' jobs are among the few research positions that entail no experimental work. Incumbents labor to rationalize facts linking chemical structures to biological activities. Their efforts might explain how a new drug binds to a known receptor, so eliciting a biological response. Computational chemists construct three-dimensional models of hypothetical or actual receptors and enzymes, with powerful computer hardware and software. Ideally their work lets them predict what small molecules would bind within protein pockets, suppress or induce biological activity, and lead to new drugs by optimization.

Protein crystallographers have greater responsibility for experimental work than computational chemists. The former scientists, for example, cocrystallize would-be drugs and the receptors to which they bind. Operating X-ray diffractometers and applying their knowledge of crystallography, they elucidate the three-dimensional structures of the drug–receptor complexes. A detailed, spatial knowledge of these elaborate structures, if it arrives early in a project lifetime, can offer welcome guidance to synthetic chemists' work.

Drug Metabolism

Chemists specializing in a remarkable variety of subdisciplines find employment in drug metabolism groups. Want ads invite applications from medicinal, organic, physical organic, and analytical chemists, and from biochemists and even computational chemists. Some employers seek chemists experienced in the work but others do not, and companies hire chemists at all degree levels.

In an exploratory phase of the work, all these scientists study biotransformations of several candidate drugs for a single project. They characterize the resulting metabolites in biological fluids, usually employing liquid chromatography in tandem with mass spectrometry. These and other efforts help pinpoint the one drug finally chosen for clinical trials by determining, comparing, and contrasting the oral bioavailability of candidate compounds. The jobs include dosing animals and collecting specimens, maintaining chromatographic instruments and mass spectrometers, or calculating physiochemical properties related to drug absorption and membrane permeability.

The developmental phase of drug metabolism entails studying the single compound slated for clinical trials in a given project and features analytical and structural aspects. The researchers devise chromatographic and spectrometric assays to detect and quantitate the drug and its metabolites in biological fluids. They validate the analytical methods developed and profile the metabolites formed *in vitro* and *in vivo*. *In vitro* experiments may include incubation of the drug with cultured hepatocytes, precisely cut liver slices, or liver microsomes. *In vivo* experiments involve administering the drug to several mammalian species, collecting specimens from them, and isolating and profiling metabolites across species. The kinds of mammals studied might include the mouse, rat, guinea pig, dog, monkey, and human. Thorough profiling requires measuring the relative proportions of biotransformation products in each species, often as a function of time after dosing. In the number and amounts of metabolites detected, scientists account for the quantity of labeled drug originally administered. They elucidate metabolite structures as a preliminary to testing the metabolites for the desired biological activity or any unwanted toxicity. A knowledge of human and animal metabolites allows toxicologists to choose the mammal most resembling the human one in terms of drug metabolism. This crucial choice assures that long-term toxicity assays begin with the best animal species for the studies.

During this development phase, researchers follow good laboratory practices. They keep written, witnessed records, which pertain to important Investigational New Drug applications and New Drug Approvals. Their notes are subject to inspection by the Food and Drug Administration.

Radiochemistry

Organic chemists knowing synthetic methods work in radiochemistry groups. These groups offer entry-level positions to inexperienced scientists holding B.S., M.S., and Ph.D. degrees. Newly hired chemists receive in-house training in making, handling, and disposing of radioactive organic and inorganic compounds. They can expect to incorporate radioisotopes into drugs slated for clinical trials. Radiolabeled drugs find use in metabolic profiling, so the tasks of making them are urgent and visible ones that must be concluded with safety for all concerned.

Radiochemists also prepare compounds containing stable isotopes like deuterium, carbon-13, and nitrogen-15. Drugs labeled with stable isotopes create no waste-disposal problem, so they are advantageous to metabolism studies. An organic chemist familiar with multistep synthesis is an asset to a radiochemistry group.

Many *in vitro* biological assays depend on the binding of radiolabeled ligands to receptors. Experimental compounds with greater or lesser binding affinity displace the radioligands in equilibria. The quantities of displaced ligands are readily measurable thanks to their radioactivity, and the measured values characterize the affinities of other compounds for the same receptors. Radiolabeled ligands do not occur naturally and so must be synthesized. Although the need for one ligand ends with the success or failure of the project using it, a new endeavor creates demand for other radioligands. This need recurs periodically, and radiochemists fill it by making labeled ligands. In a sense, their jobs can span the whole range of a discovery project, from establishing primary assays to profiling human metabolites.

SERVICE GROUPS

Analytical Chemistry

An integrated pharmaceutical company employs chemists trained in analytical techniques and deploys them throughout its research and development organization. So important is their work that the FDA licenses certain analytical chemists employed in pharmaceutical companies. Depending on corporate size, individual analysts or groups of them work in the manufacturing division, the pharmaceutical or chemical development organizations, or in chemical discovery research. In some companies, a single group of analysts serves all the foregoing units. Analytical chemists at all degree levels find entry-level employment in FIPCOs, not always as permanent employees, however. Some companies hire temporary workers (see text below).

Permanent jobs go to new graduates holding B.S. degrees who apply standard analytical techniques to organic compounds. Commonly needed techniques include elemental analysis and high-performance liquid chromatography (HPLC), optical rotation, optical rotatory dispersion, and circular dichroism. Analysts also need familiarity with infrared, ultraviolet, and carbon and proton nuclear magnetic resonance (NMR) spectroscopy and with the proliferating variants of mass spectrometry. Mass spectrometric or NMR spectroscopic measurements made in tandem with liquid chromatography figure in the work of modern analytical chemistry groups. So does automated, high-throughput mass spectrometry and HPLC. Familiarity with computers is essential because software operates modern analytical instruments. It also records and stores their measurements.

Measuring purity and establishing identity, analytical chemists in the pharmaceutical industry mostly study organic compounds. Their work serves chemical discovery and development research as well as pharmaceutical development groups responsible for drug formulation. These scientists need the strong backgrounds in practical and theoretical organic chemistry that graduate training provides. Consequently, FIPCOs hire graduates with master's and doctor's degrees who earned their highest degrees in this subject. They also employ chemists with advanced degrees in physical organic and physical chemistry.

Compound Registration

To transform a biological concept into a single innovative drug can require thousands of new organic compounds in one therapeutic project alone. A small or medium pharmaceutical company carries on several projects during 1 year, so it sees several thousand new compounds individually synthesized by traditional solution-phase reactions. With the advent of combinatorial chemistry, annual output of novel substances rises to tens or hundreds of thousands. Suitable amounts of these compounds must promptly reach the appropriate biologists. They test them for pharmacological activities and report their results. A large-enough pharmaceutical company employs a staff of registrars and distributors to carry out this interchange of samples and findings. These personnel give valuable help to both the biologists and chemists, partly by allowing the researchers to devote their efforts to science rather than administration.

Various employers designate registrars differently: as information analysts or specialists or as compound archivists, curators, or librarians. These employees register the structures of the compounds submitted and the makers' names, linking the submissions to numbered notebooks detailing the chemists' preparations. In some companies, compound registrars aided by research librarians compose chemical names for the submitted compounds, using *Chemical Abstracts* nomenclature. They assign company codes to the substances, create a computer file for each new compound, and prepare the paperwork that accompanies samples dispatched to biologists.

To plan and schedule their experiments, the biologists require certain information, which registrars transmit to them. Items include which compounds they are receiving, how many to expect, when the samples will arrive, how to identify them, and whether they are stable. The researchers need answers to several questions. Do the samples weigh enough to allow evaluation in a demanding *in vivo* screen? Why do these particular compounds deserve testing or retesting in a particular assay? What are their molecular weights? Are they salts or other water-soluble substances? Who should receive the assay results? Registrars record the whereabouts of the remaining portions of the originally submitted sample and account milligram by milligram for the amounts distributed to the biologists.

In some companies, compound registrars also receive and store submitted samples, retrieve them from storage, and distribute them. Sample handling in a compound dispensary entails clerical work: weighing quantities of samples or dissolving them to prepare solutions of specified concentrations and repackaging as well as relabeling.

Compound registrars have no responsibility for chemical research, so they work in cubicles or offices rather than laboratories. Educational qualifications are a B.S. or M.S. degree in chemistry or biology; 1 to 3 years' experience in the work can be a prerequisite. Sometimes a year's work in discovery or development research suffices to make a job hunter eligible for transfer to a registry. Candidates knowing proprietary relational databases like ISIS have an advantage; knowing how to use computers is indispensable. Any one company needs only a comparatively small number of compound archivists; a handful adequately serves 200 medicinal chemists.

Cheminformatics

Ventures in combinatorial chemistry rely heavily on computer technology. It helps produce and deal with the numerous data arising from synthesis and testing of thousands of compounds. These efforts recently created the discipline of cheminformatics. To work in this brave new world, job hunters can exploit backgrounds in pattern recognition or artificial intelligence or draw on experience in developing chemical database software. Successful candidates help design sets of compounds (called *libraries*) to be prepared by combinatorial methods, seeking to ensure adequate structural diversity and incorporate suitable pharmacophores. With computer applications, they analyze the voluminous results of biological testing, looking for relations between chemical structures and pharmacological activities (known as *data mining*). Prerequisite to employment in cheminformatics is a doctoral degree earned in chemistry, computational chemistry, or computer science; postdoctoral experience is desirable. In some cases, pharmaceutical industry experience is required.

Cheminformatics represents an emerging discipline with varying job titles. Ads, for example, sometimes call for chemometricians to work in a field known as *chemometrics*.

> *Consequently, in this field as in others, job hunters reading help wanted advertisements are well advised to skim the position titles and peruse the job descriptions.*

Patent Coordinators

Certain drug houses employ organic chemists as patent coordinators, who cooperate closely with biologists, other chemists, and attorneys. Examples include Abbott Laboratories in Illinois, Bio-Méga/Boehringer Ingelheim in Québec, Metasyn and Vertex in Massachusetts, and Bionumerik Pharmaceuticals in Texas. Patent coordinators employed there and elsewhere help to identify worthy inventions and to draft, file, and prosecute patent applications. The inventions for which patents are sought originate in other scientists' laboratories, and the coordinator effectively represents their work to the lawyer assigned to the case. He also presents to the inventors the attorney's information needs. Such an intermediary allows the scientists to direct their energies to research. The coordinator frees the attorney from gathering the patentable material and from some of the organizing and amplifying that it needs. He may help ascertain the validity of issued patents and determine whether patent infringements are taking place.

Patent coordinators' jobs are few and may be hard to find. They are not entry-level posts, but require direct experience of the work, for which employers seek B.S., M.S., or Ph.D. chemists. A qualified Ph.D. scientist achieves as a prerequisite the research successes that lead to patent applications. For a medicinal chemist employed in discovery research these successes are a series of biologically potent new compounds. For his counterpart in development research, they are novel syntheses of compounds awaiting or undergoing clinical trials. In either case, attorneys help the successful chemists patent their inventions. In this way the researchers gain the experience needed to work as coordinators. Some large pharmaceutical companies employ such liaison personnel to deal with in-house patent attorneys. Small drug houses, if they have large patent portfolios but do not employ their own patent lawyers, dispatch coordinators as envoys to external law firms. Patent coordinators do not necessarily represent their employers to the U.S. Patent and Trade Office or to other patent agencies, except as follows.

Patent Agents and Attorneys

A few chemists within pharmaceutical companies find work as registered patent agents, licensed to practice before the U.S. Patent Office. Their licenses, rather than doctor's or master's degrees, are prerequisites met by passing an examination called the patent bar. They are sometimes obtained with help from a patent department in a pharmaceutical company. There, chemists learn from attorneys how to compose and prosecute patent applications. The licenses entitle agents to practice before the Patent Office but not before any court, which is a task reserved to lawyers.

Other chemists originally employed as researchers transfer to their employer's patent departments, which may sponsor career development by providing tuition. They work as coordinators while attending law school in the evenings, aiming to become patent attorneys. Baccalaureate degrees are prerequisite for such transfers. Budgetary restraints govern transitions from lab to law, as does the supply of lawyers with chemistry Ph.D.s. Many companies prefer to employ these attorneys rather than patent agents. Chemists interested in this work attend the Chemistry and the Law symposia held during national meetings of the American Chemical Society. Symposia speakers offer career advice in their talks.

Chemical Information

Some of the largest FIPCOs hire a few chemists as information scientists or consultants, employing them in corporate libraries. These scientists search the patent and chemical literatures on behalf of attorneys and biological, chemical, and clinical researchers. They search manually or with computers, using *Chemical Abstracts, Beilstein,* or the Derwent patent database, for examples. Experience with proprietary software is often a hiring requirement, so proficiency with computers is taken for granted. Employers seek chemists at all degree levels for these posts but demand 2–9 years of prior training, especially in patent searching. Backgrounds in chemistry, biochemistry, or biotechnology are suitable, and masters' degrees in library science may be prerequisite.

Broad, practical knowledge of *Chemical Abstracts* nomenclature can help land one of these jobs. In some companies a chemical information specialist composes the names of chemical compounds from structural drawings and draws structures from names. Systematic names are essential ingredients of NDA documents, IND applications, patents, and articles in chemical and pharmacological journals. They are also important in carrying out literature searches and in understanding the results.

Synthetic Services

Some of the largest pharmaceutical companies employ groups of synthetic organic chemists whose duties fall between discovery and development research. They are employed neither to make new compounds for biological testing nor to create manufacturing processes for drugs soon to be launched. Instead, these chemists prepare substantial samples of compounds already known to possess interesting but unexplored biological activity or of intermediate compounds leading to the active ones. The required samples weigh hundreds of milligrams to hundreds of grams, so the scales on which these chemists work range from small to medium. They use flasks holding as much as 22 liters. Job hunters interested in this work look for positions in resynthesis or preparations groups or in kilo laboratories. Chemists at all degree levels find permanent, entry-level positions in these organizations.

Profiling and Identification

During drug discovery and the initial stages of development, compounds with unknown structures unexpectedly appear as process impurities, drug degradants, or metabolites. *Process impurities* are the minor by-products of reactions composing the synthesis finally adopted to make a drug. Pharmaceutical companies employ chemists to elucidate the structures of these substances and to determine the levels of each contaminant in the final drug substance. Also, the work entails preparing reference samples to validate assays and making metabolites for toxicological evaluation. This last aspect is a demanding one because the metabolite of a drug can be more difficult to synthesize than the drug itself. Liver enzymes, for example, do easily what synthetic chemists cannot readily achieve. The job therefore calls for proficiency in synthesis as well as in chromatographic techniques and analytical instrumentation, especially HPLC and mass spectrometry.

Researchers seeking this work find positions in drug metabolism, pharmaceutical or chemical development, or analytical chemistry groups. A pharmaceutical company may devote a single group of these scientists to the tasks of profiling and identifying unknown substances originating in synthesis, degradation, and metabolism. Alternatively, researchers employed in drug metabolism groups sometimes work exclusively on metabolites, while scientists in chemical or pharmaceutical development respectively study process impurities or degradants. They find work as B.S., M.S., and Ph.D. chemists and may or may not need experience.

Positions in profiling impurities and elucidating their structures can be hard to identify from job titles alone. Frequently these posts go unnamed in the help wanted ads that describe them, or they bear nonspecific names like *research investigators in technical registration resources*. Other advertisements call for degradation chemists or reference chemists. Understanding job descriptions is essential to finding one of these positions; titles can differ from one company to another.

Chromatography and Separations Science

For decades the largest pharmaceutical companies have hired chromatographers to support chemical discovery and development efforts. They offer work to chemists at all degree levels and do not uniformly demand industrial experience. These researchers are also known as separations scientists because their training can reach beyond chromatography to techniques like capillary electrophoresis and countercurrent distribution. Their other duties can include analytical and preparative HPLC, using columns with 20-cm diameters and chiral stationary phases. Since personal computers control modern HPLC instrumentation, familiarity with them is obligatory. In some cases, separations scientists make and use preparative-layer chromatography plates that reach 1 meter in width. Individual chromatographers find work in chemical discovery or development groups, within independent service organizations dedicated to separations, and in analytical chemistry, radiochemistry, or drug metabolism groups.

Regulatory Affairs

A few organic chemists, often those experienced in chemical development, find employment in regulatory affairs groups. Academic qualifications for such work are necessary but not decisive in hiring. For chemists and other scientists, they range from baccalaureate to doctoral degrees. Anyone working in a regulatory affairs group holds a firing-line job, and 3 to 8 years of industrial experience is desirable or obligatory. A working knowledge of FDA guidelines for registering new drugs can be a prerequisite. In general, members of regulatory affairs groups represent the company in its interactions with the U.S. FDA and with other countries' drug-regulatory agencies. Chemists with regulatory affairs jobs take responsibility for preparing, reviewing, or tracking the chemistry, manufacturing, and control sections (CMC) of New Drug Approval applications. Alterations to a chemical synthesis are subject to FDA approval, as are other aspects of manufacturing processes and changes in manufacturing site. Other duties may include coordinating companywide efforts to gather and compose the technical documentation needed for drug registration. Familiarity with chemical development and process research is essential, while prior employment in regulatory affairs can be necessary for certain jobs.

For more information on regulatory affairs jobs, consult the Internet: www.job-sites.com/ra/rajobs.htm.

Development Research

The importance to a pharmaceutical company of developmental efforts surpasses or equals that of discovery projects. Process chemists work exclusively on commercially promising undertakings and often on successful ones leading to product introductions. By contrast, medicinal chemists expend most of their efforts on ventures that may be scientific successes but that ultimately become commercial failures. Development, to plunder Oscar Wilde, has a future while discovery has a past. It is a little known subject, however, so this section supplements both the preceding introduction to chemical development and the coming chapter detailing the topic.

Synthetic organic chemists at all degree levels find employment in chemical development, and many but not all of their jobs are permanent ones. FIPCOs hire inexperienced B.S. chemists for development jobs, and many experienced researchers transfer from discovery to development research early in their careers.

Chemical development creates many pharmaceutical industry jobs for organic chemists. Most of these scientists prepare clinical supplies of new medicines or do process chemistry. As we saw, these researchers invent, develop, and scale up syntheses, and they optimize existing syntheses devised by discovery chemists. Process researchers also improve long-established manufacturing syntheses of approved prescription or over-the-counter drugs. Small yield increases in such cases can offer large savings in the cost of goods. If sales volumes remain high, efforts to improve syntheses can continue for decades. Process research appeals to many new graduates because it bears a greater resemblance to academic training in organic synthesis than does industrial research in medicinal chemistry.

In addition to process research positions, a variety of other jobs go to chemical developers. Some organic chemists work in kilo laboratories or pilot plants, while others labor in natural products, biocatalysis, or safety groups. Finally, anyone interested in chemical development work may consider temporary employment with a scientific staffing agency having FIPCOs among its clients (see text below).

Natural Products

Although many drugs come from syntheses begun with commercially available fine chemicals, other medicines are natural products or semisynthetic derivatives of them. To obtain supplies of these medicines, or of the naturally occurring starting materials, can require extractions of fermentation broths, or of plant or animal matter, initially on a small scale for biological screening. However, success in finding a clinical candidate requires a transfer of the isolation technology to a chemical development organization, where synthetic chemists prepare the large supplies needed for human studies. Drug houses therefore employ chemists to scale up extractions. These scientists prepare clinical supplies and, like other development chemists, devise and optimize manufacturing processes for marketed natural-product or semisynthetic medicines.

Bioorganic Catalysis

Nearly 50 years ago, organic chemists and microbiologists employed in FIPCOs began using microorganisms to introduce certain functional groups into particular drug structures through enzymatic reactions. The 11α-hydroxyl group of anti-inflammatory steroidal medicines like dexamethasone and betamethasone resulted from this practice, which generally can impart or augment a desired biological activity. Independently, microbiological fermentations can make a synthesis economic or feasible. This important practice continues, despite the last half-century's far-reaching advances in organic synthesis. Organized in biotransformation departments, groups of chemists and biologists screen microorganisms and isolated enzymes to learn which will effect a desired reaction of a specific substrate.

Often the biocatalyzed reaction sought is not merely one that introduces a functional group or transforms a preexisting one, but is also a stereospecific conversion creating or destroying one enantiomer or epimeric center. In such a case, bioorganic catalysis can dominate traditional optical resolution by selective diastereomer crystallization or chromatography. It affords means to prepare large quantities of an enantiomerically pure drug. Biotransformation groups may help discovery researchers by providing optically pure or racemic supplies of intermediate chemicals or clinical candidates. Alternatively, they use enzymes to make clinical supplies of a drug or to help develop and demonstrate a manufacturing process. The importance of microbiological fermentation to the pharmaceutical industry can hardly be overemphasized. Together with semisynthesis, it may be responsible for more drugs in tonnage or number than total synthesis.

Bioorganic catalysis groups employ organic chemists at all degree levels. Practical training in biotransformations is advantageous but not necessary for job seekers. Nevertheless, whether particular employers offer work to inexperienced applicants depends on attitudes and economics.

Safety

Development research aims to make intermediate and final chemicals on scales imposed by marketing requirements. The requisite quantities are more than large enough to pose hazards not met in discovery research.

To avoid hazardous reactions, reagents, and solvents, large FIPCOs employ chemical engineers and organic chemists. They identify and assess the risks associated with new chemical processes and changes to established syntheses of older medicines. Their positions necessarily entail experimental work, and a working knowledge of adiabatic and differential scanning calorimetry offers an advantage to job seekers. Any scientist who accepts such a position can expect to learn about the explosivity of vapors, the combustibility of solids, and the self-ignition of dusts. These jobs need B.S. and M.S. scientists as well as Ph.D. chemists who know chemical kinetics and thermodynamics. At times, some of them prepare so-called Material Safety Data Sheets (MSDS). These documents concisely state the risks associated with making, handling, storing, and shipping the new compounds that comprise the intermediates and final products of drug syntheses.

Temporary Jobs

To work in the pharmaceutical industry, you can sign on with a temporary-employment agency. Such agencies include Advanced Scientific Professionals, Integro Scientific Staffing, Kelly Scientific Resources, Lab Support, Manpower Technical, Scientific Staffing, and The Scientists Registry. Temporary workers earn an hourly wage to do assigned jobs in drug companies short of permanent staff. In 1998, B.S. scientists received a maximum hourly wage of $25, while M.S. and Ph.D. scientists' pay equaled at most $125 per hour. Eligible agency employees receive a paid vacation and holidays but no pension credit. At least half-a-dozen top-10 FIPCOs employ temporary scientific staff who work in analytical chemistry or chemical development, where the workload depends on discovery researchers' output or business developers' in-licensing. Some discovery research groups also include contingency staff. Temporary jobs last 9–24 months and, in one estimate, nearly half of all scientific agency workers convert their contingent positions to permanent ones. In part, employers use temporary staff positions to evaluate the scientists who fill them.

If you seek temporary work, beware of needlessly paying a fee to a staffing agency. According to *Saralee Woods,* author of "Executive Temping: A Guide for Professionals," enough agencies exist to find job seekers temporary work without charging them fees. Indeed, a scientist who accepts a temporary job becomes the employee of the staffing service, not the pharmaceutical company where he works. That firm pays the staffing service for his labors.

Expect to negotiate your wage when an agency offers you a temporary job. To set your expectation, consider that a FIPCO not only pays a permanent employee a yearly salary but also spends about 30% more than this amount. The extra money finances benefits, which you will pay for yourself. The hourly wage that you seek should therefore exceed by a factor of 1.3 what a FIPCO pays its permanent staff. If a permanently employed M.S. chemist earned $45,000 in 1997 and you have such a degree, you should then have sought $28 per hour for a working year of 2080 hours. (2080 hours = 40 hours $\cdot$ week^{-1} $\times$ 52 weeks $\cdot$ year^{-1}). Scientific staffing agencies pay their temporary employees wages competitive with—but less than—the salaries FIPCOs pay permanent workers.

Temporary scientific-worker agencies advertise in *Chemical & Engineering News* and in publications of local ACS divisions. Agency representatives attend the twice-yearly national ACS meetings as exhibitors, and they staff booths at the annual Eastern Analytical Symposium. They hold seminars exemplified by Kelly Scientific Resources' "How to Use a Temporary Service to Get a Permanent Job." Lab Support, Inc. sponsors an employment workshop called "Résumé Writing and Interviewing Skills" and distributes brochures on composing résumés.

SATELLITE COMPANIES, GOVERNMENT AGENCIES, AND NONPROFIT INSTITUTES

Appreciate the variety of businesses within the pharmaceutical industry. Foreknowledge may help you find an unadvertised job in a division other than pharmaceuticals or in a department other than chemical discovery research. Several of the largest drug companies engage in businesses having nothing to do with human pharmaceuticals. AstraZeneca produces agricultural chemicals, Pfizer operates an animal health division, and Bayer and Aventis make industrial chemicals.

Invited to interview for a human pharmaceuticals post, a successful candidate can unexpectedly receive a choice between that job and one in another division like animal health. Similarly, an applicant interviewed for a position in chemical discovery research sometimes becomes a candidate for a job in chemical development research and vice versa.

> *To express knowledgeable enthusiasm for both posts during an interview, a diligent job seeker learns beforehand about the possibility of an alternative offer and the nature of the work.*

Pharmaceutical industry employers include many organizations whose operations exclude the sale, registration, clinical development, and discovery of human medicines. In a sense, these institutions are satellites orbiting the FIPCOs that form the nucleus of the industry (Figure 3-1), and all of them hire organic chemists.

As outsourcers, many satellite companies serve not only the pharmaceutical industry but other chemical industries as well. They sell services or chemicals for research or manufacturing and can be small specialized firms employing only a few people or large, long-established companies carrying on many businesses. The Eastman Chemical Company exemplifies such a seller from the latter category. It offers to do custom synthesis, manufacture pharmaceutical intermediates, sell fine organic chemicals, or carry out process research and development. All these businesses broaden the job market for organic chemists.

Service Firms

Other firms specified in Appendix B engage in several businesses of a fully integrated pharmaceutical company, but not in all of them. In the northeastern United States there is Albany Molecular Research; in the Midwest near Chicago there is MediChem Research. Both companies offer traditional and modern medicinal chemistry, custom synthesis, and process research and development, among other services. Abbott Laboratories, itself a FIPCO, also acts as an outsourcer. It offers its services in chemical development and in making advanced intermediates, custom organic chemicals, and peptides.

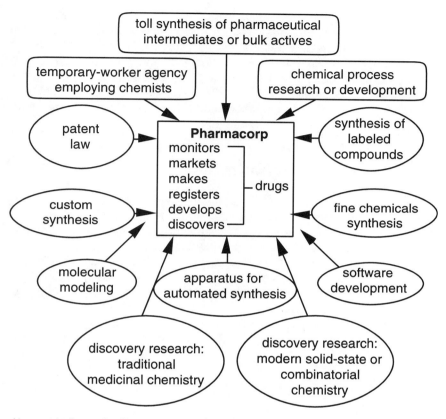

(Arrows indicate the flow of goods or services)

Figure 3.1 Idealized functions of selected satellite companies orbiting an imaginary FIPCO and employing organic chemists

Some service firms may restrict themselves to a single business, perhaps temporarily. For examples, the North Carolina firm SARCO offers combinatorial chemistry services as do Pharmacopeia in New Jersey and NéoKimia in Québec. Discovery research services generally include medicinal chemistry by traditional or modern means, custom synthesis, preparation of radiochemicals and stable-isotope-labeled drugs, isolation of drug metabolites, or patent law.

In Québec, C/D/N Isotopes specializes in custom synthesis of compounds labeled with stable isotopes as does Mass Trace in Massachusetts. Amersham Pharmacia Biotech in Wisconsin makes radiolabeled nucleotides.

Other companies offering radiolabeling include Advanced Chem Tech (Kentucky), Cambridge Isotope Laboratories (Massachusetts), ChemSyn Laboratories (Kansas), Greenfield Laboratories (Indiana), PPD Pharmaco (Wisconsin), and Ricerca, Inc. (Ohio). Chemists find work in these companies and in medical centers, where they make labeled organic compounds that help diagnose, monitor, and treat diseases. The Kettering Medical Center in Ohio, for example, hires radiochemists to make labeled pharmaceuticals useful in positron emission tomography.

Scientific Advising

Certain law firms specialize in intellectual property matters like copyrights, trademarks, and patents. Large ones hire small numbers of Ph.D. chemists and biologists as scientific advisers. One such firm employs 22 attorneys and two scientists. Patent law offices employing chemists include Campbell and Flores in California and Washington State; Fish and Neave and Pennie and Edmonds, both in New York State; and Needle and Rosenberg in Georgia. Fish and Neave, a Manhattan firm of 120 attorneys, is large enough to operate a patent-agent trainee program for its scientific advisors. It hires scientists with Ph.D.s and pays their tuition to law school, which they attend at night. "$63,000/per year to start," reads its November, 1997, advertisement.

Chemical Development and Manufacturing

Developmental research services comprise process chemistry and *toll synthesis*. The latter phrase refers to large-scale, contract manufacture of pharmaceutical intermediates and drugs, which are called *bulk actives*. A chemist operating a toll synthesis needs to carry out the workable procedures it comprises, but he or she is not asked to improve them. To provide an intermediate or bulk active to a drug house may represent a one-time effort serving clinical development or a continuing arrangement aiding manufacturing operations.

A sales figure makes the importance of this business evident. Pharmaceutical companies spent $3 billion in 1996 to buy large quantities of pharmaceutical intermediate chemicals and bulk actives (*Stinson, 1997*). Ordinarily neither kind of chemical is available from other commercial sources because each represents a new compound. Contractual work provides both kinds of compounds and employs chemists. Several companies specialize in making chiral intermediates on large scales. They include Catalytica in California, Degussa in New Jersey, Synthon in Michigan, and NSC Technologies in Illinois, but are not limited to these examples.

Fine Chemicals

Organic chemists also find work within fine-chemical supply houses, where drug companies spent more than $18 billion during 1996 (*Stinson,* 1998). These suppliers offer comparatively expensive, pure substances in relatively small amounts. Fine chemicals serve discovery researchers as starting materials that synthesis transforms to target compounds. They may themselves represent compounds worthy of biological testing. For development chemists, fine chemicals provide the starting materials needed to explore new syntheses of clinical candidates. Supply houses do not necessarily manufacture all the substances they sell; and some manufacturers sell only large quantities of chemicals yet identify their businesses as fine chemicals. Appendix B lists more than two dozen fine-chemical companies of both kinds.

Automated Synthesis

The advent of combinatorial chemistry and the opportunity to automate organic synthesis created a need for new apparatus in which multiple chemical reactions take place simultaneously. Special equipment manufacturers arose in response to this demand, offering employment to organic chemists. Some of these scientists devise new ways of using the apparatus to solve synthetic problems posed by discovery research in the pharmaceutical and other industries. Others demonstrate and sell the equipment. Companies that engage in this business are, for example, Argonaut Technologies (California), Bohdan (Illinois), Gilson (Wisconsin), PE Applied Biosystems (California), RoboSynthon (California), and PerSeptive Biosystems (Massachusetts).

Software Development and Molecular Modeling

"Being interested enough to find our Web page and respond to our listing is a strong qualification [for employment]." So states CambridgeSoft, a software development company in Cambridge, Massachusetts, appealing over the Internet to job seekers. Other software developers include Chemical Design in New Jersey and, in California, IRORI, MDL Informations Systems, Molecular Simulations, Simulations Plus, and Tripos. These and similar companies employ chemists to develop, demonstrate, and sell software useful in discovery and development research.

With computational chemistry and chemometrics among your qualifications, a position with an independent molecular modeling company may appeal to you. FIPCOs devote some of their research budgets to this work, whether in chemical discovery or development, bioinformatics, or in drug metabolism and pharmacokinetics groups. They can outsource some of it because satellite companies specialize in it. Such companies, some of them included by Appendix B, serve not only the pharmaceutical industry but other industries as well.

The Food and Drug Administration

Several agencies of the United States government hire organic chemists. One of them is the Food and Drug Administration, which needs both researchers and reviewers. It employs 1000 chemists, many of them organikers. The researchers find work doing organic synthesis, especially of drug degradants and metabolites. Reviewers evaluate Investigational New Drug applications and the chemistry, manufacturing, and control (CMC) sections of NDA applications. As an FDA want-ad makes clear, theirs are responsible positions. "The chemist reviews . . . the adequacy of the methods, facilities and controls used for manufacture of drugs; reviews proposed labels; summarizes findings; and makes recommendations for approval or nonapproval of drug applications." Qualifications include U.S. citizenship and a degree in the physical sciences that includes 30 semester hours in the study of chemistry. The administration regards as highly desirable postgraduate education or experience in synthetic medicinal chemistry, scale-up, or drug manufacturing.

The U.S. Patent and Trade Office

Another agency of the federal government that employs organic chemists is the Patent and Trade Office, which distributes ballpoint pens as recruiting aids. The barrels and pocket clips give the toll-free job line number (1-800-642-5670) and Web address (www.uspto.gov).

Patent examiners who are engineers, chemists, or other types of scientists evaluate applications for utility, novelty, nonobviousness, enablement, and unity. The evaluation of a single application takes place over several years and entails much written correspondence between the examiner and the attorney or agent working on the application. Examiners ascertain the scope of the inventions claimed in a patent application. They search the scientific and patent literature to discover whether the claims of a given application find precedent in earlier inventions. They also compose a final opinion concerning the allowability of the inventor's claims. New examiners learn the needed legal and analytical skills from Patent Office mentors and from classroom and on-the-job training. To work as a chemical examiner of U.S. patent applications requires minimal academic qualifications: a bachelor's degree from an accredited school and 30 semester hours in the study of chemistry. In practice, examiners' qualifications reach as high as doctoral degrees.

The Walter Reed Army Institute of Research

Among the nonprofit research institutions employing chemists (Table B-1 in Appendix B) is the Walter Reed Army Institute of Research (WRAIR). Its Department of Medicinal Chemistry, which forms part of a Division of Experimental Therapeutics, engages in discovery research seeking drugs to treat or prevent malaria and leishmaniasis. The department employs a handful of Ph.D. chemists as civilians or soldiers. The former work under contract or as permanent government employees. Soldier-chemists at WRAIR hold the rank of U.S. Army captain and may work there until they reach retirement age. They are trained as synthetic organic or medicinal chemists. Depending on budget, the medicinal chemistry department also employs postdoctoral researchers supported by the National Research Council.

The National Institutes of Health

Medicinal chemistry research also takes place in three of the National Institutes of Health (NIH). "It's a great place to work at the interface between chemistry and biology," said *Dr. Kenneth Kirk,* Acting Chief of the Laboratory of Bioorganic Chemistry. His institute, which is the National Institute of Diabetes and Digestive Kidney Diseases, and two others employ 20 to 30 organic chemists in tenured research jobs. The other two are the National Cancer Institute (NCI), which operates the Laboratory of Medicinal Chemistry, and the National Heart, Lung, and Blood Institute.

Those Ph.D. chemists who are U.S. citizens employed in the NIH receive permanent positions after a probationary term. During it they demonstrate their ability to carry out independent research. A search committee finds suitable candidates for tenured positions, some of whom come from the ranks of intramural postdoctoral researchers. "It's tough to get a permanent position," said *Kirk* (personal communication), "but the experience of working here as a post-doc can be extremely useful." United States citizens who hold Ph.D. degrees win Intramural Training Awards, while foreign nationals receive Visiting Fellowships. However, few positions either temporary or permanent go to B.S. or M.S. chemists.

The NIH and NCI also employ chemists in technology transfer positions. The scientists work not as researchers but as coordinators who gather intelligence concerning patentable inventions made by NIH and NCI investigators.

They convey this news from the researchers to administrators who decide whether to patent the inventions, disclose them publicly, or to release the rights to their inventors.

Miscellaneous Posts

Merck Research Laboratories hires organic chemists to edit *The Merck Index*, a concise encyclopedia of organic name reactions, chemicals, drugs, and other topics. Its editors have bachelor's or master's degrees and bring laboratory experience to their jobs.

A few other entry-level jobs entail no laboratory work. They track an orbit remote from the pharmaceutical industry nucleus, but make indispensable contributions to this and other chemical industries and to the community of researchers. For example, the Chemical Abstracts Service employs organic chemists to summarize the world's scientific literature. Academic qualifications span the range from bachelor's to doctor's degrees; strong candidates know languages like Japanese, Russian, or German.

Finally, in 1996, a start-up pharmaceutical company sought a specialist through a terse want-ad offering no more information than what follows in the next two sentences. The successful applicant's duties were to help publish or document scientific research. Discharging them required a Ph.D. in chemistry or biochemistry but no laboratory work. To understand some jobs, you have to ask about them.

SOURCES AND SUPPLEMENTS

Anonymous; "Resident Research Associateships, Postdoctoral and Senior Research Awards, tenable at the US Army Medical Research and Material Command, administered by the National Research Council," Washington, DC, 1999.

Borchardt, J. K.; "Out of work? Try temping or contracting out your skills," *Today's Chemist at Work*, April, 58 (1996).

Brown, F. K.; "Chemoinformatics: What is it and how does it impact drug discovery," *Annual Reports in Medicinal Chemistry* **33**, 375–384 (1998).

Ember, L. R.; "With little fanfare or public note, research is thriving at FDA," *Chemical and Engineering News*, June 17, 20–29 (1996).

Fish and Neave; advertisement in *Chemical and Engineering News*, November 24, 82 (1997).

Food and Drug Administration; advertisement in *Chemical and Engineering News,* July 28 (1997).

Kirk, K. E.; private communication; May 13, 1998.

Ringrose, P. S.; Bristol–Myers Squibb press release, 9:18 am, February 9, 1998; Reported by PRNewswire.

Stinson, S. C.; "Custom chemicals," *Chemical and Engineering News,* Feb. 3, 38–64 (1997).

Stinson, S. C.; "Fine chemicals' healthy ferment," *Chemical and Engineering News,* July 13, 57–73 (1998).

Woods, S. T.; "Executive Temping," Wiley, New York, 1998.

chapter

4

DISCOVERY AND DEVELOPMENTAL CHEMICAL RESEARCH

Common Features

". . . in organic chemical work, one's failures are the firm basis upon which one's successes are built."—R. B. Woodward

INTRODUCTION: WHAT YOU SHOULD KNOW ABOUT THE JOB YOU SEEK

What work organic chemists do in a pharmaceutical company informs this and the next two chapters. Chapter 4 surveys the requirements of jobs common to discovery and development research but foreign to chemistry students (Table 4-1). Specifically, this chapter details performance reviews, aids to productivity, goal setting, budgeting, and corporate meetings. It discusses the importance and the means of communicating research results. Other topics include the possibilities for publishing scientific research and the matter of personal visibility.

Much of the chapter concerns patents because of their lasting importance to pharmaceutical companies and their utter unfamiliarity to incoming chemists. Chapter 4 tells you which chemists need to know something about patents and why. It defines patents, presents the fundamentals of patentability, and describes a patent of a pharmaceutical invention, showing how to

Table 4-1 Duties of Discovery and Development Chemists

Chemistry
 Make organic compounds
  Conceive syntheses of the chosen, final, and intermediate compounds
  Select appropriate starting materials, reagents, reactions, and conditions
  Monitor the progress of reactions, employing suitable analytical methods
  Isolate, purify, characterize, and identify organic compounds
 Operate standard items of laboratory equipment
  Spectrometers
   Infrared
   Nuclear magnetic resonance
   Polarimeters
   Ultraviolet
   Visible
  Other
   Automatic synthesizers
   Balances
   Gas and liquid chromatographic instruments
   Melting-point equipment
   Microcomputers
   Rotary evaporators
   Vacuum pumps
  Execute or arrange for searches of the chemical and patent literatures

Laboratory management
 Arrange for repairs and maintenance of equipment and fixtures
 Monitor progress of ordered work
 Select and order goods and services

Communication
 Report results, problems, and plans to supervisor
 Write weekly, monthly, semiannual, or yearly reports
 Share work experiences, procedures, and chemicals with coworkers
 Present own research at regularly scheduled group meetings
 Describe own experiments for patent applications or manuscripts of scientific articles
 Publish in chemical journals

Recordkeeping
 Sign and keep a witnessed, dated, legible laboratory notebook
 Witness coworkers' notebooks
 Submit new compounds, registering them with the research archive

Safety
 Work safely
 Learn
  Escape routes
  Locations and use of fire extinguishers, eyewash fountains, and related equipment
 Keep laboratory free of clutter and other hazards

(continues)

Table 4-1 Continued

Personal development
  Keep abreast of technical and scientific developments
    Attend seminars and conferences
    Read chemical, patent, and pharmaceutical industry literatures
    Take courses in chemistry, biology, or nontechnical subjects

---

read the document. This chapter thereby complements other writings covering patent strategy or chemical process patents. It deals with inventorship and concludes discussing laboratory notebooks. Good notes establish firm grounds for patent attorneys' judgments of who invented what.

Chapter 4 confers three benefits. (1) It provides wherewithal to decide whether job hunters care to work as organic chemists in the pharmaceutical industry. (2) It offers applicants confidence that they can project in the interviews they attend. Not always a dangerous thing, a little knowledge can dispel the air of desperation sometimes associated with willingness to do any offered job. (3) This chapter empowers job seekers to display a distinctive sign of intent to work and initiative to learn, which is prior knowledge of what available jobs entail. Acquired by preparations, this familiarity can favorably impress interviewers who are chemists.

THE BASICS

Look over Table 4-1 to appreciate what most research chemists do who work in pharmaceutical companies. But, as you read the entries, reserve final judgment concerning what you will do until you inquire about the specific job offered. No single job requires experience in all the tabulated areas, so the table represents a sketch more than a blueprint. Academic degrees and research experience determine how much responsibility you receive for some of the tabulated duties. Many of them pertain largely to synthetic chemists rather than to experts in radiochemistry, biotransformations, drug metabolism, or natural products. Tasks change from time to time and differ from one company to another. Finally, the table provides no idea of the hours per day spent on any given activity, in some of which even old-timers participate only infrequently. If you contribute experimental write-ups to a patent draft, for example, you can expect to work intensely or even exclusively on this task for a few hours or days. But the need to contribute such material may recur at intervals spaced years apart. For as long as a decade after filing, however, follow-up work on a single patent application can call for several short-lasting efforts.

Performance and Productivity

Whatever their duties, chemists employed in a drug house must work productively to thrive as researchers. They should expect their labors to receive objective and quantitative evaluation, regardless of whether they belong to discovery or development groups. However, these groups apply different criteria in judging performance.

Discovery Research

Productivity in a discovery group requires that recently hired chemists make target compounds. This work is largely individualistic, and success in doing it contributes to corporate progress. Not surprisingly, the number of these substances prepared each year offers an easily measured and widely adopted standard used to evaluate chemists' contributions. The intermediate substances of a synthesis, if they are not themselves biologically active, count less than the final ones. They may not count at all, although they outnumber the final ones and are indispensable to the work as well as laborious to make. Consequently, ease of synthesis remains an important principle of medicinal chemistry. In minimizing the required number of sequential chemical reactions, a concise synthesis maximizes the ratio of final to intermediate compounds.

> *A high ratio concerns newcomers and old-timers alike because a chemist's annual output of target compounds influences his yearly salary. His relative productivity ranking largely determines his pay increase, regardless of whether he holds a doctor's, master's, or bachelor's degree.*

He and his coworkers in discovery research compete with one another for unequal shares of the fixed dollar amount allotted to raises. So, how a medicinal chemist synthesizes target compounds and what structures they bear are less important than their numbers and pharmacological activities. Weighing most heavily in a chemist's favor is a wealth of biologically active and desirable target compounds made within an assigned project. A serendipitous finding of desirable but extracurricular biological activity may tip the scales only slightly.

Development Research

Two aspects of discovery research, structure and synthesis, play a crucial role in drug development. They define the goals and thereby direct the efforts of process researchers. Early in development these chemists must produce pure samples exclusively of the drug chosen for clinical trials, not of any other compound except for intermediate ones. They have to do so on a schedule specifying weights and dates. Their problem can be largely logistical, if the original drug synthesis can meet the material demands made by impending clinical trials. A solution then requires the researchers to begin by buying reagents and starting compounds in adequate quantities and purities and with suitable delivery times. Process researchers demonstrate the usefulness of the original preparation, usually by incrementally increasing the scale of each reaction that composes the synthesis. They improve or circumvent any troublesome step in the existing preparation. If necessary, they abandon the synthesis for a new one that they devise and demonstrate, still keeping to schedule. Early-stage development may in one sense surpass medieval drawing and quartering by horses. It creates tension in several directions—among them speed, timing, safety, quality, and quantity. Later in development, the cost of a synthesis becomes important.

Chemists participating in a narrowly defined, initial development campaign engage in teamwork more than their colleagues in medicinal chemistry. As a result of this and other factors, individual contributions can be more difficult to quantify than in discovery research. Process chemists are partly judged by how completely and promptly they carry out assigned tasks that represent important elements of the whole effort.

Other achievements used to assess performance include improvements in environmental friendliness, reductions in the number of isolations and purifications, and boosts in throughput. The synthesis of a clinical drug is friendly if, for example, it does not entail large volumes of organic solvents or aqueous solutions. Properly disposing of large liquid volumes costs money. Fewer isolations and purifications lower the costs and speed the work. These improvements are especially welcome when clinical trials require large weights of the drug. More concentrated solutions increase throughput because they use large plant apparatus more efficiently. Throughput is important since stationary, glass-lined steel equipment is less in supply and more in demand than small pieces of laboratory glassware.

During the later stages of chemical development, when clinicians' and toxicologists' material demands are satisfied, researchers seek a suitable manufacturing process capable of yielding tons of drug. The processes examined for this purpose may be unrelated to the synthesis originally devised in discovery research. Individual process chemists or groups of them consequently compete to demonstrate the usefulness of one or more reactions or even of entire syntheses. In these circumstances, they strive to reduce the expense of manufacture, which the cost per kilo measures. This figure quantitatively gauges performance. Groups and individuals whose work lowers the per-kilo drug cost receive rewards proportional to their successes.

Aids to Productivity

A newcomer wisely tracks what he accomplishes and attempts. Doing so spurs him, and provides a good account of work that receives others' scrutiny. Each day should see new experiments begun and old ones pursued. Satisfactory experiments have definite outcomes, regardless of whether they succeed or fail. It helps productivity to review progress weekly; to count the experiments begun, ended, and progressing; and to set and meet a daily standard for their number. Working this way entails multitasking, which is commonplace among discovery and development researchers. Among them, the minimum number of daily experiments exceeds or equals two.

Seeking Tempi

"A [chess] player who makes two moves to accomplish the same result he could with one move is said . . . to have lost a tempo" (*Horton*, 1959). Like a contestant moving efficiently, a chemical researcher gains tempi by allowing reactions to run overnight or during weekends and holidays. In much of industrial discovery and development research, it is unnecessary to know how long a reaction takes to maximize the product yield. And, few other reasons forbid starting an experiment toward the day's end. While researchers rest, their molecules work.

Embracing Services

Chemical researchers can boost their own productivity by seeking and accepting the many valuable services that pharmaceutical companies offer them. Expert information scientists search the literature using computers, combing much of *Chemical Abstracts* or *Beilstein* to learn how to make or where to buy needed compounds.

Pharmaceutical companies can afford what many university research groups cannot: glassware washing services, which veterans gladly use. Often, however, newcomers to industrial chemical research insist on washing their own flasks, as if these items were irreplaceable or precious. Their habit understandably arises during undergraduate or graduate research. But in industry, it sacrifices valuable time. Whether its practitioners or their employers realize genuine benefits from any savings achieved is questionable, and most chemists soon shed the habit to aid productivity.

Analytical services are also available, and exploiting them fully can increase a chemist's output. He need not record any but urgent spectra, leaving the spectroscopists to acquire routine or technically difficult ones. The chemist thereby frees himself to carry out reactions for which he would otherwise find less time or meet some delay. By contrast, industrial spectroscopists, although they have chemists' training, never conduct chemical reactions. Restricting their work to their own duties, they offer an exemplary lesson to synthetic chemists. They teach another economic lesson, which is to discard inexpensive nuclear magnetic resonance tubes used once.

Handling Paper

Soon after you begin work, your mailboxes will overflow. Mail demands judicious treatment since not all of it offers value to you. Merely reading the subject of an e-mail message and its sender's address is a helpful guide to purging your electronic in-basket. To throw away worthless, hard-copy items without ever bringing them to your desk saves time; otherwise, you needlessly handle the material twice. The losses count little in a day or week, but mount over a working lifetime.

In some companies, cumulative time losses come not from what researchers receive but from what they send. Chemists repeatedly complete several different kinds of forms to order supplies, register compounds, or obtain in-house spectra. Some forms require information that changes infrequently, like a chemist's name or an internal company mailing address. Using photocopies of such a form can conserve time, if the copies already bear the repetitive data. To take another example, printing a chemist's name saves writing it on the labels used for vials containing compound samples. In other companies, however, researchers make some of these submissions electronically, which saves time. Computer applications fill in the repetitive information.

Goal Setting

Some drug houses insist that employees formally write goals each year. Individuals seek goals that implement or complement those adopted by their bosses, groups, or corporate divisions. A good goal features other characteristics: it is objective, feasible, specific, and measurable. It can lend impetus and direction to scientists' work, if achieving the objective brings a commensurate reward that is immediate, tangible, and valued by the recipient. Despite the merits of goal setting, this practice can wane in one pharmaceutical company or division while it waxes in another.

More often than once a year, of course, individuals set their own goals to organize and occupy their working days. Because they strive to discover or develop drugs, they evaluate each proposed experiment in these terms. Chemists ask not only which trial they are to make next but how a successful outcome advances a practical project. Do the advance and the reward for success justify the risk of failure? Some avoidable victories are Pyrrhic, like a series of successful experiments that culminate in understanding of reaction mechanism but yield no samples for biological screening. Consequently, it is as important to set a time to end an experiment, or a series of them, as it is to choose a time to begin. Although persistence helps to achieve feasible goals, difficult ones are perhaps better taken by indirection than by tenacity, a flaw that can dog a researcher throughout his career. Progress in industrial research can depend less on understanding and solving problems and more upon circumventing them. Moreover, a scientist's attitude seems flexible to managers when they see him go around obstacles.

Medicinal discovery and development remain the ultimate goals of the pharmaceutical enterprise. Defining goals, however, brings no guarantee of success to any particular research project. "As in prospecting for gold, a scientist may dig with skill, courage, energy, and intelligence just a few feet away from a rich vein—but always unsuccessfully. Success or failure... may be almost accidental, with chance a major factor in determining not *what* is discovered, but when and by whom" (*Kubie*, 1954). Industrial chemists recognize the uncertainty of research and the incomplete control they exert over projects and outcomes as distinct from efforts.

To implement a new business strategy, a chemist's assigned project occasionally changes abruptly. Even large groups of researchers sometimes switch overnight from seeking dermatological drugs, say, to discovering central nervous system medicines. Such changes take place in development research, too. They frustrate researchers but challenge them as well. So, it is crucial in these circumstances to down old flasks and raise new ones. A chemist, in other words, resembles a thoroughbred called back to the post and exhorted to run again, although its last race, because of a false start, was neither won nor lost. Lingering at the gate or facing backward in the stall makes gamblers anxious; whereas a galloping start thrills onlookers.

Budgeting

Newly hired chemists taking their first permanent jobs disburse research funds without controlling budgets. At the outset, they can look forward to spending a certain amount on equipment, particularly to furnish an empty laboratory that they are to occupy. Every year afterward they take responsibility for identifying and filling needs for instruments, service contracts, or other purchases. Preparedness and promptness count in requesting costly purchases, so it helps to write a wish list, compare prices, and procure timely price quotations. Depending on the expense, the needed written approvals may issue from several management levels, perhaps in different departments. Gathering signatures takes time, and each request competes with other demands for the attention of managers authorized to approve it. During the time taken to obtain approvals, the corresponding quotations may approach expiration, so scheduling can be critical. When approval does arrive, it is wise to spend the appropriation immediately, lest unexpected budget cuts made late in the fiscal year delay or prevent the purchase. Appropriated money unspent in 1 year can spark budget cuts in the next, and such a situation can needlessly embarrass the manager who originally championed the appropriation. For these reasons, and to keep important commitments, the year's end sometimes sees a surge in budgeted spending.

Internal Meetings

Industrial scientists attend in a year's time many corporate meetings, some of which concern administrative or project-related matters or combinations thereof. All the researchers go to certain administrative meetings dealing with Occupational Safety and Hazard Administration (OSHA) regulations, chemical hygiene plans, right-to-know law, and with other safety and environmental matters. Turning up is obligatory, and participants sign attendance records, which demonstrate corporate compliance with governmental regulations.

Other administrative matters inform monthly or weekly meetings of research groups at which individual members also present their work. Although the talks pertain to current projects, they are not necessarily restricted to problems needing solutions; so they can deal largely with recent progress. Each chemist may speak briefly during every meeting or talk once or twice a year at greater length. Sometimes senior and junior scientists share the podium; at other times, only the more experienced researchers speak. The best talks are unified, coherent, and concise, and they keep Michael Faraday's Rule; the eminent English chemist said, "One hour is enough for anyone."

Some researchers also meet academic consultants in problem-solving sessions convened a few times yearly. The problems attacked pertain to individual scientists' work, so these meetings are project related. They elicit helpful suggestions, which can productively redirect research. Depending on the consultant's preferences or other factors, these sessions can be informal one-on-one meetings taking place in individual offices. They can also be get-togethers formally convened in conference rooms and attended by dozens of researchers and research managers. An ambitious and prudent chemist prepares his presentation soberly, for the consultant's job may require him to evaluate each speaker. Good work presented in such a meeting brings recognition to its author from a respected independent source. An adviser worth his wage lends intellectual excitement to the discussions, so many chemists eagerly look forward to his visits despite the preparations they entail.

In other project-related meetings, chemists discuss, and assume responsibility for, various aspects of work to be done. These team meetings occur when a new project begins and later when its direction changes. They are successful if the team members agree on what needs to be done, in which order, and by whom. Agreement tends to ensure that important tasks are done promptly and that unintentional duplication wastes no efforts. A team meeting like this needs a leader, an agenda, a record of proceedings, and follow-up. As much as possible, a good leader lets his coworkers choose the aspects of the project on which they are to work.

Certain project-related meetings involve representatives from more than one discipline. Medicinal chemists, for example, may meet with biologists or patent attorneys or with experts in radiochemistry, drug metabolism, biotransformations, natural products, or process chemistry. However, not all chemists take part in the decisive team meetings or other cross-disciplinary encounters that advance projects. A chemist's participation depends partly on his managerial duties, with greater responsibilities bringing more of these interactions (Figure 4-1). In large FIPCOs, for example, vice-presidents and directors spend much of their time in meetings, while entry-level chemists employed as experimentalists have little contact with their counterparts in biology, clinical medicine, or regulatory affairs. These chemists do meet many salespeople, who give seminars or visit laboratories.

Some meetings bring together fellow employees from all branches of the company. They convene to learn useful, transferable skills like managing time, negotiating agreements, and reviewing performances. They study other nonacademic subjects, such as the nature of the pharmaceutical enterprise or speed reading. In other gatherings, employees learn the state of their employer's business, especially the prospects (in a FIPCO) for launching new drugs in the coming year.

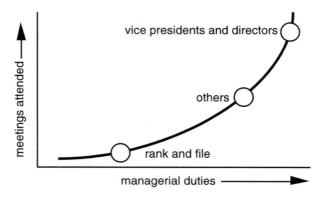

Figure 4-1 Who participates in crucial meetings?

The sheer number of required meetings, as well as their length and diversity, differs drastically from student days. Meetings, because of their topics and participants, lend variety to a working life. In them chemists interact with other employees ranging from executive vice-presidents to maintenance workers. So, in taking part, a scientist effectively represents his whole department and his immediate boss as well as himself.

Communicating Results, Plans, Problems, and Persona

Act early and often to establish your reliability and productivity as a researcher, making both evident to your coworkers. If you are to thrive as an employee, you'll need to make your good works known. Do so yourself to ensure effective communication, for no one else knows your work as well or stands to benefit as much as you. Admittedly, your boss may have a stake; but certainly not all supervisors are equally successful in spreading news or are concerned to do so.

Reporting Results

Throughout the working year, opportunities repeatedly arise to share results in conversation or writing. Bosses welcome unsolicited reports, particularly of good news that they can pass upward. They also feel gratitude for alerts concerning potential problems or early warnings of real difficulties. A chemist employed in his supervisor's laboratory enjoys ample time on any working day to discuss results, to plan, and to mention in an informal setting any problems arising.

> *Newly graduated chemists, especially those with M.S. degrees, accustom themselves to illustrating their conversations with structural drawings. Professionalism demands it, for these ideograms impart specificity to discussions that would suffer vagueness without them.*

In drawing structures, practice helps in acquiring facility and liberality aids in communicating accurately and clearly, particularly in casual conversations. Discussions move too quickly to benefit from the help of computer-generated structural drawings—on which many inexperienced job seekers hope to depend during interviews.

Random chances to communicate your results come when you meet coworkers in the hallways, seminar rooms, or cafeteria, so take care to seize the moment. The coworker whom you encounter may be the director of your department or the vice-president to whom he reports. Either may ask how and what you are doing, and the question lets you give a good account of yourself if you prepared. To do so, you need to relate not only your own recent results and immediate plans, but also to mention how they serve the larger project in which you are working. In such a discussion, it is not wise to assume your questioner knows or remembers your assignment, role, accomplishments, or plans; you'll need to voice them tactfully and coherently.

Monthly, quarterly, semiannually, or yearly, you may have to compose or contribute to a written research report. A well organized one effectively indexes your work, so it makes retrieving information easy. You'll find yourself consulting your own reports long afterward when you seek data for a patent application or a scientific publication. An old report will refresh your memory and so speed your efforts to contribute to the application or publication. Composing a research report sharpens your appreciation of your own work, even if you recently completed the job. It can point out omissions that you can rectify, foster fresh interpretations of old data, and thereby suggest new directions to take. Reviewing work done several weeks or months ago prepares you to appraise your own performance, which you may do once or twice a year.

Publishing

Pharmaceutical companies encourage employees to publish the results of their scientific research. Industrial chemists write articles for journals, give lectures at universities and conferences, or present posters at scientific congresses. Some of them serve as journal referees or editors and as conference organizers. Their employers also sponsor scientific publishing in a variety of ways. They cover the page charges levied by certain journals, pay for splendid posters and colorful slides, reimburse employees for travel expenses, or subsidize conferences.

By encouraging scientific publishing, pharmaceutical and other companies help develop their scientists' careers. Employees who learn to deal with editors and referees benefit from the interplay that article writing brings and do themselves a favor by compiling a list of publications. Such a list represents a work history interesting to another employer because it is evidence of accomplishment. As few as one or two publications can help to obtain a significant promotion. A long list of scientific publications, however, does not necessarily bring any advantage, and the question of whether such a list is detrimental can rouse controversy. The answer depends heavily on the circumstances of employer, time, and supervisor. Several business advantages can accrue to a pharmaceutical company that encourages publication of chemical research. By disclosing work done, a published article can be advantageous to an employer if it deflects a competitor from a particular research area. Publication can prevent a competitor from obtaining a patent, help recruit and retain talented scientists, and open doors to leading academic researchers.

Chemists who seek to publish their work incur two reasonable delays in doing so. Both spring from a commonsense requirement that manuscripts, posters, and slides betray no valuable proprietary information. Much industrial research goes into U.S. and foreign patent applications, which hold up publication. About 12 months after the initial U.S. filing, patent attorneys make an international application to patent the same invention. Patent offices abroad publish the application 18 months after the original filing, which occurs 6 months after the international filing (Figure 4-2). From submission to foreign publication, however, the contents are secret, and an inventor loses valuable foreign patent protection if he publicly discloses the contents before the 18 months end. A U.S. filing precedes a foreign submission by 12 months, so at least 18 months elapse between the first filing and foreign publication of the application. This $1\frac{1}{2}$-year period, which represents the first delay, is minimal. It may take 2 or more years of research to warrant applying for a patent at all; in that period, no disclosure of commercially valuable information is advisable. So, $3\frac{1}{2}$ years may pass before publication can gain approval from the patent attorneys.

The final judgment concerning what data are valuable usually rests not with the author or patent attorneys but with a hierarchy of research managers. An author needs their written approvals for what he hopes to publish. Gathering the requisite signatures consumes time and creates the second and shorter delay, which amounts only to a few weeks. However, to

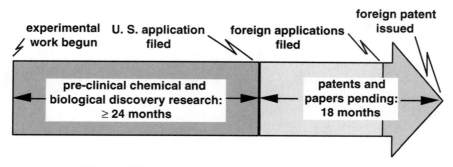

Figure 4-2 Approximate time course of research and patenting: delay of publishing

obtain approvals in time to meet submission deadlines takes planning, which is less troublesome to all concerned than unexpected appeals for speedy review.

Provided that the author's manuscript wins the approvals that patent protection brings, he may publish his work, subject to any restrictions on which approval was contingent. For example, a medicinal chemist might write extensively about the chemistry and biology leading to the choice of a clinical candidate, but say nothing about when the substance is to enter clinical trials or what progress it had made in them. A process chemist might relate several attempts to devise a suitable manufacturing synthesis but withhold from publication the route finally chosen or keep as a trade secret crucial procedural details. Organic chemists employed in pharmaceutical companies write articles or reviews on many topics. Their subjects include, beside those already mentioned, the invention or discovery of new reactions or novel uses of known reagents. Elucidation of reaction mechanisms and labeling of drugs with radioactive or stable isotopes make suitable articles. Other subjects include syntheses of natural products, combinatorial libraries, and new heterocyclic ring systems. They also cover determination of drug metabolite and other natural product structures. Employment in the pharmaceutical industry lets chemists do and publish much engaging research.

The Importance of Being Visible

In a large FIPCO, service on committees brings visibility to newcomers that they might not otherwise receive. It gives them an opportunity to observe, acquire, and demonstrate people skills. Research groups have continuing needs for representatives to serve on safety, chemical disposal, library, seminar, and recruitment committees. Committees decide which of several pieces of expensive equipment to buy and interview salespeople representing the manufacturers. Members may be appointees or volunteers. Research groups also need individual chemists to take charge of communal apparatus like infrared spectrophotometers, cameras, or high-performance liquid chromatography instruments. Duties entail showing others how to use the equipment, maintaining it with the help of technicians, and buying and keeping supplies. Common rooms also need upkeep for which researchers assume responsibility. These rooms include, for example, chemical and solvent storage areas, instrument rooms, and walk-in freezers. Chemists who take on these jobs give valuable service to their coworkers and company, but do not allow community service to reduce research productivity, which takes priority.

Patenting Your Work

Learning about patents hardly represents a decisive step in a job search, so a successful effort won't bring an offer of employment to a new graduate. Why read or write about patents, then, when doing either invites sleep? For three reasons.

1. Prior knowledge of patents can help a job seeker decide whether he wants to do the work patents entail. Although not all industrial chemists take on extensive responsibilities for composing and prosecuting patent applications, many organikers contribute experimental procedures sometime during their careers. They also consult patents to find procedures useful in their own research.
2. Hence, a familiarity with patent matters prepares a researcher to do what his employer needs.
3. An early acquaintance with the concept of inventorship can prevent the personal disappointments that sometimes arise when cold air dispels the myths trailing patents.

To stay awake, absorb this section in small doses. Its topics are independent, and their importance to you depends on whether you are seeking or doing an industrial chemist's job (Table 4-2). Topics of concern to job seekers hold less interest for job holders and vice versa.

If a marketed drug profits the pharmaceutical company that originated it, then generic drug houses begin selling cheaper, bioequivalent samples of the medicine hours after the originator's patent term ends. The originator's sales of the drug sales can then plummet, so important are the monopolies that patents afford.

Table 4-2 What Chemists Should Understand About Patents

Topic	Relative Importance of Topic	
	To Job Seekers	To Job Holders
Inventorship	+++	+
Function	++	+
Structure	+	++
Reading claims	+	+++

Medicinal chemists study their competitors' patents for several reasons. Perhaps chief among their incentives is a desire to improve upon other researchers' lead compounds, when doing so is relevant to their own projects. Another reason is to prepare their competitors' patented compounds as standards for their own biologists' assays. Therefore, they seek to learn from a patent the structures of the substances having the best biological activities of all those compounds protected. The chemists peruse the sections of this patent that afford any insight to potency and to exploitable facts associating structures with activities. Researchers working in chemical development help compose the process patents that protect marketed drugs. So, a knowledge of how to read and write patent applications forms an essential part of an industrial chemist's working life, regardless of whether he labors in development or discovery research.

Patent applications come from both kinds of groups. They create the basis of an employer's future business, give insight into competitors' research, and express some of a scientist's accomplishments. For an organic chemist seeking to change industrial employers, granted patents are the currency of résumés as much as or more than scientific papers are.

Patents Defined

A patent represents a mutually beneficial, written agreement between a government and an inventor. It is a social contract that brings different obligations and benefits to both parties. The government's obligations are to maintain legislative and judicial systems that help define and enforce rights to inventions, and they occasion the inventor's benefits. The government offers legal recourse to the patent holder, who may be the inventor or his assignee, and is often his employer. It allows the owner—assisted when necessary by an attorney and a court—legally to prohibit unauthorized parties from practicing the invention. In the case of a drug, *practicing the invention* means making, using, or selling the medicine. Unauthorized practice would diminish the patent holder's business or detract from his financial interest.

Governments, however, take no responsibility or action to ensure that businesses fostered by patents break even or make money. Patents do *not* bestow on their inventors or assignees *any* right to practice their own inventions—only to exclude others from doing so. Furthermore, the grant of a patent does not imply that the patent is valid; findings of validity are the province of a special court, not of the examiners who allow claims.

Term

The dates of filing and issue are important. They specify the time when patent protection began, and indicate when the invention will enter the public domain, if no extension to the patent term is allowed. United States patents filed on or after June 8, 1995, run 20 years from the filing date. Patents granted before June 8, 1995, go on for 17 years from the issue date. The Congress sought to make the term long enough for the originator to recover its research and development costs and to benefit from the invention. If the protected invention is a drug, then development can last some years before the originator can secure regulatory approval. When a drug company's composition-of-matter patent expires, a competitor may manufacture the active ingredient for sale, if doing so does not infringe existing process or formulation patents.

The 20-year patent term and the variable development period determine how long a pharmaceutical company enjoys the sales monopoly brought by a composition-of-matter patent. Let us estimate this time with three typical premises. First, chemical and clinical development begin 2 years after biological and chemical discovery research start. The company files a U.S. patent application when development begins and, finally, development lasts 12 years. In these circumstances, the drug wins FDA approval 14 years after discovery efforts began, but only 8 years remain in the patent term. With fewer years remaining to the patent life, a pharmaceutical company may stop development if the expected sales revenue fails to ensure profitability.

Disclosure

An inventor's obligation requires him to describe his conception completely, saying what it is, why it's useful, and how to implement it. Only then can a valuable invention ultimately benefit the people of the country issuing the patent. When its term expires, anyone there becomes free to use it without paying royalties and likely to gain without financing development. National governments issue patents to bring the advantages of valuable inventions to their citizens. The enabling particulars presented by patents do not remain company secrets, as they might otherwise do, but enter and augment the public knowledge. To disseminate this knowledge, the U.S. Patent and Trade Office mails copies of its patents for a small fee, currently $3. Readers may consult patents in certain public and university libraries found in 43 states and the District of Columbia, and patents are also available over the Internet. Many of them, incidentally, make entertaining light reading found, for example, in the Gallery of Obscure Patents (http://www.patents.ibm.com).

Costs

The expenses of obtaining patents, however, are not light at all. Depending on circumstances, the Patent and Trade Office imposes as many as 24 fees amounting to approximately $8500. (Qualified individual inventors pay less than this amount, thanks to a two-tier fee system.) An outside patent counsel's services can add as much as $15,000 to the filing cost. And, according to *H. Jackson Knight,* an author and a DuPont patent specialist, expenditures for procuring and maintaining a patent in only a few foreign countries can exceed $100,000. Costs aside, so critical is patent protection that over 100 grants were issued to each of several FIPCOs in 1997. For example, about 110 patents went to Pfizer, 140 to Bristol-Myers Squibb, and 370 to Merck.

Types of Patents

The universe of U.S. patents currently comprises three types, which are plant, design, and utility. Plant patents protect asexually reproduced plants, while design patents cover ornaments like wallpaper patterns. Utility patents claim devices, processes, and compositions of matter (Figure 4-3). Pharmaceutical patents are directed to any or all of these three subtypes, but industrial chemists employed in FIPCOs mainly deal with patents of certain processes or compositions of matter. In such patents, compositions of matter may include mixtures like drug formulations or pure compounds like the active ingredients themselves. The components of such formulations are themselves known compounds. The processes claimed by a pharmaceutical patent comprise chemical reactions for making drugs and methods for treating or preventing disease. These reactions show how to make the compounds of the invention, while the methods teach how to use them. Both the reactions and methods are themselves inventions deserving protection. A pharmaceutical patent emerging from a discovery research effort ordinarily contains three kinds of claims: to methods of treatment or prevention, to single compounds as medicines, and to pharmaceutical compositions. Such patents may include chemical process claims, which also form the substance of independent patents emerging from chemical development research.

A Fortress of Claims

As a newly identified drug progresses through chemical, pharmaceutical, and clinical development, it becomes the subject of more and more patents. They restrict themselves to newly invented chemical processes and drug formulations and to additional claims of methods of medical use. The drug would not have entered development if it were not already protected by a

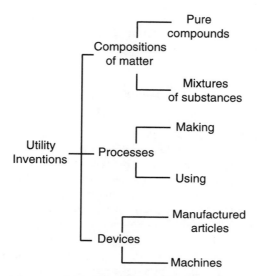

Figure 4-3 Patentable utility inventions

composition-of-matter patent. Not discovered or sought during the short discovery phase, the newer inventions are purposely devised or serendipitously found during the longer development stage. Sometimes development work leads to a patent protecting a drug hydrate or other solvate not known or protected during the discovery phase. Such a patent can advantageously extend protection. In any case, by the time the sponsoring company introduces the drug to the marketplace, a fortress of claims surrounds the rights to the invention. Its construction and architecture pay tribute to the skills and training of patent attorneys, examiners, and agents.

Patentability

Acceptable applications and valid patents must present inventions that are novel, unobvious, and useful. They show unity of invention, teach physical means to realize the inventions and, in several respects, disclose what attorneys call *the best mode* of practicing them. In a typical pharmaceutical patent, the best mode may refer to the compound most suitable as a drug, to the best way of making it, and to the disease it treats most effectively. Patent law requires inventors to reveal the best modes without, however, necessarily emphasizing or even identifying them. Although a patent relates the best mode of, say, making a compound of the invention, patent law does not require the inventor to ascertain which method is superior. The law merely obliges the inventor to disclose this information if he possesses it and otherwise to name a best mode.

Unpatentable Ideas

Mere ideas lack patentability, as do scientific principles and physical phenomena. In a sense, patents are analogous to copyrights: the former protect the physical embodiments, while the latter cover novel expressions of ideas. Patents and copyrights differ because patented ideas must be novel and useful, whereas copyrighted expressions need not protect new or useful ideas. Both patents and copyrights foster commerce; neither exclusively rewards nor records creativity. Consider an inventor who conceives a new chemical reaction. To preserve his rights, he keeps his invention secret if he does not realize his idea through experimentation. If he can realize his invention, two alternatives arise. The inventor can develop and publish his work. Or, he can keep his results quiet until he finds or devises a commercially valuable use for the compounds that the reaction forms. Itself an idea, such a reaction alone is not a patentable invention, and its success in creating the expected products does not impart credible usefulness. Thoughtful or even brilliant, this inventor's creative idea is not patentable without implementation and extrinsic utility.

Novelty and Nonobviousness

Patent Office examiners in the United States scrutinize applications to ensure they meet the foregoing criteria. They read the patent and scientific literature themselves, searching for precedents, which are known as *prior art*. To help examiners determine the novelty of their claims, inventors must disclose in their applications any relevant precedents. Prior art might include information from textbooks, from articles in scientific journals, or from patents current or expired, but would not be limited to facts in these publications. Items offered for sale through catalogs or stores can constitute precedents. So, any inventor who devises a variant of, say, the wood screw can expect to contend with rich prior art to obtain a patent for his creation.

Like rival bull elks, examiners and attorneys lock antlers over the obviousness or nonobviousness of claims in patent applications. Mightily do they struggle, sometimes prolonging their engagements for years. Consequently, an adequate treatment of their contests may lie beyond the span of chemists' attention and the scope of this book. It can merely offer an alert to readers with two instances.

(1) To begin with a vital example, minor structural changes within a chemical compound claim may fail to qualify it for a patent. Trivial efforts to

distinguish one's own compound claim from a competitor's do not escape the notice of sharp-eyed examiners. Consequently, for one inventor to invoke an n-heptyl group instead of his competitor's n-hexyl side chain invites a finding of obviousness, unless the change leads to a surprising or unexpected result.

Between n-hexyl and n-heptyl side chains there lies a difference of homology, which represents one of several shoals of obviousness. Another is the choice of alkali metals or halogens lying immediately above or below one another in the periodic table. No new patent application necessarily escapes obviousness by claiming, say, a chlorophenyl substituent instead of a fluorophenyl or bromophenyl group that an older patent protects. Other obstacles are salts and isomers of various kinds.

If compounds of the invention can form salts or exist as isomers, any ordinary inventor contemplates protecting them. His obvious thought is worthy, and patent practice respects it. He need not claim these derivatives specifically or generically. To cover them, it suffices if he discloses them broadly in the patent specification, not individually in the special section devoted to claims. A competitor, consequently, cannot secure his own protection of the same compound merely by claiming a different salt or isomer of it.

(2) However, certain evidence might entitle your application to an allowable claim if it covers a salt with surprisingly and unexpectedly good activity. This evidence would support a so-called *selection invention*. Such an invention arises when a subclass of compounds chosen from a patented class possesses a demonstrable advantage. The advantage must be one that the other compounds lack. In this case, a claim to the subclass is allowable in a patent.

Varying physical forms of neutral compounds may also be patentable and can create the basis for selection inventions. So, an amorphous solid may show better oral absorption and thereby higher blood levels than crystalline forms of the same drug. The amorphous substance would therefore offer a distinct therapeutic advantage. Similarly, a particular crystalline form can also bring such an advantage over other crystalline or amorphous states.

Utility

An examiner demands a statement of relevant, credible utility in a patent application directed toward chemical compounds as drugs. The compounds claimed should demonstrate some ability—as indirect as may be—to detect, monitor, prevent, treat, or cure disease. That they can stop doors, weigh papers, or balance boats is irrelevant. However, the standard of acceptable evidence is minimal, a point that eludes some inventors for years.

For example, the evidence of medical utility that a patent offers may be slight. It suffices if some compounds of the invention prove weakly active merely in a particular *in vitro* biochemical assay using isolated animal cells.

The assay itself may have no demonstrated relevance to the discovery of human medicines, having never before been used to find a clinically effective drug. Potency and dose responsiveness in this or a proven assay are as inessential to patent applications as is statistical significance of the measurements. Efficacy in a live-animal model of a human disease and safety and effectiveness in treating human patients, important as they are, are merely helpful to support claims to using the compounds as drugs. Furthermore, the inventor need not synthesize all the compounds he should claim, although his patent must teach means to make them. Nor must he completely purify, fully characterize, or adequately identify those substances that he does make and claim. Patent practice does not oblige his bioanalysts to test all the compounds he submits or to him submit for testing all those substances he prepares.

The biological findings that demonstrate utility need not appear in the patent or patent application at all. Instead, they may be submitted to the examiner before the patent issues, yet will not appear in the patent. Such 11th-hour changes are tactics that keep competitors guessing as long as possible, which is a business strategy that some pharmaceutical companies adopt. In some respects, acceptable patent applications meet neither the standards of academic chemistry and biology nor the criteria of regulatory agencies. They need not do so because valid patents represent commercial tools, not scientific masterpieces or medical triumphs.

Reduction to Practice

A patent must teach its reader how to implement the invention it protects. So, in the case of one directed to chemical compounds as drugs, the inventor discloses the best method he knows for making the claimed compounds. This method is not one he must identify as such, but may be one of several described in the patent. It reveals selected details of the chemical synthesis adopted, including the starting materials.

When its term expires, a patent confers on the public the benefits of a chemical invention. It must therefore enable an interested party successfully to repeat the experiments it describes. To ascertain whether these efforts succeed, a helpful patent provides means to recognize compounds of the invention by their characteristic physical properties like melting points and mass-to-charge ratios of molecular ions. Other physical data may replace or supplement the foregoing ones, so many strongly supported patents include details of infrared, ultraviolet, proton, and carbon nuclear magnetic resonance spectra.

An inventor may squander his working life who musters in his patent a great body of experimental evidence. A wealth of supporting detail is wasted in patent applications and only needed to persuade a referee evaluating the manuscript of a scientific article. Most patent offices do not require this level of evidence, and many patent attorneys think it superfluous. Consequently, the decision to include or omit such minutiae becomes a matter of business strategy, not one of good science or legal obligations.

Remarkable Claims and Paper Patents

Patent law treats chemical inventors generously. To obtain protection, it does not demand they elucidate the structures of the compounds they claim. A so-called *product-by-process* claim lets inventors secure a patent for compounds of unknown structure. An allowable claim of this kind relates the starting material(s), solvent(s), reaction(s), and procedure(s) needed to produce the compound of interest. Effectively it specifies that the object of the invention is whatever remains when the inventor's instructions have been scrupulously followed. Such a claim offers uncertain protection, for a competitor can perhaps circumvent the patent by making the compound through a different synthesis.

Fingerprint claims are also available to chemical inventors who do not disclose the structures of the compounds that they invent. In such a claim, the inventor provides chemical, physical, and perhaps biological data characterizing the compounds but not identifying them. The claim is stronger the more data it provides. Enough data effectively define the invention and make allowable a claim to thoroughly characterized compounds of unknown structure.

Filing a U.S. patent application without having done any chemical or biological experiments at all is permissible. So is adding to it examples of experiments that were never done. The U.S. patent law presumes that individual inventors cannot afford to do all the experiments needed to realize their conceptions and thereby substantiate their claims (*Maynard and Peters*, 1991). It lets their applications include such examples, effectively to increase public benefits. Here is a European patent attorney's rationale for such examples, with italics added. "The inclusion of a paper example in a patent is *not* a false representation that the compound has been made; it is an honest representation that the compounds *can* be made in that way" (*Grubb*, 1982).

The mere act of filing an application containing such examples constitutes what patent lawyers call a *constructive* reduction to practice. The resulting grants are known as *paper patents*, which an inventor may later have to defend with the results of authentic experiments. Well composed paper examples therefore do not include numerical values of physical properties that are difficult or impossible to predict: melting points, for example, or optical rotations. They use the present tense, and reserve the past tense for genuine examples, a convention that not all patents honor.

The need to defend paper examples arises rarely because few patents interfere with other grants. According to the U.S. Patent and Trade Office, "about one per cent of the applications filed become involved in an interference proceeding." Pharmaceutical corporations, although they precautiously treat all patents as valuable intellectual property, do not bring suit against one another without a monetary stake. Moreover, their attorneys need a strong case to prove allegations of interference by another patent or infringement by a competitor. Examiners, although they scrutinize applications, exercise no responsibility for judging or ascertaining the operability of working examples.

A valid patent presents an invention capable of use. The inventor's instructions for implementing his idea must therefore be operable. Not surprisingly, the procedures furnished for making compounds of the invention must provide those substances. But patent law does not require the reactions employed to be efficient; it suffices if they are merely effective. Indeed, the reactions employed may not even form part of the patent claims. Patent practice tolerates minute chemical yields if they fall within certain limits. Acceptable yields may be low enough to inspire disdain yet high enough to attract notice. Such scant yields do not invalidate claims.

Organizational Structure of Patents

A patent directed (for example) to biologically active compounds as drugs can fill nearly 200 square feet of paper because it runs for 300 pages. To extract specific information from such a volume, its reader must know what to look for and where to find it. Selectivity in reading the sections of a patent is not only helpful, therefore, but also essential. It aids in the efficient use of working time.

Like any compound, a patent has a certain structure. This organization recurs from one patent to another and rewards study because it eases reading. Noteworthy structural elements of a utility patent summarize bibliographic facts, present any drawings necessary to understand the invention, and give a background. They summarize the invention, describe it in detail, stake claims, and abstract the patent. Next-to-last in appearance, the claims are first in importance. They define the protection sought by the inventor(s) and granted by the patent office. Each of the major sections of a patent has its own substructure.

Bibliography

The bibliography appears on the cover page. It presents the title, inventors, assignee, attorney, filing and issue dates, other information, and, in some cases, the *Markush* structure (see text below). Patent titles may or may not explicate the inventions covered, depending on the business strategy adopted by the agent or attorney responsible for the application. A title can be broad enough to reveal nothing about the nature of the invention. What it's all about a reader must learn elsewhere in the document.

United States patents list their inventors exactly, neither excluding anyone who contributed to the invention nor including anyone who did not do so. Because corporate inventors usually assign their rights in any workplace invention to their employers, the roster of inventors often differs from the list of assignees. The latter list ordinarily includes only the name of the organization sponsoring the research. Patents tell which company took an interest in what area of research about 2 years before the inventors filed their application. (Incidentally, U.S. patents offer means to identify and contact chemists, biologists, and other scientists all employed by the same pharmaceutical company. They do not, however, specify which inventor practices which science, so a reader cannot identify the chemical inventor(s) from the patent alone.)

Markush Structures

Named after a patent attorney, a *Markush structure* is a generalized drawing. It succinctly conveys the structures of all the compounds that the patent protects, which are known as *compounds of the invention.* Inspecting this structure can quickly introduce the subject matter of the patent to the reader, although some of them are so generic that they are uninformative. An understanding of these structures derives from examining them and reading the accompanying text. In a long claim, it demands to-ing-and-fro-ing within the document.

Figure 4-4 presents an imaginary but otherwise typical formula (**1**). A patent citing this structure defines the letters that compose it, using standard phrases in a conventional format. The definitions read like this:

m and *n* are independently chosen integers ranging from 1 to 3;
A represents NH or $N(CH_2)_m R^1$; and
R^1 and R^2 are independently chosen from hydrogen, aryl, substituted aryl, or lower alkyl.

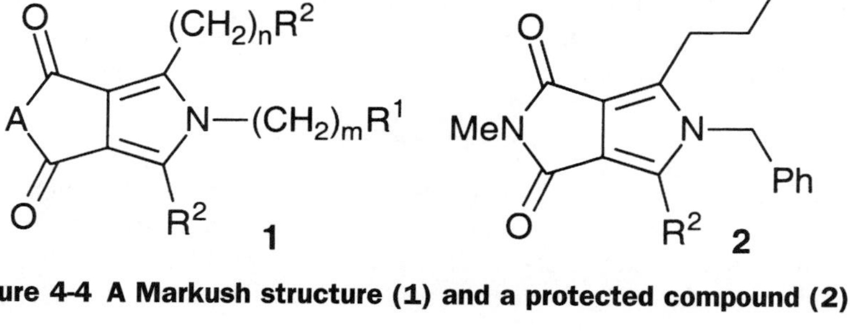

Figure 4-4 A Markush structure (**1**) and a protected compound (**2**)

The patent specifications define the terms *aryl, substituted,* and *lower alkyl. Aryl,* for example, might be restricted to benzenoid aromatic rings. Typically, the specifications but not the general structure would cover any geometrical isomers or enantiomers formed by the compounds of the invention.

Structure **1** covers the bicyclic lactam **2** (Figure 4-4), which meets the following conditions: m equals 1, n equals 2, A is MeN, R^1 is phenyl, and R^2 represents methyl in one instance and phenyl in the other.

Background and Summary

Following the bibliography appears a section stating the problem that the invention solves and showing that the problem merits a solution. For example, the background might simply state that arthritis sufferers need a drug that relieves, say, joint inflammation. The background relates the history of the problem, and the attorney who composes this section will ask the inventor to prepare him with a disclosure as complete and candid as possible. Suppose, for example, that some aspect of a competitor's patent is inoperable and that the present invention overcomes that obstacle. Then, in the background section, the attorney calls the inoperability to the patent reader's attention.

For example, imagine the inventor suspected that a competitor's claim to certain compounds is deficient. His patent fails to teach any workable method for making them and instead alleges to have prepared them by a chemically unreasonable reaction. These compounds would therefore not have existed before the present invention, which would be patentable. To succeed in patenting his own inventions, a chemist must find, read, and understand his competitors' patents. Attorneys and agents without training in organic synthesis cannot be expected to spot unworkable synthetic methods.

The *summary of the invention* presents the kind of patent that lies in hand, for example, a utility one dealing with compositions of matter, and shows that the invention possesses novelty. It indicates that the patent is directed to chemical compounds as drugs and names their broad structural class. The summary also specifies the diseases that the compounds prevent or treat and mentions methods of making and using them. Compared to summaries from the chemical literature, patent summaries can be insubstantial: many convey few particulars of the invention, which appear in the detailed description instead.

Detailed Description

There follows next a detailed description of the invention, which usually forms the largest section of the patent. The description distinguishes the invention from any other workable solution to the problem. To persuade the examiner that the invention is novel, it cites any relevant prior art. The description shows how the present invention represents an advance over any previous attempts to solve the problem of interest. It argues that they did not anticipate the solution offered now. An application covering antihypertensive drugs would show how the claimed phenethanolamines, for instance, differ structurally from those already known to lower blood pressure. Large structural differences ensure that patent claims possess novelty and lack obviousness. If the novel compounds of the present invention act by an innovative biological or biochemical mechanism, the specification points this out to strengthen the application.

In a long, minutely detailed passage, the description sets out the scope and limitations of the invention. This section defines the variable features that compose the *Markush* structure. It includes all the claims appearing within the separate claims section in the draft submitted to the patent office. The scope of the *specification*, however, may exceed what the examiner finally allows in the issued patent. Anyone reading an issued patent to find out whether a particular compound is claimed should therefore base an understanding of the coverage on a study of the claims, not the specification. The original specification usually undergoes no alteration to reflect the changed scope of the claims finally permitted.

To define the scope of the invention, the specifications may include the kinds of hydrates, solvates, salts, and isomers that the compounds of the invention might usefully form. Continuing to set boundaries, the description defines the terms in which the invention is couched. For example, it might specify that the phrase *lower alkyl* means side chains containing, say, one to seven carbon atoms, including branched chains and cyclopropyl groups. It might ostensibly define the term *aromatic heterocyclic,* furnishing several pages of structural drawings that depict such groups. In composing these and other definitions, attorneys often exercise more freedom than chemists might do. But, to obtain coverage, they must define in writing any extraordinary usages and do so without offending any scientific principle. Such a usage might, for example, exclude iodine from the elements that correspond to the term *halogen.*

Finally the detailed description teaches the reader how to use the invention. It offers the preferred means of doing so, which represent the best available to the inventor at the time of filing. In a patent directed toward chemicals as drugs, the application names the diseases that the drugs are to monitor, prevent, or treat. These diseases may be broadly or narrowly specified, for example, as inflammatory conditions or, within that category, arthritis. Drugs are administered differently depending on their chemical and physical properties and on the diseases they treat. Consequently, the patent application may state how the inventor expects patients to take the drugs. It lists some or many of the usual routes, which are oral, rectal, intravenous, intramuscular, topical, inhaled, or sublingual.

Topical administration, for instance, is the preferred route for certain drugs to treat skin diseases. Treatment therefore demands a cream that clings to the affected area. How to prepare this cream and what ingredients it contains are sometimes related by the patent: knowledge of them seems essential to practicing the invention. The description details the percentage composition of the active ingredient in the cream, for example, as a 10,000-fold range. Other ingredients and methods of formulating them—pressing them into tablets, for example—may also appear in this section. However, some patents of new chemical entities do not disclose formulation details. The responsible attorney declines to make what he regards as a limiting, strategic revelation.

To obtain protection of chemical compounds as human drugs, a patent offers evidence of medical utility. Such a patent is sometimes issued and ordinarily sought before any of the compounds claimed even enters clinical trials. Consequently, these testing results are not available to the inventor and his attorney when they write their application. Therefore, they depend for their evidence on the lesser showings of live-animal models of human disease or upon functional or biochemical assays *in vitro*. Their patent application includes a procedural description of the assay used. It offers enough information to let any suitably trained and equipped pharmacologist or biochemist replicate the work. Often the patent gives the biological findings. They take the form of a table linking four essential elements: the assay adopted, the compounds tested, the doses or concentrations used, and the results obtained. The simplest kinds of numerical results frequently suffice.

The detailed description discusses the chemical reactions used to prepare the compounds of interest. It illustrates these changes with structural drawings of starting materials, intermediate compounds, and final products, compiling the drawings into schemes. These reaction schemes, unlike many counterparts in chemical textbooks and journals, employ generalized structures and reagents rather than individual ones. The accompanying text puts broad limits on reaction conditions, solvents, and times, not specific boundaries. For example, it might call for temperatures ranging from 0 to 150°C, saying that the preferred range for a given reaction equals 25 to 50°C, for instance. The description offers ranges even though the inventors used exactly one temperature to carry out the reaction. Although they may never investigate the effects of higher or lower temperatures or ascertain any preferential conditions, patent practice insists they speculate. Chemical inventors therefore choose ranges that they believe to be workable. To identify the conditions and reagents actually employed by the inventors, a knowledgeable reader seeks the phrases *most preferred* or *more preferably*.

Patent attorneys generally encourage inventors to make reasonable conjectures to expand their conceptions. Their patent applications should bring the broadest coverage that the inventors envision rather than the narrowest scope that their experiments validate. No thoughtful miner stakes a claim exclusively to a blanket-sized area that yielded gold-bearing ore.

Next the specification may introduce *working examples* of the invention, which concern the compounds claimed as drugs. Examples—although they are *not* required by patent laws—help demonstrate the operability of the invention, provide physical constants for the claimed compounds, and give detailed instructions for making them. Such compounds are ordinarily the pure products of deliberately chosen and run chemical reactions, not items of commerce that can be purchased, mixed, repackaged, and sold. Consequently, the inventor must furnish his procedure for making them. This recipe enables someone skilled in the art of synthesis to practice the invention. So, it may relate facts mundane to a chemist but crucial to an examiner—for example, the names of a supplier selling commonplace equipment or a well known reagent.

Working examples take the familiar form of standard operating procedures that a chemist records in a scientific notebook as he carries out chemical reactions. They state concisely which compounds of the invention arise from these reactions. The accompanying list of compounds gives their complete chemical names or sometimes structural drawings. It identifies the compounds, which in either case can be recognized by the physical property or properties that the example furnishes.

Minimizing details, a single working example usually includes all the relevant products realized from only one chemical reaction. It does so even though a given compound of the invention results only from the last of a sequence of many essential reactions. Descriptions of these other reactions appear in a separate section of the specification devoted to so-called *preparative examples*. Each of these examples lists the chemical names and physical constants of the intermediate compounds. Like working examples, preparative ones also describe standard operating procedures. The difference between these and working examples is that the latter treat only compounds of the invention, while the former present only intermediate compounds that are not the subject of claims.

Some patents shun specific structural drawings, which instantly and unambiguously convey meaning to anyone skilled in the art of organic chemistry. Often they do not use Arabic numerals to designate individual compounds. Such numbers, or even letters or Roman numerals, are second only to drawings as identifiers. In contrast, unnumbered systematic chemical names—on which so many patents exclusively depend—finish dead last. To establish from such a patent the exact synthetic route that forms a given compound demands work if the synthesis is long.

Tracing the steps of a synthesis repeatedly entails searching for a particular chemical name, for example, of an essential intermediate compound. Such a job resembles the task of walking a reluctant cat backward through its day. The name appears within a list, but even a short patent can contain half a dozen or more of these lists, each referring to scores of organic compounds. So, to find the compound requires the reader to locate the proper list. Because all the compounds within a given list represent minor structural variants, the names differ only slightly from one another. An exemplary difference between two distinct but easily confused names appears in 10-phenyl-2,3-dihydroimidazo[1,2-*a*]pyrido[2,3]pyrimidin-5(10**H**)-one and 10-phenyl-2,3-dihydroimidazo[1,2-*a*]pyrido[2,3]pyrimidin-5(10**H**)-thione (Claim 11 in U.S. Patent 4,725,596 (Feb. 16, 1988)). Only three characters of 66 distinguish these names from one another.

To understand a patent usually requires the reader to draw some of the many structures of the compounds cited. Complete structural drawings express knowledge of the long systematic chemical names compiled in the claims and in the preparative and working examples. Often, to compose names from structures, or draw structures from names, calls for the help of a librarian or another specialist trained in nomenclature. A cornucopia of nomenclature is "The Ring System Handbook."

The headings or subheadings within a patent may not clearly distinguish working from preparative examples. So, it is helpful to know that often the last reaction of a multistep synthesis appears first because that reaction makes the strongest claim to a patent reader's attention. With such a nonchronological organization, the inventor relates the synthesis by reversing the sequence in which he executed its steps. He presents last the structures of the compounds used first, which are those of the starting materials.

Claims

To read a patent, it is wise to begin with the material that ends the document, namely the claims (Table 4-3). This section usually begins with the phrase, "What is claimed is. . . ." Containing the most important information, it receives the attorney's and inventor's greatest efforts. The claims define the invention and specify the rights that the patent office grants, often after prolonged negotiations. They may run for hundreds of words, all composing a single sentence. So, it helps understanding to read such a sentence one phrase at a time. This pace makes for slow going, but a reader takes comfort knowing that the claims abjure boilerplate and present essentials. A patient patent reader pauses at semicolons to await comprehension as a bicyclist dismounts at stoplights to wait for opportunity.

Patents of pharmaceutical inventions contain one or more of five types of claims. They pertain to individual compounds, other compositions of matter, chemical processes, methods of use, or medical devices. Formulations like creams or tablets exemplify the subject matter of some composition claims, which may include the active ingredient as well as other substances

Table 4-3 Speed Readers' Guide to Composition-of-Matter Patents

Industrial organic chemists often consult a pharmaceutical composition-of-matter patent only to answer the following two questions. To do so, they need to know both the structure of the compound and its systematic chemical name.

Does this patent protect this compound?

Search for the numbered claims, opening the patent at the back and turning toward the front.

Look for a typical opening phrase, "What is claimed is...."

Find the composition-of-matter claims, which usually open the section.

Search the specific claims first, looking for the compound of interest among the particular structural drawings, taking the next step if you do not find what you seek.

Compare chemical names to match the compound of interest to a substance in the specific claims. Move on if the compound of interest is neither drawn nor named in the specific claims.

Find the broadest generic claim, which ordinarily contains a general formula and begins the set of composition-of-matter claims.

Substitute the structural components of the compound of interest into the formula.

How did the inventors make this compound?

Find the working examples, which compose a section of the document marked by the word "Examples."

Search each example for the claimed compound of interest, looking among the particular structural drawings. Move on if the examples give names rather than structures. Go to the next-to-last step if you find a drawing of the structure that interests you.

Compare chemical names to match the protected compound of interest to a substance in the examples. Continue if you do match the name.

To find a method for making an intermediate substance, look in the preparative examples.

Repeat the preceding step to find all the intermediate compounds of a synthesis.

---

in specified proportions. Binders, for example, help hold tablets together. Chemical process claims concern methods for making the active ingredients by specific reactions, reagents, and starting materials.

To foresee that a compound acting by a particular pharmacological mechanism can treat a specific human disease reflects a certain inventiveness. Patents therefore include method-of-use claims in which, for example, an anti-inflammatory substance relieves arthritis. Method claims may also extend to the use of compounds for diagnosing, monitoring, and preventing diseases. The compounds selected need not be novel, nor must the disease have been hitherto untreatable. New treatments by old compounds justify patents, which need not contain any other kind of claim.

Pharmaceutical patents helpfully identify the kinds of claims they contain. Each claim opens with a typical phrase such as "A compound having structural formula I. . . ." It often continues by defining the substance with a general formula, sometimes taking one or more pages to do so.

Another example concerns a method of use. This claim starts with "A method of treating arthritis in a mammal. . . ." It goes on to relate specifics of the method: ". . . comprising administering to said mammal an antiarthritis effective amount. . . ." It ends specifying the compounds useful in this treatment: ". . . of an anti-inflammatory compound of structural formula I. . . ." The claim would repeat all the material of the preceding compound claim, again defining *Markush* structure I. Staking claims is as repetitious as digging postholes. Repetitions, however, avoid gaps in the metaphorical fence with which a patent surrounds intellectual property.

There are two other points to be made concerning the method-of-use claim in the preceding example. It is broad in one sense and narrow in another. Breadth comes from the utility of the compounds in treating mammals rather than humans only. One of these compounds might become an animal health drug, and the wording of the claim preserves the right to such a use. The claim is narrow because it deals only with arthritis, which is one of several inflammatory diseases.

Breadth or narrowness also characterizes other kinds of claims, particularly to compounds. An attorney skilled in composing compound claims begins with one having the greatest structural scope. Succeeding claims progressively restrict the breadth of coverage by including fewer and fewer of the substances covered by the preceding claims. Narrower claims continue until one of them lists individual compounds. The narrowest one identifies these compounds by unambiguous chemical names or structural drawings. Broad structural claims are known as *generic* ones, while the narrowest one is called *specific*.

This ultimate claim to individual compounds can be important, even though it is the narrowest. It contains the name or structure of the best compound of the invention. If the research culminated in a clinical candidate or a drug in clinical development, then that valuable compound appears in this list. The list, however, does not identify the substance chosen as the drug for marketing or the compound for development, nor does it indicate that any compound was selected. Nevertheless, the list includes the substance and thereby satisfies the best mode requirement.

Progressive narrowing of claims provides many fallbacks, which are desirable to retain as much coverage as an examiner allows. All is lost, obviously, if a patent contains a single broad claim that he denies. But much is saved with a succession of increasingly narrower claims, which an accomplished patent attorney writes with fluency and an examiner dismisses with difficulty. An all-encompassing claim resembles the outermost layer of an onion. It can therefore be sacrificed, just as a cook discards the crackling brown skin.

Inventorship

Myths swarm to a patent like bees to a hive. Respectively, the prestige and extra compensation supposedly associated with inventorship exemplify two popular myth-understandings. Recurring often, these conceits threaten to sting newcomers with the acid of reality. Another myth pertains to inventorship as a just and due reward for efforts to realize a patentable invention. This last misunderstanding can persist throughout decades-long careers, embittering some believers who do yeoman work but make no inventive contribution. Learning the meaning of inventorship as it pertains to U.S. patents dispels misunderstandings among industrial chemists at all degree levels.

Definition

Invention is a mental act in which the originator conceives the patentable idea and provides a workable means to realize it. His status as an inventor does not require or forbid him to actualize the idea himself, which coworkers may do instead. Although they elevate the idea to practice, their efforts—prolonged, extensive, or successful as they may be—do not entitle them to share inventorship. Merely making the compounds of the invention does not suffice to earn this status, even if one of these substances becomes a best-selling medicine. Joint inventorship of patents is commonplace, however, but because each of the inventors substantively contributes to at least one claim.

Controversy

"... [P]atent departments look forward with dismay to ... disputes about inventorship" (*Grubb,* 1982). Disputes arise for several reasons, and, in large companies, they make the task of ascertaining inventorship one of the thorniest aspects of patent practice. The duty falls to patent attorneys, who rely on the signed, dated, and witnessed documents known as disclosures or records of invention. These records, kept preferably in bound, numbered notebooks, are succinct histories prepared by some inventors when they conceive their ideas. Correct inventorship is crucial to the validity of U.S. patents, which differ in this way from grants issued by other countries.

Glory

Although the chance of inventing a best-selling drug may be small, being named an inventor of such a medicine is a source of enduring satisfaction and not a little prestige. However, less prestige may attach to a patent claiming no compound that becomes a marketed drug, undergoes extensive clinical development, or even enters human trials. Alas, not all research projects culminate in the choice of a compound for a Phase I human study, although many efforts do evolve compound patents. Some companies file patents as territorial defenses. They mean their filings to repel competitors who might otherwise enter or overrun a seemingly fertile field of research. If the field proves barren because no clinical candidate emerges, the patent fails to justify its filing and maintenance costs. Alternatively, a pharmaceutical company understandably loses interest in certain compounds acting by a particular pharmacological mechanism. The loss arises when its drugs or its competitors' fail clinical trials. For one or more of a variety of reasons, therefore, FIPCOs abandon many patents that claim compounds as drugs.

Rewards

Nevertheless, some pharmaceutical companies do recognize inventors' contributions to their businesses. With in-house newsletters and ceremonies, some of them celebrate the scientists whose discoveries led to valuable patents. To reward success and spur innovation, a Japanese pharmaceutical company offers extra compensation to its researchers. Takeda awards a bonus of as much as $427,000 to a scientist whose drug brings sales of more than $60 million. No more than four researchers may share the award.

In the United States, knowledgeable inventors abandon hope of compensation beyond ordinary salary and benefits. Nothing extra is due them under U.S. law or in patent practice, although German and British firms must offer extra compensation linked to the financial success of patented drugs. Here, however, newly hired chemists begin work by signing certain agreements. The scientists exchange their rights to employment-driven inventions for regular salaries, not extra payment. The agreements oblige them to transfer the rights to their employers by executing separate documents called assignments. Chemical inventors sign these documents shortly before their patent attorneys file their applications with the Patent and Trade Office. And, in doing so, the chemists waive all monetary claims to their pharmaceutical inventions.

Laboratory Notebooks within Industry

Who invented what, and when did he do so? These questions are among those that well-kept notebooks answer in an industrial laboratory. Like a university research group, an industrial one depends on notebooks—still largely handwritten—to describe experiments and to record research failures and successes. Notebooks in both places contain the raw materials of finished scientific articles. Corporate laboratories, however, make additional demands on their chemists' recordkeeping. The contents of their notebooks must inform patents, help in prosecuting applications, and provide credible documentary evidence when necessary. Proper notes of experimental work can be especially important in obtaining U.S. patents.

In prosecuting applications, good notes can help to establish the novelty, utility, nonobviousness, or operability of an invention. Few applications emerge from prosecution unscathed by examiners. So, for example, a dated entry can show that one chemist's invention preceded another's publication of the same idea and thus aid in procuring patent coverage. Biological or chemical data recorded in a notebook, but additional to what a patent application presents, can help establish the utility or operability of an invention. Even a record of a failed experiment can have value. It may convince an examiner that the invention claimed in a patent application lacks obviousness; it is arguable that the first attempt would have succeeded if the invention were obvious. A handwritten entry identifies its author and thereby gives evidence of inventorship. Such questions sometimes arise after a patent issues, if a competitor infringes it, or before, if two competitors make the same claim. In the latter case, the Patent and Trade Office declares a so-called *interference*.

When two U.S. patent applications interfere with one another, a notebook entry can decide the matter of which applicant receives the patent sought. Such an entry must be signed, dated, and witnessed. In U.S. law, the party first to invent wins the patent, not the one first to file. To establish first inventorship calls for credible evidence, which is most convenient in the documentary form that authenticated notebooks offer. Patent attorneys can use notebook entries to show that company scientists were diligent in implementing a given invention or in attempting to do so. Abandonment of an invention can justify awarding the pending patent to a competitor in case of an interference.

Six crucial features of uncompromised records include binding, inking, crossing out, witnessing, dating, and filling in.

1. Notebooks should contain only bound pages, not loose ones, because the latter can be (dishonestly) inserted out of chronological sequence. The possibility of such an insertion weakens a loose-leaf record.
2. Notebooks demand ink because penned notes are difficult to alter inconspicuously.
3. Mistaken entries should be crossed out legibly, at the expense of a sloppy-looking record. Such a record is preferable to an otherwise perfect one

showing erasure abrasions, which conceivably correspond to dishonest alterations of the description or date of an important experiment.
4. Notebook entries demand frequent witnessing by scientists. The witnesses chosen must be noninventors capable of understanding the experiments described, so the best ones are other chemists.
5. Both witnesses and experimentalists must date their signatures, which should be contemporaneous with the experimental descriptions.
6. Diagonal or vertical lines should fill in blank pages or half-pages between entries. Later, therefore, no one can undetectably add to the existing record in a fraudulent attempt to establish priority of invention.

SOURCES AND SUPPLEMENTS

Anonymous; "Drugs Under Patent: the Comprehensive Guide to FDA-Approved Pharmaceuticals Under Patent & Market Exclusivity," FOI Services, Gaithersburg, 1996.

Anonymous; "The Ring System Handbook," The American Chemical Society, Washington, DC, 1993.

Bailey, P.; "The art of lecturing," *Chemistry & Industry,* 7 March, 90 (1994).

Bryant, J. L.; "Protecting your Ideas: The Inventor's Guide to Patents," Academic Press, San Diego, 1997.

Committee on Patents and Related Matters; "What Every Chemist Should Know About Patents," American Chemical Society, Washington, DC, 1997.

Committee on Patents and Related Matters; "Electronic Record-Keeping for Patent Purposes: Cautions and Pitfalls," American Chemical Society, Washington, DC, 1990.

Committee on Patents and Related Matters; "Record-Keeping Fact Sheet," American Chemical Society, Washington, DC, 1988.

Grubb, P. W.; "Patents for Chemists," Clarendon, Oxford, 1982.

Grubb, P. W.; "Patents in Chemistry and Biotechnology," Clarendon, Oxford, 1986.

Hargreaves, M. K.; "Hints for lecturers," *Chemistry in Britain* **16**(10), 552–555 (1980).

Horton, B. J.; "Dictionary of Modern Chess," Philosophical Library, New York, 1959.

Kanare, H. M.; "Writing the Laboratory Notebook," American Chemical Society, Washington, DC, 1985.

Knight, H. J.; "Patent Strategy for Researchers and Research Managers," Wiley, New York, 1996.

Kubie, L. S.; "Some unsolved problems of the scientific career," *American Scientist* **42**, 111 (1954).

Maynard, J. T., and Peters, H. M.; "Understanding Chemical Patents: A Guide for the Inventor," American Chemical Society, Washington, DC, 1991.

Maynard, J. T.; "Understanding Chemical Patents," American Chemical Society, Washington, DC, 1978.

Saliwanchik, D. R., and Saliwanchik, R.; "Effective documentation of research results," *Drug News and Perspectives* **5**(5), 306–313 (1992).

Souleau, M.; "Legal aspects of product protection—What a medicinal chemist should know about patent protection," Chapter 41 in "The Practice of Medicinal Chemistry," C. G. Wermuth (ed.), Academic Press, San Diego, 1996.

U.S. Patent and Trade Office; http://www.uspto.gov/web/offices/pac/doc/general/interfer.htm, 1999.

Woodward, R. B.; "The Total Synthesis of Strychnine," XIVth Intl. Congress of Pure & Applied Chemistry, Zürich, 21–27.VII.1955, Birkhäuser Verlag, Basel u. Stuttgart, 1955, p. 222.

chapter

5

DISCOVERY RESEARCH

Medicinal Chemistry

"The most fundamental and lasting objective of synthesis is not production of new compounds, but production of properties."—George S. Hammond

INTRODUCTION: ONE DAY IN THE WORK OF A MEDICINAL CHEMIST— A FICTIONAL ACCOUNT

Eight-fifty on Monday morning. The medicinal chemist sits before a computer, issuing what she prays is the next-to-last command to print. Unable to finish her report before now, she needs to proof it again to satisfy her professionalism. It is too easy to make confusing mistakes in more than 100 citations of 75 numbered items including literature references and new compounds. Tapping the print key, she glumly foresees doing several hours' work checking three tables of numerical data and 55 numbered drawings. They depict chemical structures and compose 10 reaction schemes. She rises, leaves her office, crosses her laboratory, and enters a corridor to collect her two-dozen laser-printed pages.

This chemist is 29 years old. She embarked on her Pharmacorp career only 11 months ago. She is 3 years removed from the beginning of her postdoctoral research and 7 years distant from the start of her Ph.D. studies. Enrollment in the college that conferred her baccalaureate degree seems as remote as the Cambrian age, 600 million years before the present era.

After 2 hours of reading and correcting, she is nearly done. Her readers, however, will little note nor long remember this report. Despite its length and complexity, it describes no compound entering clinical trials. Her current project, which employs seven chemists and two biologists, has yet to yield a promising molecule. Neither a failure nor a success, the nascent project is to vanish like dirt beneath a broom, if whispers about a new executive are credible. "Why write these chronicles of wasted time," she asks herself, "when composing them defers the experiments that would advance new research?" Recognizing that bureaucracy pervades any large organization, she grouses good-humoredly. "At Pharmacorp," she says aloud, "presenting your results is more important than getting any."

Delaying experiments planned on Friday, her computer screen pulses at 11 AM. One of her colleagues, who is a biologist, is sending her a message that discloses the latest result of screening for inhibitors of an enzyme implicated in a particular illness. Were a drug to inhibit the enzymatic activity, physicians might treat or cure a hitherto intractable disease that is widespread and devastating. Discovering and developing such an inhibitor, and introducing it to medicine, form the goals of her emerging project.

Several months ago Pharmacorp embarked on this venture after a management meeting. Representing chemistry, biology, clinical medicine, sales, and marketing, the executives convened to judge the biologists' research proposal. They decided that the existing market for such a drug warranted the $600 million that discovery and development cost. The biological foundations of the therapeutic approach were sound, their corporate competitors had not already marketed such a drug, and Pharmacorp's entry could dominate the market by reaching it first.

Nevertheless, persistent tension attends the decision. A competing company might already have begun the same research. Perhaps it would devote more in budget, time, and manpower than Pharmacorp proposes to invest. Its researchers might have begun 1 to 2 years ago. If they keep this lead, they will bring their drug to market first. Pharmacorp scientists would not learn they were racing or losing until the winner seized the prize money.

Initially the new project requires no synthetic chemists, partly because no one knows the structure of the enzyme substrate. Moreover, no existing evidence suggests that any naturally occurring compound suppresses the enzyme, let alone one with a known chemical structure. Without that crucial structural knowledge, chemists are powerless to mimic in another molecule the important features of any substance that inhibits this or any other enzyme. Drugs do not arise through inventing molecules, but largely by modifying existing structures. Synthetic chemists who create drugs require knowledge of active structures as automobile mechanics need defective cars to work on. "For the future," thinks the medicinal chemist, "drug design will remain more a conceit than a reality." Instances of successful drug design, which introduce new chemical entities to medicine, resemble second marriages, for they are triumphs of hope over experience.

She reviews the biologists' progress. To begin the approved project, they import, invent, and validate the needed assays. As one of two primary screens, they choose a high-throughput assay that detects inhibition by components of mixtures prepared artificially. They also devise a lower capacity test to quantify inhibition by isolated samples. Pharmacorp biologists create a battery of other enzymatic assays for profiling the selectivity of the compounds tested. Selectivity is essential, lest an inhibitor undesirably halt a normal physiological process. Other newly emplaced tests later measure the effects of enzyme suppression on experimental animals challenged with the disease. With a restricted number of newly made compounds the pharmacologists plan to correlate these effects to the influences of dosing orally and intravenously. Giving the inhibitor orally, they plan to investigate the time course of drug action, periodically noting duration and measuring intensity. In parallel with the development work, and 2 months ago, the biologists started screening the many samples in Pharmacorp's 50-year-old collection of proprietary organic compounds. Once daunting, this still formidable task is now commonplace throughout the international pharmaceutical industry.

*According to her counterpart's message on the screen before her, one compound from the latest batch tested in the primary assay shows the activity desired. It is also more potent than those substances turned up initially. Only a few weeks ago, her colleagues in biology surveyed nearly 100,000 compounds, merely finding the few weakly active ones she tabulated in her report. Recently they began to test a second set of 100,000 compounds, soon discovering the activity of this one substance. She may have a solid lead in Phrmco **133997**.*

It is an old compound, which she infers from the low number, but one that belongs to her employer, as the simple code name implies. A more complex name would have meant that Pharmacorp bought certain rights to the compound from a university researcher or a drug discovery company. But, in this instance, all the rights belong to Pharmacorp, where one of her predecessors made the compound for a purpose unrelated to her present effort. Only 18 months ago did academic biochemists elucidate the enzymatic function that Pharmacorp seeks to interrupt. So recent is their work that no one in industry or academe has crystallized the enzyme. Without crystals, no X-ray analysis of the enzyme is possible, so its three-dimensional structure remains unknown. "No help there," she notes.

*She wonders what structure Phrmco **133997** has, thinking that her bosses will be inquiring soon. They are also chemists, whose names appear on the biologist's e-mailing list. All these chemists have to guess or ask because his note lists no structures, conveying only code names and numerical measurements. So she searches a computerized database, soon finding the structure of **133997**, which a call to the compound registry would also have done.*

*The woman arranges by a telephone call to take lunch with a veteran coworker. In the cafeteria she mentions the structural class to which Phrmco **133997** belongs. Her remark elicits an estimate of its age: 25 years and old enough to have been repeatedly studied in various assays at different times. However, this substance never progressed to clinical trials or preclinical animal toxicology, her graybeard colleague says firmly. It and analogous compounds were inactive in the assay for which they were made. Subsequent studies revived no interest in these compounds until now. As far as her coworker knows, no adverse biological activity—nor any threat of it—taints the compound despite its age. Such a prospect perfuses drug discoverers with anxiety. Tarring with the same brush might prevent clinical development of a safe successor from the same structural class of compounds. Her chat with the old biologist—who recalls 40 years of corporate history—is therefore helpful. His knowledge prepares her, relieves her of any duty to bear unwelcome news, and frees her to pursue a promising lead.*

*Back in her office, she looks again at this morning's e-mailed message. In it the bioanalyst says he retested **133997**, confirming its inhibitory activity. Its structure therefore represents a series of active compounds with only one known member. His finding exhilarates the medicinal chemist because, in the new assay, this compound is the first to show any high activity at a low concentration. Other compounds in the tested group exhibit low activity at a high concentration. So impotent are these others that mak-*

ing new ones in the same structural series might imperil careers. She can still hear the naive but powerful vice-president for research demanding of one of her colleagues, "Why do you make inactive compounds?" However, Phrmco **133997**, which unites structural novelty with the minimum activity for a promising series of compounds, offers her first real chance to create a winner. Even if it never goes to clinical trials, its structure may point to other, structurally related compounds, among which lurks a billion-dollar drug.

The discovery challenges her, too. Not surprisingly, the potency demonstrated by the concentration–response study of Phrmco **133997** is insufficient to bring the compound into development as a clinical candidate. "Chance favors the prepared mind," said the stereochemist Louis Pasteur, but regardless of preparations, rarely is anyone's luck so good that the first active compound becomes the clinical candidate! As high as the **133997** potency is, it is still too low by a factor of 100. Research management declared that a strong candidate for human trials would inhibit 50% of the enzymatic activity at a one-nanomolar concentration in the primary assay. Her task is to make quickly a new compound with this potency but a different structure. Cycles of synthesis and testing lie ahead.

Her enthusiasm overcomes a recurring doubt. The e-mailed report gives no indication of the error associated with the inhibition measurements. Results and time pressure allowed only one determination at each concentration. So, the chemist doesn't know and won't learn how certain the result for Phrmco **133997** is. She has to believe the assay neither yields false positive results, nor overlooks highly potent compounds. The test does not discriminate among substances of low or modest potency. Nevertheless, her colleague's somewhat uncertain finding demands what Winston Churchill, wartime Prime Minister of Britain, wrote on important memoranda called to his attention, "Action this day!"

Next the chemist poses obvious questions with answers that help her advance the project. What weight of the compound remains on the company shelves? And what are the biologists' material and timing requirements? They need to profile Phrmco **133997** to secure proof of concept, even though this compound isn't slated for clinical trials. Material requirements increase as it passes successfully through their assays, while scheduling of assays becomes more demanding. Requirements initially amount only to milligram quantities of the compound. Given time and success, however, live-animal studies demand tens of grams and perhaps even hundreds. Delays in furnishing quantities of the experimental drug to the pharmacologists become visible and intolerable under these favorable circumstances.

"From now on," she says, "I'll have to supply our biologists." The substance is to serve as a standard in every project-related assay they carry out. Each time they test a batch of new compounds, they have to measure the inhibitory activity of **133997** as a control experiment. Any significant rise or fall in its activity warns of trouble with their assay procedures, which, absent the measurement, might go unnoticed. Under adverse circumstances, but without such controls, testing of new compounds can yield grossly inaccurate results, misleading attempts to create more potent compounds. The medicinal chemist, to reveal any disturbing trends, promises herself to plot the **133997** activity against time and to keep the graph current. Otherwise, her research, like an addled driver, might pass a throughway to enter a dead-end. Looking around her small office, she also resolves to fill the room with the lead compound if stocking up helps the biologists.

"However, Pharmacorp shelves may already store ample material," she thinks, "so I should get a sample of **133997** and check its purity." An old compound may have suffered photolysis in the fluorescent light illuminating the compound storage rooms. Such light, emitted by standard industrial lamps, contains energetic and perhaps destructive ultraviolet wavelengths. During long storage, the sample may have undergone hydrolysis or oxidation in moist air. To judge from its structure, however, **133997** should be as inert as brick dust. Hence, the likelihood that it suffered degradation seems small, although the consequences would be important. The observed inhibitory activity might have arisen not from the compound originally tested but from a potent, unknown impurity formed in a minute proportion. She needs to know the sample is pure or to discover any biologically active contaminant.

She visits the compound storage rooms, seeking the same batch of Phrmco **133997** that her colleague in biology assayed. Successful in finding the sample, she sees that the chemist's notebook number on the **133997** jar matches the number in the biologist's e-mailed note. The match helps establish that the active sample has the originally assigned structure, suggesting that no mistaken switch of samples occurred. Ready to determine thin-layer or high-performance liquid chromatograms and to record nuclear magnetic resonance spectra, the chemist departs holding a small vial. She'll soon know the purity and identity of the specimen that the bioanalyst tested.

Now the company's stock of **133997** is nearly exhausted. Can she buy the material that her biological colleagues need and, if so, how soon? There is no excuse for making what she can buy promptly. In time and money, a purchase costs less than a synthesis when overhead expenses are counted, as they must be. Her colleagues, she realizes, may need more than the few milligrams contained in this vial and remaining in the storage room. "It would be smart to look into this," she thinks.

Ascertaining commercial availability necessitates a manual or computer search through the catalogs of fine-chemical suppliers. At the larger companies, research librarians gladly assist with these searches, so the chemist visits the library to request help. She completes a form by entering the structure and, as a precaution, the empirical formula. Afterward, the information scientist in the library assures her that he can begin the work right away. His networked personal computer lets the scientist search the "ChemSources" and "CHEMCATS" databases and the Available-Chemicals Directory.

Leaving the librarians' suite of offices, the chemist enters the spacious glass-walled reading room. She chooses a copy of a medicinal chemistry journal, which lies on shelves displaying hundreds of new numbers of other scientific and medical periodicals. Seated at a long table, she spends 10 minutes leafing through the issue, viewing the structures and reading the titles. With this survey begins a 3-year-long effort to track competing companies' attempts to discover an inhibitor of the same enzyme that now engages her and her colleagues. Before studying any article, however, her attention wanes and she resumes planning.

"Who," she asks, "is to make Phrmco **133997** if no supplier carries it?" Well, the duty of furnishing as much as 100 grams may fall to her and her coworkers, even though they all work in a chemical discovery group. Chemical development efforts in her company often are reserved for drugs in clinical trials or production. More advanced projects than hers preoccupy the in-house scale-up scientists who work in chemical discovery research, as she does. However, with her good luck and her bosses' planning, the departmental budget permits her to buy help from an external custom-synthesis firm. But regardless of whose duty the task becomes, the medicinal chemist responsible for the project needs to know how to make the compound. If she doesn't prepare it herself, she must still furnish to other chemists written procedures for doing so. And giving them a sample from the vial in hand helps them validate their own experiments.

She wonders how Phrmco **133997** was originally synthesized. What reactions and starting materials did its makers adopt? Did running through the whole synthesis take a few days or a week or more? Answering these questions calls for another literature search to begin in-house. There one or more of her predecessors first made the compound, describing the work in witnessed notebooks. The scientific records department loans these notebooks to researchers. Also, it encourages the scientists to read and photocopy microfilmed records.

"Wait," she says to herself, "think again." She draws the structure and examines it. The best preparation of the compound, because of its structural size, surely requires many chemical steps. One or several chemists would have described these steps, possibly in many notebooks but almost certainly within pages scattered through any one book. Looking through the notebooks for a preparation of **133997** could be a labyrinthine task. Success in quickly finding procedural descriptions depends on whether several time-pressed chemists also designated the best synthetic route, even though they indexed and cross-referenced their many notebooks.

So, she might be wiser to look first for a published preparation of **133997**. An article in the chemical literature, if one exists at all, is easy to find. Published in any of numerous chemical journals, it furnishes procedures for carrying out the synthesis. An article also provides a wealth of physical constants identifying the compound of interest, which patents or other journals might neglect. Indeed, the main value of chemical journals to working chemists lies in the sections describing the experiments done. These write-ups preserve invaluable preparative details and characteristic properties for coming generations of experimentalists. "Berichte," she says aloud, naming an exemplary German journal. Nineteenth-century issues of Berichte der Deutschen Chemischen Gesellschaft contain procedures published nowhere else and never bettered in a hundred-and-fifty years.

"I'll look at the patent literature, too," she thinks, recalling a talk she heard. Attorneys from Pharmacorp's intellectual property department recently taught an intramural course entitled "Patents for Medicinal Chemists." And she remembers what one of the lawyers said during her lecture. A valid patent presents workable methods for making and using the compounds of the protected invention. It does so to meet a requirement that it enable its reader to practice the invention. Such a patent obligatorily offers the best methods known to the inventors when they file their application, even though its examples relate few procedural details and scant physical data. Consequently, agricultural, chemical, and pharmaceutical patents describe reliable preparations and reactions of many synthetically useful and easily procured molecules. Not reported elsewhere in the primary chemical literature, these compounds are not necessarily covered by the patent claims. To consult the patent literature is essential lest advantageous starting materials be overlooked.

Replacing the journal on its shelf, she returns to the librarian's cubicle to request a search for articles or patents dealing with the synthesis of the lead compound. The search may also illuminate the patentable novelty of Phrmco **133997** and its analogs. Although the novel finding that this substance markedly reduces the enzyme activity justifies several therapy claims, Pharmacorp wants more from its U.S. patents. Each patent costs a total of $8000–9000 to file, issue, and maintain. To justify expenses and exclude competition, the company seeks composition-of-matter coverage of the active compounds claimed in its applications. Consequently, it requires structural novelty for these materials, which a thorough literature search establishes. Pharmacorp needs active compounds with structures not publicly divulged in print or in speech.

She continues to think about the literature search, which may answer another question: what structural analogs of Phrmco **133997** have other chemists already made and published? Such substances may be items of commerce. If any are, then the chemist would be wise to buy and submit them for testing in the primary assay. Knowledge of their structure-activity relation directs a quest for novel compounds with high potency. Testing commercially available, structurally known substances can quickly and inexpensively illuminate this relation.

But the thought of this literature search indirectly raises a second query. What are the analogs from Pharmacorp files that the bioanalyst tested and found wanting? An acquaintance with their structures would prevent her and her coworkers from unwittingly making inactive or impotent target compounds. Because her colleague in the biology department has already given her the company code numbers for the assayed compounds, she can solve this problem easily and returns to her office to do so.

Exiting the library, she walks a corridor narrowed by chemists' supplies: cylinders of argon, drums of silica gel, chests of dry ice, and cabinets of glassware. To enter her office she skirts shoulder-high steel carts bearing vacuum manifolds mounted over oil pumps. Seated again before her computer, the chemist hunts through a company database correlating code numbers and inhibitory activities with chemical structures. This effort can be genuinely helpful rather than merely historical. Combined with a knowledge of the structures tested, even negative assay results can indirectly guide efforts to make new substances with increased potency. "Maybe I can draft a molecular activity map," she thinks. Such a map is a two-dimensional structural drawing showing atoms and bonds within a lead compound and annotated with the effects on biological activity wrought by molecular modifications. It would show her where in the structure chemical changes abolish the inhibitory activity that the modifications were meant to augment. By concisely recording these changes in a

> *graphic format, molecular activity maps prevent medicinal chemists from unknowingly making new compounds incorporating only minor structural modifications. Such compounds tend to possess the same inactivity or activity as one another. The maps therefore tell a chemist what not to make and show where the lead structure invites drastic new changes; yet they do not specify which compound to make next, but instead evoke creative answers to a perennial question. And, in the present case, such a map would usefully illustrate the research presentation she has to give next month.*
>
> *Having done what she could to advance her new project, she goes into her lab to do a job growing old. This is the column chromatography of a crude reaction product, a task put off last week and postponed again this morning. If successful, it would afford a pure sample of one of the last compounds she made for biological testing in her current project. Securing such a sample is not worth doing for completeness—a quality rarely desirable in discovery research—but submitting the stuff for testing would let her score again. "Another target compound," she thinks. It boosts her productivity and may be a barn burner like **133997**, if not in today's assays then in tomorrow's.*
>
> *Six-ten PM. Finished with her chromatography, the chemist sets aside a solution of the purified product to crystallize overnight, reflecting on Pharmacorp's emerging project and asking how next to exploit the biologist's finding of enzyme inhibition. At once her leisure reading floods her mind, and she recalls* Robert Robinson's *remark about a chemist synthesizing new compounds as anesthetics. Now as well as 60 years ago, "It [is] obviously necessary to extend the series . . . to find the best member."*

One day in a medicinal chemist's working life can evoke the exhilaration and urgency portrayed here. In some respects, however, such a day is atypical. It represents a transition between research projects, so it comes rarely. A summons to a researchers' meeting more probably heralds the change than does a biologist's report. Moving the new project from planning to action may demand the labors of several scientists, not one alone. To work effectively, they need understanding and experience as much as—or more than—advanced degrees. Several researchers may cooperate to complete the initial jobs or work independently to do so. Nevertheless, the tasks depicted reveal some of the underpinnings of a project in chemical discovery research.

Writing for an online career center (www.wetfeet.com), *Bonnie Wasserstein,* a B.S. scientist employed by a FIPCO, describes a typical day in the working life of any medicinal chemist. Her portrayal, presented as a "Real

People Profile" and found under "Pharmaceuticals," depicts the multitasking that is commonplace in industrial chemical research. Other aspects of the work are introduced in the remainder of Chapter 5.

Organization, or Who Chooses What

Underlying any piece of chemical discovery research is a wealth of decisions necessary to begin and advance the endeavor. Responsibility for making these choices is unevenly distributed throughout a hierarchy of employees. They range from newly hired B.S. scientists through veteran researchers and directors to vice-presidents for chemistry and for biology. Representatives of other parts of the company take part in the more momentous decisions. Each participant labors under certain restrictions but also acts with some autonomy. Job hunters, to understand the freedoms and fetters of the positions that they seek, need to know who makes which decisions. Although no universal answer to that question appears here, for none is independent of time, place, and other circumstances, a summary may help in posing the query (Table 5-1). Chapter 6 details the matrix organizational structure.

Getting Started

Chemists can expect to begin their careers with individual projects conceived by their bosses. Newcomers' projects typically form part of a larger effort and offer easy acclimatization to research in a corporate environment. For instance, a newly hired chemist may be asked to make a series of target compounds, include some substances with structures specified by her boss, and suggest relevant structural variations. The synthesis she adopts is usually a reliable one, having already been successfully applied to structurally related examples. To operate the synthesis may require her only to use existing supplies of intermediate substances to prepare the target compounds and to photocopy and follow procedures described in her colleagues' notebooks. The job may otherwise call for buying or locating starting materials and for making her own supplies of those intermediate compounds. In either case, the researcher's initial experiments take place within a framework familiar to chemists who conduct research in universities.

To make the foregoing target compounds, a newly hired scientist soon finds herself devising new synthetic routes or looking up old ones. Which responsibility she receives depends on her training, experience, ability, and knowledge of organic synthesis. Depending on these and other factors, she can (or cannot) look forward to deciding herself what target compounds to

Table 5-1 Preparations for a Discovery Research Project: Who Decides What?

Choices	Choosers
Entering a therapeutic arena	Managers from sales, marketing, finance, biology, chemistry, and clinical medicine
Making a biological approach	Biologists and research managers
Allocating chemists and biologists	Research managers
Inventing biological assays and importing known ones	Biologists
Finding a lead compound of known structure	Biologists and chemists
Identifying the defect presented by the lead compound	Biologists and chemists
Planning a remedy	Biologists and chemists
Choosing individual target compounds to represent the structural type	Chemists
Devising synthetic routes	Chemists
Assigning or assuming responsibility for synthesizing target compounds	Chemists
Assigning or assuming responsibility for testing target compounds	Biologists

make next and how and why to do so. These important choices determine the direction and outcome of many months' work. Such suggestions need approval in a hierarchical organization. Well-founded, far-reaching suggestions, regardless of whether they come from newcomers or veterans, win approval in the form of managerial consensus. The originator, to gain approval, persuades her boss that the notion merits action, the boss, in turn, convinces her supervisor, and so on. It is a rare idea that is self-explanatory. Even good ideas demand persuasive champions, which partly explains why an ability to communicate is so valuable to employers and so desirable among job candidates.

Ambitious chemists may note that FIPCOs do not lightly give supervisory responsibility to newly graduated and hired employees, if they do so at all. To demonstrate competence, attain some success, and thereby create a need for help, Ph.D. chemists can work alone several years after starting their careers. Policy, which varies across the industry, determines this period as well as the necessary qualifications and the number of chemists reporting to a newly appointed supervisor.

DISCOVERY RESEARCH

Within a pharmaceutical company, a certain progression of events and decisions typifies a successful project in discovery research. The medicinal-chemical aspects of such an effort begin with the finding that a particular compound possesses the desired biological activity. Usually and preferably the chemical structure of this compound is already known. The substance becomes the lead compound, and its structure guides chemists' attempts to modify the molecule. Structural modifications are to improve one or more chemical, physical, or biological properties or to impart or abolish some other property or activity. The project advances if it yields one or more series of target compounds synthesized during 2 to 4 years. Each worthy series contains many active members and not a few potent ones. Further advancement brings newer target compounds with smaller IC_{50} values, each number representing the drug concentration needed to inhibit 50% of the desired biological effect. The lower this concentration is, the more potent is the drug. Plotted against project lifetime, declining IC_{50} values show that the potencies of the newer compounds increase, thereby confirming the chemists' choices of targets and intensifying their hopes for success. Activity in the desired assays justifies the expense of patenting, so, series by series, the active compounds become the subject of patent applications. The biologists extensively investigate a small number of potent substances emerging from the series. Their studies spotlight one or two substances as candidates for further development. Satisfactory biological, toxicological, and chemical development wins Investigational New Drug status for the candidate compound(s), which may then enter human clinical trials. To appreciate this progression calls for familiarity with its stages, which this section describes, focusing on what chemists contribute.

Common Goals of Discovery Chemists

Veteran chemists and tyros alike strive to create structurally diverse series of patentably novel compounds possessing the desired biological activity and improving upon the properties of the lead compound. One of these new compounds, and preferably more than one, must outdo the lead compound in one respect at least. Lest success be impossible to attain, however, competent research managers keep the number of required improvements to a minimum. Advances can draw on many possibilities, among them oral activity, greater potency or selectivity, or lessened toxicity. Whatever the improvement sought, chemists seek to link it to changes in chemical structure and ask their biological colleagues to assay quantitatively for it. Structural changes to improve upon the potency of the lead compound rank among those alterations most often sought. This frequency is understandable; there is, for example, no need even to measure the toxicity of an impotent or inactive compound. Other problems that arise often concern the pharmacokinetic behavior of new compounds that are potent *in vitro*. In live-animal assays, some of these compounds may exhibit too low a bioavailability, too short a duration of action, or too great a susceptibility to metabolism. A plethora of metabolites increases the risk that one of them may be toxic, while a short-acting medicine may require too frequent dosing or too elaborate a means of drug delivery. Inadequate bioavailability, as measured by low blood concentrations of the experimental drug, prevents clinical development altogether, making it impossible to establish a safe dosing regimen. Medicinal chemists must make so many potent, selective compounds that some of them leap these hurdles.

Once a clinical candidate presents itself, or shortly before, the chemists begin to make yet more new compounds from which the biologists can choose a back-up drug. This compound may begin development only when its predecessor meets delay or difficulty. Alternatively, development can begin immediately. In any case, the back-up candidate should offer some advantage over the original clinical drug.

Lead Compounds

Finding a lead compound is one of the crucial tasks in a successful discovery project. Until the arrival of high-throughput screening, it ranked among the most frustrating problems in medicinal chemistry as practiced in pharmaceutical companies. To understand the importance of a lead compound, consider a definition amplified beyond what Chapter 1 offers.

Lead Compounds Defined

Combining small molecular size with a known chemical structure rich in functional groups, a lead compound possesses the biological activity sought to treat a human ailment of commercial interest. In the primary biological assay, it may be potent and dose responsive but must also exhibit in this test or another some flaw that prevents developing it as a drug. The substance might lack patentable novelty, activity in an animal disease model, bioavailability or clinical efficacy in humans, or another essential property. Harmful side effects, toxic metabolites, or short duration of action because of rapid metabolism or fast clearance can be disabling faults of the lead compound. Varied as its faults may be, they should be few. A viable successor, often drawn from a series of structurally related substances, makes up the deficiency of the lead compound but introduces no crippling failing of its own.

The molecular size of a lead compound plays an important but limiting role in drug discovery. High molecular weights tend to reduce or abolish bioavailability. The upper limit of suitable weights of a lead compound and its successors apparently falls within an approximate range of 350–500. Beginning a program of molecular modification with a large lead compound, if its size restricts the scope of structural changes, can exacerbate the medicinal chemists' task.

Prerequisites

To make any rational molecular modifications at all demands a lead compound that has a known structure, shows the desired biological activity, and is available for testing.

One standard approach to making modifications entails preparing new substances analogous to the lead but using a synthesis independent of it. An analog, for example, might contain a chain of atoms where the lead compound bore a ring of them. It might incorporate a heterocycle instead of a carbocyclic ring, or include an amide where there was an olefin or vice versa. Gross structural differences necessitate a wholly new synthesis, so this approach obviates need for some but not all samples of the lead compound. In the new route, it serves neither as a starting material nor as an intermediate. But, this avenue obviously requires structural knowledge as a prerequisite to conceiving the original analogy to the lead compound.

In another approach, structural modifications result from chemical reactions of the lead compound itself, and making these changes requires it to contain reactive groupings of atoms. Unless the functional groups are known, choices of reactions and therefore of product structures and properties become random. Particular reactions demand specific functional groups in the reactants. The famous *Diels–Alder* reaction, for example, requires two components, one of them containing a double bond. The other partner must hold two double bonds conjugated to one another. So, a compound lacking the one group is inert to a substance containing the other. No chemist can expect to induce the *Diels–Alder* reaction by presenting a diene to a saturated lead compound. With hundreds or thousands of reactions and groups to choose from, random synthesis offers scant hope of successful drug discovery:

> ... *chemical syntheses, at least if the target is at all complicated, never conform to the popular picture, which has it that though the chemist may have to perform many hundreds of experiments before he finds the magic formula, he finally discovers that if he mixes the right things together, in the right kind of pot, he succeeds in producing, say, synthetic quinine.* (Woodward, *1963;* italics added*)*

Taking either approach, the chemists provide the biologists with a quantity of the lead compound, in amounts ranging from a few milligrams initially to many grams later on. Samples may come from a corporate compound collection, from isolation or fermentation, or from synthesis. Medicinal chemists or their coworkers in preparations groups may make the lead compound in-house or contract the work to a custom-synthesis house.

Samples of the lead compound are indispensable to biologists and chemists. Biological researchers need to confirm the activity found in, say, a biochemical screening assay, profile the dose–response relationship, and establish selectivity in related tests. Ultimately they seek to discover the therapeutic activity of the lead compound using some animal model of the human disease for which they seek a new drug. All these investigations furnish evidence that their therapeutic hypothesis has validity, which permits chemists to begin making derivatives of the lead compound or encourages them to continue. In testing compounds with structures analogous to the lead, the biologists include the latter to control their experiments. The chemists, too, may need samples of the lead compound, usually to establish the purity and confirm the identity of the specimen that originally showed the interesting biological activity. For all these reasons, a paper lead is inadequate. Mere reading knowledge of structure and activity does not transform an organic compound into a desirable lead, unless a sample of such a substance can be procured and its activity confirmed by biological testing.

Origins

The traditional sources of information concerning lead compounds are diverse. One source is publications reporting interesting biological activity and the responsible structures. These publications include patents, patent abstracts like the *World Patent Alert,* and industry newsmagazines and newsletters. Examples of the latter comprise *Pharmascope, Drug News and Perspectives, Emerging Pharmaceuticals, Drugs in Development,* and SCRIP. Another source of competitive intelligence concerning drug discovery and development is *Pharmaprojects,* which is supplied on compact disks. News of chemical structures producing biological activity also issues from scientific conferences. In these meetings, chemists, biologists, and drug-metabolism experts publish abstracts of their presentations, give lectures, and show posters of their work. Scientific journals and annuals are a rich source of structure–activity information. Chemists regularly consult the *Journal of Medicinal Chemistry,* the *European Journal of Medicinal Chemistry, Bioorganic and Medicinal Chemistry, Artzneimittelforschung, Medicinal Chemistry Reviews, Bioorganic and Medicinal Chemistry Letters,* and *Annual Reports in Medicinal Chemistry.* Journals devoted to the life sciences—for example, *Nature, Science,* and the *Journal of Pharmacology and Experimental Therapeutics*—also bear news of potential lead compounds. In all the foregoing sources, the quantity of information pertaining to lead compounds varies, as does the timeliness and completeness of revelations. None of these publications represents an unfailing source of useful lead compounds, but all make valuable contributions to the task of finding them.

Empiricism remains important to drug discovery, so the results of experiments necessarily supplement published information concerning useful leads. Pharmaceutical companies continue to seek lead compounds by high throughput, rational screening of compound collections, which can be somewhat randomly assembled. They enter reciprocal licensing agreements with other firms, which furnish collections of organic compounds made, say, as agricultural chemicals, and test these compounds for pharmacological activity. Also, they arrange to test compounds in collections made by university laboratories and synthesized by graduate students and postdoctoral researchers. Industrial medicinal chemistry groups buy large sets of compounds from fine-chemicals suppliers anywhere in the world and arrange with combinatorial chemistry companies to prepare and test large numbers of new compounds. Plant extracts and microbiological fermentations are also sources of new lead compounds and marketed medicines, as they have been for decades. FIPCOs periodically test all the compounds in their own archives, updating the batteries of assays employed whenever biologists and biochemists devise or import new tests. Casting its net broadly, a large pharmaceutical company typically tests several million compounds each year.

Lead Compounds and Medicinal Chemists: A Summary

In helping to find a lead compound for a new project, medicinal chemists play a defined role that they assume or accept. They survey the relevant literature to discover the active compounds of known structures studied by any competing companies. These chemists determine the purity of lead-compound samples tested in house or scheduled for testing, purifying the samples if necessary and arranging for retesting. They confirm the assignment of structure to the substance. Finally, they make or otherwise procure adequately sized samples of the lead compound, maintain a stock of it, and monitor its quality. In any given project, all this work may require the services of only one chemist working part-time. She devotes her other efforts to different aspects of this or other concurrent projects.

Making Target Compounds

Baccalaureate chemists who never took part in research familiarize themselves with its rudiments, particularly the elements of traditional solution-phase synthesis. As pharmaceutical company researchers, they conduct organic reactions, learning to do benchwork as competently as M.S. chemists. Although this book cannot impart an M.S. chemist's experimental skills, it portrays here the basics of discovery and development research in synthetic organic chemistry, which B.S. scientists acquire on the job. Besides running reactions and monitoring them for completion, the fundamentals comprise isolating a new compound, showing that it is pure, purifying it if necessary, characterizing it, and assigning its structure. Within a discovery group, organic chemists work to make new target compounds for pharmacological testing. Their basic tasks therefore include registering these substances in the company archive and submitting them for testing.

Suitable Target Compounds

In addition to their structures and activities, suitable target compounds present certain chemical and physical characteristics helpful to drug discoverers and developers. They show reasonable stability toward laboratory air and heat and to visible light and atmospheric moisture. For ease of handling in discovery research, solids are preferable to viscous oils and flowing liquids as long as their melting points exceed the physiological temperature of 37°C. Solids need not be crystalline and may be amorphous, and crystalline solids may contain more than one polymorph. Discovery chemists deal with oils, liquids, and low-melting solids by weighing and dissolving them before they undergo testing. Water-soluble target compounds help biologists prepare samples for dosing and so do compounds forming water-soluble salts. To prepare a target compound reproducibly as, for example, a specific crystalline form is a task that lies in the realm of chemical development, which becomes important only when the substance is to begin clinical trials.

Running Reactions

In executing a synthesis, a chemist changes a starting material, or a mixture of them, to one or more intermediate substances; transforms all but one of the intermediates to others; and converts the last one to target compounds. The stages of such a synthesis call for different chemical reactions, not all of which are already familiar to the researcher, regardless of her corporate experience and academic training. She needs to answer basic questions in five areas before starting any reaction:

1. Is the starting material for the first reaction of the sequence an item of commerce? If so, do we possess it now; if not, how soon can I obtain it and how much do I need?
2. Does the appropriate functional group of the available reactant form the desired atomic grouping in the product sought and do so safely?

3. What is the stoichiometry of the planned reaction, and what by-products does it form, if any?

4. What reagent(s), catalyst(s), solvent(s), atmosphere, and conditions effect the change, and, if a choice exists, which maximizes the purity and yield of the product?

5. Where can I learn what I need to know about the reaction planned?

Starting materials. Elevating a planned synthesis to practice requires a commercially available starting material. In contrast to their counterparts in academe, discovery and development chemists almost invariably buy such substances.

> *To make oneself any compound that the company stockroom possesses or a fine-chemical house sells is to blunder; and, in considering what compounds are affordable, self-indulgence proves superior to self-denial.*

Searching in-house for such a substance often reaps rewards if one's employer keeps an up-to-date computer inventory. A successful in-house search lets a scientist begin a synthesis during the hour it was conceived and thereby nurtures enthusiasm for the work.

To locate external sources of commercially available starting materials, chemists consult vendors' printed catalogs. Other valuable sources of similar information are proprietary databases available internally, through the Internet, or on compact disks. Databases like "CHEMCATS," the Available-Chemicals Directory, and ChemSources present many vendors' wares. Not only can a chemist find different suppliers of a single chemical, but she can compile lists of related substances like primary aliphatic amines by substructure searching of one supplier's database. Such a list helps extend a series of compounds analogous to a lead structure. Manifested in commercial databases and networked microcomputers, computer technology makes searching for starting materials quick, broad, and versatile.

Resources. Three pragmatic principles help many veteran chemists engaged in the organic synthesis that drug discovery projects require. (1) They *select target compounds that no colleague is already making.* To do so calls for awareness of other researchers' future plans and current tasks. Oral and written reports supply this information, as do informal inquiries.

> *(2) Experienced scientists avoid inventing experimental procedures that are already known. (3) They make no effort to improve workable procedures or effective syntheses.*

To begin work that demands familiarity with the chemical reactions planned, they need certain resources to answer the questions enumerated above. Information sources include their own stores of knowledge, their colleagues', and the appropriate chemical literature.

It takes only a few coworkers for their cumulative experience to span many decades. So, such scientists retain vast stocks of diverse chemical lore, while their shelves hold relevant books. Fellow chemists, particularly those working in the same project, are therefore likely sources of practicable suggestions, theoretical explanations, and useful procedures. Colleagues within one research group or company meet an obligation to help one another and take satisfaction from doing so. Even a researcher unable directly to assist can usually name another chemist who can. If a newcomer welcomes help she can expect her coworkers to advise on a variety of matters, from learning to use unfamiliar equipment to getting a haircut nearby.

Planning a synthesis, or learning about one of the reactions composing it, often means visiting a chemical library. There a chemist can browse as needed the hundreds of volumes that survey the primary literature of organic synthesis. These volumes provide journal or patent references to specific procedures yielding particular compounds. Also they direct a reader to tried procedures applicable to a variety of related substrates and can indirectly suggest alternative starting materials, intermediate compounds, and synthetic routes. The survey literature, much of it priced beyond an individual chemist's purse, includes handbooks, dictionaries, annuals, series, an encyclopedia, textbooks, monographs, and review journals (Table 5-2). Its great store of relevant knowledge means that an hour in the library can be worth days in the laboratory. However, to draw promptly from this store and the primary journal and patent literature still requires a reading knowledge of certain European languages.

In doing the library work needed to develop plans for a synthesis, it helps to view the proposed reaction sequence from different standpoints. Suppose, for example, that the sequence conceived initially is $A \rightarrow B \rightarrow C \rightarrow D \rightarrow E \rightarrow F$. If so, the planner obviously needs to find out where to buy or how to make the starting material A. Unexpectedly, but often enough to be exploited, it happens that an intermediate compound like D is commercially available. Or, it is made as readily by another route as substance B or C is prepared by the planned synthesis. Much synthetic organic chemistry has been done and published in the past 15 decades. It is thus essential to consider each intermediate substance of one synthesis as a starting material for another, shorter reaction sequence yielding the same product. A knowledgeable chemist looks for references to such a substance, if its structure is relatively simple. Discovery of a journal article showing how to make this intermediate material can, by preventing unnecessary benchwork, speed progress and avoid embarrassment. The resources needed to search com-

Table 5-2 Selections from the Literature Surveying Organic Synthesis

"Comprehensive Organic Functional Group Transformations," Katritzky, Meth-Cohn, and Rees (eds.), Elsevier Science Ltd., Oxford, 1995.

"The Chemistry of Functional Groups," S. Patai et al. (eds.), Wiley, New York, 1966–.

"Encyclopedia of Reagents for Organic Synthesis," L. Paquette (ed.), Wiley, New York, 1995.

"Reagents for Organic Synthesis," L. Fieser and M. Fieser (eds.), Wiley, New York, 1967–.

"Annual Reports in Organic Synthesis," various editors, Academic Press, San Diego, 1970–.

"Comprehensive Organic Chemistry," Barton and Ollis (eds.), Pergamon, Oxford, 1979.

"Comprehensive Organic Synthesis," Trost and Fleming (eds.), Pergamon, Oxford, 1991.

"Methoden der Organischen Chemie," Houben-Weyl, Vierte Auflage, George Thieme, Stuttgart, 1957.

"Rodd's Chemistry of Carbon Compounds," 2nd ed. S. Coffey (ed.), Elsevier, Amsterdam, 1964.

"Organic Reactions," various editors, Wiley, New York, 1942–.

"Organic Syntheses," various editors, Wiley, New York, 1921–.

"Synthetisch Methoden der Organischen Chemie," W. Theilheimer, S. Karger, Basel, 1946.

"Chemical Reviews," American Chemical Society, Washington, DC, 1922–.

"Chemical Society Reviews," The Chemical Society, London, 1970–.

"Advances in Heterocyclic Chemistry," Katritzky et al. (eds.), Academic Press, New York, 1963–.

"Comprehensive Heterocyclic Chemistry," Katritzky (ed.), Pergamon, Oxford, 1984.

"Heterocyclic Compounds," various editors, Wiley, New York, 1950–.

"Comprehensive Organometallic Chemistry," Abel, Stone, and Wilkinson (eds.), Pergamon, 1995.

prehensively for particular compounds are *Beilstein* and the printed edition of *Chemical Abstracts*, both of which are searchable online.

These two works, as well as the application "REACCS," make possible efficient searches ranging over much or all of the primary chemical literature. *Beilstein*, for example, covers organic chemistry into the 18th century. Computer searches also facilitate asking open-ended questions. What, for example, are the reactions, reagents, and conditions that create a pyrrolidine incorporating at its C-(3) and C-(4) positions the atoms of an olefinic starting material (Figure 5-1)? Each suitable reply comprises structural

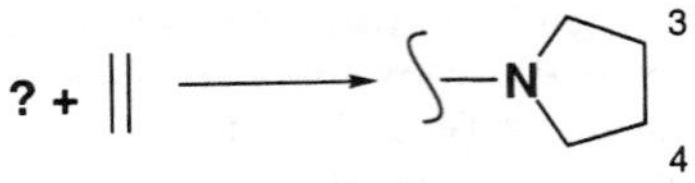

Figure 5-1 Typical search query

drawings, textual material, and one or more journal citations including experimental procedures. Now answered in minutes, such a question would have consumed days or weeks of manual searching as recently as 15 years ago. Fast, broad literature searches like this are among the valuable benefits that the Silicon Age confers upon computer-literate organic chemists.

Persistent efforts and satisfactory experiments. "There are *no* general reactions" (*Woodward*, 1969). For this or other reasons, not all attempts to effect a known reaction succeed with a particular substrate, even in the hands of veteran experimentalists. Success in the synthesis of organic compounds demands a certain persistence, even though much labor can be fruitlessly expended to induce desirable chemical changes to take place and run to completion. Intense heating, for example, meant to force such a reaction to occur, can unexpectedly change colorless crystals to an opaque and intractable tar. Such a result, unsuccessful as it is, nonetheless represents a satisfactory experiment, if other, less brutal but more numerous bids merely demonstrate the inadequacy of the reaction conditions used.

> *On the whole, destruction of the starting materials is preferable to dishonor of the chemist who cannot make molecules react.*

Persistence in research, however, is an ambivalent trait, sometimes justified by success but often condemned by failure. Slow progress, furthermore, draws attention to the opportunities to engage in more promising research, and these prospects reinforce a decision to halt in mid-project. Uncertainty always attends the work, for the hard-won product of a difficult synthesis may exert no desirable biological activity whatsoever. To drop a troublesome line of research is therefore commonplace in drug discovery and development as well as in academic research. Even the Nobel laureate *Robert Robinson*, a great organic chemist remembered for brilliant achievements attained with simple apparatus (*Williams*, 1990), reevaluated his method and goal:

> *His instinct when confronted with a difficulty in experimental work was to seek at once an alternative route to his objective or even, at times, to change the objective itself. This contributed positively to the discovery of new reactions. . . . (*Todd and Conforth, 1976; italics added*)*

Although no single analytical technique offers universal applicability, microcombustion analysis closely approaches this ideal. The technique applies to a variety of structural classes and can require little sample preparation. To analyze crude products, which is sometimes helpful, requires no preparation beyond labeling, filling, and dispatching a vial. For pure compounds, microanalysis gives the percentage composition within 0.1–0.4% for each element that the sample contains. Such accuracy demands painstaking sample purification like repeated crystallizations, vacuum distillations or sublimations, and prolonged drying of solid samples. It is desirable to monitor the progress of these purifications, often by TLC, lest successive attempts waste time by failing to remove a contaminant. Accurate microanalytical values not only establish the sample purity but, by providing an empirical formula, afford valuable structural information. This information confirms that the reaction forming the compound took the expected course. Usually done by the analytical departments of large FIPCOs, complete microanalyses take 1 to 3 days, from submitting the sample to receiving the results. External analytical houses, which smaller pharmaceutical companies use, are equally fast, thanks partly to courier services, e-mail, and facsimile transmissions.

Characterizing and Identifying Samples

An organic chemist characterizes the pure compounds she makes, often identifies samples of them with one another, and sometimes deduces their structures. These are recurring tasks that demand skill and judgment and are commonplace throughout discovery and development research. The chemists who do them work in medicinal or process chemistry, drug metabolism, radiochemistry, pharmaceutical analysis, and synthetic services. These tasks occupy much of the working life of any chemist who remains an experimentalist and reward sometimes protracted efforts to purify samples. A chemist who likes these aspects of her job, which this section introduces, enjoys her career.

Characterization. Measuring the physical properties of a pure sample characterizes a new compound. Thorough and reliable characterization is important for four reasons. It lets any chemist distinguish samples of different substances, permits identifying one sample with another specimen of the same compound, and creates a basis for deducing structures and seeking a patent. The properties usually measured comprise melting or boiling point, optical rotation, or chromatographic profile. Microanalytical values usefully characterize samples. Characterizing also includes measuring infrared and ultraviolet spectra as well as proton or carbon nuclear magnetic resonance or mass spectra. The job entails keeping a sample of the new compound, storing the spectra measured, and describing spectrometric and other properties in a laboratory notebook.

Keeping an original sample of the compound and a record of its physical characteristics permits later comparisons and contrasts. A sample permits direct physical comparisons, like a mixed melting point or a thin-layer or high-performance liquid chromatogram of an artificial mixture of two specimens. The record includes a note of microanalytical values; printed spectra or detailed descriptions of them; prints of HPLC traces; and photographs, photocopies, or drawings of TLC plates. Comparisons and contrasts unequivocally tell a later researcher that she does or does not possess the same substance as its originator did. This knowledge represents a milestone.

Identification. Organic chemists use *identify* and related words in two senses, which deserve recognition. In one sense, which this section adopts, they mean comparing two (pure) samples to establish that the specimens do or do not contain the same substance. Such a comparison does nothing to elucidate the structures of the compounds studied, if neither structure is already known. In the other sense, *identifying* refers to assigning chemical structures, a task that does not necessarily involve any direct comparison of samples. Figure 5-3 illustrates the distinction with a *Euler* diagram. Careful sample identifications yield sound structure assignments in some cases but

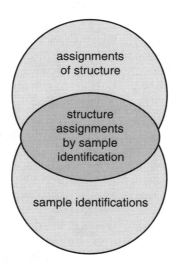

Figure 5-3 Relations between sample identifications and structure assignments

not all. They lead to assignments when one of the two samples compared has a known structure. Some assignments, however, do not depend on identifications at all. Instead, they follow logically from, say, measurements of spectrometric properties carried out with a single specimen.

Lest self-deception halt progress in making supplies, a careful researcher conclusively identifies the product of her current experiment with the substance issuing from an earlier trial of the same reaction and substrate. To continue the work without identifying the product represents a wishful Panglossian attitude that the outcome attained necessarily corresponds to the result desired. Not so. Even well thought-out experiments go awry, and the flaw may not be apparent until the result lies in hand.

To identify one sample with another calls for measuring and comparing one or more physical properties including spectrometric ones. Reliable identifications are direct: the researcher mixes the two samples and makes the measurement using the mixture. For example, a mixed melting point is a traditional means of identification, albeit one limited to crystalline samples containing the same polymorph. To complete such an identification calls for three melting point determinations, one of each sample and one of the artificial mixture. Other direct comparisons analogously use three measurements. Reliable identifications also benefit from independent confirmatory evidence, perhaps including comparisons of PMR or infrared spectra. Peaks present in the spectrum of one pure sample but absent from the spectrum of the other show the samples represent different compounds.

Indirect identifications are commonplace, and basing them on comparisons of profusely detailed spectra makes them trustworthy. Coincidence, however, sets infamous traps for researchers making indirect chromatographic comparisons. To compare mere HPLC retention times or TLC flow rates invites misidentifications because disparate compounds sometimes have identical values. As a rule, credible chromatographic identifications are always direct and even the best ones are never exclusive.

Chemists employed in different institutions sometimes exchange samples to compare them, for example, to verify that a new synthesis of an old compound succeeded. These exchanges may be more common in academia than in industry. There, precaution against giving away valuable proprietary information understandably restricts—but does not abolish—the privilege of dispatching samples. Company policy may require the requester to justify the proposed mailing and gather signatures granting permission to disclose the structures.

Tips and traps. Not all methods of characterizing a substance are equally useful in elucidating structures or in identifying samples. Consider two examples. First, the fingerprint region of an infrared spectrum may contain a wealth of absorptions that serve well to characterize the compound studied. However, fingerprint wavelengths and intensities do little to establish structure, for they lack definitive correlations to specific functional groups.

Second, think of a mass spectrum, perhaps one recorded after chemical ionization or fast-atom bombardment energized the sample. We suppose it shows only the peak of a protonated molecular ion, as expected. Such a spectrum helps assign structure because it furnishes the molecular weight. In the case of a high-resolution mass spectrum, it also defines the empirical formula. But a one-peak mass spectrum, although it affords structural information, fails to distinguish isomers because they have identical empirical formulas and molecular weights. An electron-impact mass spectrum offers a better chance for adequate characterization because such a spectrum is usually rich in fragment ion peaks of differing intensities. In gathering data, therefore, it is important to know what purpose the information is to serve—characterization, sample identification, or structure assignment—and which kinds of information the different techniques provide. The knowledge needed and the equipment available determine the method chosen. It is also advantageous to adopt versatile methods that accomplish more tasks than one, but without developing unreasonable expectations.

Such expectations typically involve attempts to infer what spectrometric data do not imply. Few conclusions concerning sample purity, for example, follow from a routine mass spectrum. The intensity of a molecular ion peak in the spectrum observed is not necessarily correlated to the concentrations of components in the sample investigated. A volatile but stable minor component can produce an intense molecular ion peak. The most abundant component, however, if it is comparatively involatile or insensitive to the ionization method, may yield a relatively weak peak. Alternatively, the molecular ion formed by the major component may fragment into smaller ions faster than it can be detected. In these circumstances, it outrages logic to conclude that a compound is absent from a sample because a mass spectrum failed to show the corresponding molecular ion. Such negative evidence deserves no trust.

Elucidating Structures

The 20th century saw the development and spread of a battery of powerful physical methods used to discover the structures of organic compounds. These methods include ultraviolet, infrared, and nuclear magnetic resonance (NMR) spectroscopy as well as mass spectrometry and X-ray crystallography. So formidable is this array that, in many cases, the time needed to deduce a structure now occupies minutes rather than months. Techniques that, for the first century of organic chemistry, had been standards of structure assignment became superfluous. They included the investigation of the chemistry of a new compound and the confirmatory total synthesis of a natural product. Deducing a structure no longer requires directly comparing a synthetic sample of a naturally occurring compound with a specimen isolated from natural sources. Such confirmations once crowned the chemical researches of many decades. The profound changes in methods of structure assignment are perhaps comparable to the recent revolution in navigation wrought by development of the satellite-based global positioning system. Over much of the globe, techniques of dead reckoning including use of the sextant are obsolescent.

Four principles. Progress aside, certain timeless principles still govern structure determinations, which newcomers to discovery and development research learn and apply. (1) Sound assignments include evidence from a variety of sources. The demands and importance of the task are too great to entrust to one method, with the exception of X-ray analysis (see text below). (2) The sample employed to gather this evidence should be a single one shown pure by at least two independent methods. A single sample avoids the extra labor of identifying one specimen with another and precludes the risk that two unidentified samples differ in structure. To use an impure sample can waste the labor of gathering and interpreting the data and delay the assignment of structure. (3) A complete assignment includes a determination both of molecular weight and empirical formula.

Finally, (4) a convincing structural assignment demands compelling positive evidence, not merely characteristic or consistent data or negative evidence. In the reduction of, say, a nitroacetophenone to a hitherto unknown nitrobenzyl alcohol, an infrared spectrum of the product can constitute convincing evidence if it shows a hydroxyl group. Positive confirmatory evidence of a different kind arises from a proton nuclear magnetic resonance spectrum. It presents (among other features) the doublet peak expected for a methyl group within the $CH_3CH(OH)$ substructure, which the starting nitroacetophenone lacks.

In this and other cases, some physical constants like chromatographic flow rates (TLC) and retention times (HPLC) provide no decisive structural knowledge. They merely characterize the sample without elucidating its structure. If no authentic sample of the expected nitrobenzyl alcohol exists, no direct chromatographic comparison of specimens is possible. Because the alcohol represents a new compound, indirect comparison of flow rates or retention times is also impossible. Lack of these numerical data and an authentic specimen therefore prevent any structure assignment based on sample identification.

The infrared spectrum, furthermore, also shows nitro-group absorption, which is consistent with the expected structure. The absorption is unhelpful, however, because it enforces no conclusion regarding the success of the reaction attempted. The starting acetophenone itself contains a nitro group that absorbs infrared waves at the same frequency, making it possible to mistake the starting material for the product. Similarly, the absence of carbonyl group absorption in the spectrum is consistent with formation of the desired alcohol, but does not require this conclusion. It is unhelpful because it is negative.

A chemist assigning the structure of a new synthetic substance routinely looks for certain spectrometric peaks. They are those appearing in the spectrum of the product but not of the starting material. These peaks typify the functional group change that occurred. The task of assigning structure to the product is simplified if the starting material itself has a known structure, which in medicinal and process chemistry is commonplace.

X-ray analysis. In the pantheon of methods employed to elucidate structures, single-crystal X-ray analysis reigns supreme. It deserves every chemist's respect for it unites speed, completeness, and dependability. Using a modern diffractometer for measurements and a computer for calculations, this method unequivocally establishes within a few days or weeks the composition and solid-state conformation of a small molecule. By contrast, 150 years separated the discovery of the notorious alkaloid poison strychnine and the confirmation of its complex structure by synthesis, which in 1947 culminated a half-century's chemical research.

X-ray analysis, however, lacks generality because it requires a single crystal. It provides no indication of the purity of the corresponding bulk sample. Because it is labor intensive, requiring training in crystallography, the method is not applied to large numbers of samples. For chemists who are not crystallographers, it precludes the exquisite intellectual pleasure provided by other kinds of evidence used to deduce structures. Nonetheless, its power is awesome. In persuading a researcher to seek an X-ray analysis, the stereochemist and Nobel laureate *Derek H. R. Barton* said, "Talking to crystallographers is like talking to God."

Nuclear magnetic resonance spectroscopy. Perhaps the most popular methods for establishing structures comprise carbon nuclear magnetic resonance (CMR) and PMR spectroscopy. Proton magnetic resonance spectra are especially quick to record. Available for the first time in the 1950s, PMR spectrometers soon became indispensable to the labor of assigning structures to organic compounds. They ease and speed what until then had sometimes been an arduous and long-lasting task. Nuclear magnetic resonance (NMR) methods, because they are versatile, quickly show the purity of the sample and characterize it. They can identify one sample of a substance with another specimen, making possible a peak-by-peak comparison of printed spectra of each sample. Alternatively, a chemist can record and inspect the spectrum of an artificial mixture of samples. Nuclear magnetic resonance methods furnish positive, detailed evidence of the functional group changes that make up chemical reactions. Moreover, a certain spare elegance ornaments the proton-decoupled CMR spectra of many compounds. Each carbon atom in the substance studied corresponds to a single sharp line in the spectrum recorded. Counting the lines simply reveals the number of different carbon atoms in the molecule. If, because of molecular symmetry, the carbon atoms outnumber the lines, merely counting the latter still offers incisive structural insights. So pervasive is the practice of NMR spectroscopy by individual organic chemists employed in discovery and development research that the method merits some discussion here.

Tricks of the spectroscopist's trade. To exploit NMR spectroscopy, experienced chemists constantly add to their working knowledge of the subject. They learn the chemical shifts and other related particulars of the compounds that they are currently using. Also, they familiarize themselves with special techniques like two-dimensional NMR spectroscopy and phenomena like the nuclear *Overhauser* effect.

Throughout their working lives researchers also follow certain guidelines in recording and interpreting spectra.

1. Proton magnetic resonance spectra, to permit valid comparisons, use solutions prepared with the same solvent. Otherwise, large (± 0.5 ppm) confusing solvent-induced changes of chemical shift can occur.
2. Small chemical shift changes (± 0.2 ppm or less), however, may not merit any interpretation. Alert to this possibility here and elsewhere, the wary chemist shaves hypotheses with *Occam's* razor, eliminating unneeded postulates.
3. Informative PMR spectra display sharp peaks, so chemists rerecord any spectrum showing line broadening. Sometimes refiltering the solution or repurifying the sample sharpens the lines, while a change of solvent, temperature, or pH helps with other specimens. Some samples lie beyond help.
4. Acceptable PMR spectra routinely show the integrated areas of all the various peaks because the areas convey important structural information. They tell how many protons of each kind are present in the molecule.
5. Prerequisite to convincing structural conclusions are CMR spectra displaying the signals of all carbon atoms composing the molecules investigated. It takes more time to record a complete CMR spectrum than to acquire a whole PMR spectrum, so more patience is required. So is an understanding that the areas of CMR signals are uninformative, unlike those of PMR signals.

6. An understanding of two recurring NMR phenomena, exchange coalescence and magnetic nonequivalence, is indispensable. These complex and sometimes baffling phenomena can mislead even contemporary efforts to assign structures. A grasp of coalescence and nonequivalence is perhaps best acquired from an NMR course or textbook or from a knowledgeable coworker. Repeated encounters with the phenomena do little to explain them, so, in this case, direct experience is a poor teacher.

7. Numbering, cross-referencing, and annotating spectra help in writing reports, patent applications, and publications. A chemist's laboratory notebook indicates that she recorded, say, a CMR spectrum of a certain sample, gives its number, and leads the reader to find the printed or electronically filed spectrum. In turn, the spectrum takes the reader to the page of the notebook stating the provenance of the sample used for the measurement. To tease structural information from an NMR spectrum demands effort and takes time. A record of the destination reached and the journey made is therefore valuable. Comments written directly on a stored, printed copy of the spectrum find use long afterward, when details of the interpretation might otherwise be lost or forgotten. Helpful notes comprise labeling of peaks with assignments, an explanation of any molecular symmetry, the conclusions drawn, and the arguments leading to them. Good notes also include references to spectra of model compounds drawn perhaps from commercial catalogs or from personal or corporate files. To compute and study the ratios of different kinds of protons in a molecule are essential to interpreting a PMR spectrum. Initially to note the findings and conclusions directly on the printed spectrum prevents duplicating the work later.

Devising Syntheses

Extending a series of compounds to find the best member obviously requires a synthesis, which may be either acceptable or good. In discovery research, an acceptable synthesis produces patentably novel structural analogs of the lead compound possessing the desired biological activity and forming in quantities and purities adequate for testing. No project can long endure if its products lack these properties, however efficient or elegant the synthesis used to make them and regardless of their structural complexity or beauty. Because most target compounds are impotent or inactive, great sample purity is initially insignificant, as long as the biological assay used does detect highly potent substances. (Not surprisingly, the quality of *clinical* samples, as distinct from research samples, assumes great importance during chemical development.)

Criteria. Ease of synthesis may be the most important determinant of successful drug discovery. A good synthesis meets many related criteria, among them simplicity, versatility, brevity, robustness, adaptability, and convergency. This list, however, conspicuously lacks one quality, efficiency, that is less highly valued in medicinal chemistry laboratories than elsewhere. An efficient synthesis has a high overall yield, thanks to a high average yield per step or a small number of steps. It unquestionably benefits the discovery project of which it forms a part and gratifies its designer.

> *High chemical yields, however, represent more a bonus than a necessity, and a merely effective synthesis suffices to discover a drug, albeit not to manufacture one.*

So, a job worth doing in discovery research is worth doing acceptably. For a discovery chemist to perfect a research synthesis by maximizing its overall yield may represent an act of folly, especially when the work yields no clinical candidate.

Simplicity. Technical simplicity of operation in each step allows the greatest number of chemists to use the synthesis without special training and thereby increases their output of target compounds. In 3 to 5 years, a dozen chemists operating a simple synthesis can produce several thousand target compounds. They need no help from combinatorial chemistry to do so, but may receive some assistance from parallel synthesis.

Versatility. In making target compounds, a synthesis that yields diverse structures is welcome. Structural diversity lends breadth to patent coverage and tends to ensure that no billion-dollar molecule goes unrealized. A single synthesis that creates a wealth of different compounds brings a desirable kind of efficiency to discovery research. Its designer need not devise and demonstrate other syntheses to make the same subsets of target compounds. Both output and diversity of target compounds get a boost from a common intermediate occurring late in a synthesis.

Brevity. However, a good synthesis is short, needing no more than five to six chemical steps to complete. It should proceed without protecting groups or ensure that one of the two reactions needed to introduce and remove such a group accomplishes some other valuable task. Longer sequences maximize the time and labor required. They minimize the number of target compounds resulting per week. A long synthesis, if it forms a target compound of no previously known value, can provoke anxiety when a chemist begins it and prolong uneasiness until she finishes.

Robustness. Also typical of a good synthesis is robustness. In no reaction of such a synthesis are the circumstances—time, temperature, concentration,

pH, and atmosphere, for example—stringent. Instead, every reaction tolerates some variability in each condition without going awry by failing to occur, forming by-products, or lowering the yield of the desired product. A synthesis comprising robust reactions promises the greatest output of compounds for the least investment of effort and demands only minimum experience to operate successfully.

Adaptability. Adaptability to scale-up lends value to any synthesis that yields potent target compounds. Invariably, the biologists need relatively large quantities of several target compounds, each amounting to, say, 250 g or less. A good synthesis produces these amounts, regardless of whether they represent racemic or enantiomerically pure substances.

Discovery research in its initial stages sometimes offers little opportunity to recycle unwanted enantiomers and few chances to devise means to do so. This source of the desired stereoisomer can be thus cut off until chemical development begins. A satisfactory preparation directly yields adequate quantities of both enantiomers, by asymmetric synthesis or optical resolution. It thereby allows the biologists readily to determine which optical antipode makes the better drug, if either does. Indeed, the racemate may prove superior to either enantiomer, but the inferiority of any enantiomer is unknowable until its biological properties are separately established. (Some racemates, of course, comprise rapidly interconverting pairs of enantiomers. In such cases, any enantiospecific differences in pharmacological properties are inconsequential on the time scale of drug development.)

Linearity versus convergency. Convergent traditional syntheses are preferable to linear ones for two reasons (Figure 5-4). They allow chemists to join large components of the target molecule to one another in few stages, much as builders assemble a two-story structure from prefabricated modules comprising rooms and corridors. Second, intermediate compounds from the two arms of a convergent synthesis can be prepared simultaneously, which saves time. A linear synthesis lacks this advantage because making any intermediate substance requires a supply of the previous one. Necessary and effective as it may be, proceeding in steps makes pedestrian progress.

Irrelevant criteria. In conceiving a discovery synthesis, which often entails choosing among alternative routes, it is important not only to adopt relevant criteria but also to spurn extraneous ones. The latter category includes two chestnuts, which are manufacturing efficiency and plant safety. Admittedly, a particular research synthesis may be uneconomic to operate on a plant scale because of low overall yield expected during manufacturing. For discovery chemists to reject it for this reason, however, is to rely—perhaps unknowingly—on an unsound argument. It neglects the likelihood that pro-

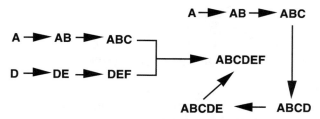

Figure 5-4 Convergent (left) and linear (right) syntheses of structure ABCDEF

ficient process chemists can circumvent or solve the yield problems. An omission similarly flaws the analogous plant-hazard analysis: development chemists skillfully handle large quantities of dangerous reagents like the pyrophoric *n*-butyl lithium and effectively substitute safe reactions for hazardous ones. Weak arguments concerning efficiency and safety can originate in discovery chemists' unfamiliarity with process research and development.

Submitting Compounds for Biological Testing

Submitting a sample for in-house biological testing culminates the labor of synthesizing a new compound. Within a FIPCO, it comprises two basic tasks, which are registering the substance in the archive and delivering it to the dispensary. Registration calls for completing one or more electronic or printed forms that are stored indefinitely. These forms request certain biological tests for the sample and instruct the biologists and dispensers in storing and distributing it. Directly or indirectly, they also furnish an experimental procedure for making the new compound. The forms include the procedure or lead a reader to a notebook recording it.

Registration Forms

Permanent records of submitted new compounds are also important on the long term. Years after the original submission, they allow searches to find proprietary substances that, if tested, might be potent in other, recently developed biological assays. Finding old compounds active in new tests is an indispensable source of lead substances and has two consequences. First, within a single large pharmaceutical company, tens of thousands of file compounds—some of them many decades old—undergo periodic testing in batteries of high throughput assays. Second, continued testing requires chemists to make and submit relatively massive samples. While a few milligrams may suffice to begin with, the initial testing can soon exhaust the supply, which prevents discovering an unexpected activity years later. Although an interested chemist certainly can replenish a supply long after it was exhausted, experience dictates that he or she needs a mighty reason to do so. It is wiser to seize the moment and make plenty at the outset.

Content. Registration forms typically ask for the information itemized in Table 5-3. The vials containing samples bear numbers referring to a notebook number and page. Each test sought requires a minimum sample weight, so the quantity available partly determines the number and kinds of tests requested. The biologists and dispensers who handle the sample need to know what precautions they should take with it, if any. So, the submitter gives some indication of stability, perhaps noting that the substance needs protection from air or light. Experienced submitters thoughtfully compose such indications lest they needlessly alarm or misdirect the nonchemists who dispense or test the samples.

Certain physical and chemical properties of the samples are useful to the biologists who test them, so the registration forms provide room to note them. Water solubility is an example. Measuring this solubility, however, may not be required if the biologists test the samples as solutions in other solvents or as suspensions. However, the biologists' practices and requirements vary from time to time.

If the sample is a salt, then submitters make sure to note this point. For the addition salts that amines form, they state the acid used. Similarly they indicate the cation in carboxylate and other acid salts. For biologists preparing solutions of known molarity, it is essential to know which acids and cations the salt contains.

Spectrometric properties allow succeeding chemists to recognize the sample and provide an evidentiary trail that patent lawyers can easily follow in the case of a patent dispute. In the registration form, therefore, submitters note the kinds and numbers of the spectra they or others recorded. A structural drawing and the chemical name complete the form.

Table 5-3 Content of Compound Registration Forms

Biological tests desired
Sample number and weight, both referred to a traceable notebook entry
Handling and storing precautions, if any
Selected physical and chemical properties
Submitter's name
Molecular weight, empirical formula, structural drawing, and chemical name

What to do about impurities. When an impurity contaminates a sample, the submitter indicates what structure this component has, if she knows. She estimates, if possible, how much of it is present. Admittedly, no estimate may be possible and structural knowledge lacking; nevertheless, professionalism demands a note revealing the presence of an impurity.

Newcomers soon learn that to submit a compound for biological testing does not demand ultrahigh sample purity.

> On the contrary, successful drug discovery so voraciously demands large numbers of compounds for testing that the presence of impurities does not delay or prevent submissions of samples.

Only a sought-after biological activity retrospectively lends importance to the sample submitted—not its purity, synthesis, or structure. Ample opportunity to prepare and submit a satisfactorily pure sample always follows the finding of potent biological activity.

Company Code Names

Registrars provide an electronic or printed copy of the submission form. They or information scientists may also compose the chemical name, and they assign a short company code name, like "Phrmco **133397**." Comprising letters that typically abbreviate the company name as well as a string of numbers, this code makes it easy to talk about, say, the biology of the compound. No one has to memorize a long and often complex chemical name, nor must anyone draw the corresponding structure. The code name therefore makes intelligible telephone conversations feasible. It also aids in tracking the sample, reporting testing results, and drafting patent applications. If the compound becomes a marketed drug, its trade and generic names soon displace from usage the company code name, which is invaluable in the meantime.

Satisfactions and Successes

A career in discovery research can bring a variety of personal satisfactions and corporate successes, which complement the salary, benefits, and publication opportunities that the work entails. Some of these achievements come to chemists employed in or out of the pharmaceutical industry, while others are peculiar to medicinal chemistry. Even veteran chemists take pleasure in applying a known reaction or synthetic method to the preparation of new compounds. To conceive a novel synthesis or invent a new reaction is gratifying and so is implementing either of these tasks through benchwork. It remains a source of satisfaction to muster evidence demonstrating that a reaction product bears the expected structure and to determine a reaction mechanism. When a fugitive insight arrives suddenly, a moment of intense exhilaration greets the solution to an unexpected problem of structure elucidation.

Medicinal chemists find satisfaction and success in dutifully producing target compounds in abundance. Through traditional methods of organic synthesis, a single researcher makes 45–65 new compounds yearly and, aided by techniques of parallel synthesis or combinatorial chemistry, prepares as many as 200. For a researcher to attribute a desired biological activity to particular chemical structures is a source of satisfaction as well as inventorship. Regardless of who makes the attribution, it is a pleasure to prepare the first member of a series of potent compounds. Subsequently, to overcome (by written argument rather than mortal combat) a patent examiner's challenges to a patent application is also gratifying. On a rare and welcome occasion, a chemist learns that one of his or her compounds is to begin clinical trials. This announcement can foreshadow rewarding collaborations with researchers from varying scientific disciplines, who are charged with chemical, pharmaceutical, and clinical development of the substance. Introduction of a safe, effective drug to medical practice culminates development and represents ultimate success.

METAMORPHOSIS: SYNTHETIC ORGANIC CHEMIST TO MEDICINAL CHEMIST

A chemist who enters a discovery group starts her career working in synthetic organic chemistry. Her tasks, outlined in the preceding section, then largely comprise the familiar ones done as a student or a postdoctoral researcher. To participate fully in discovery research, however, demands a metamorphosis completed early in her career and begun on the first day. It changes a synthetic organic chemist to a medicinal chemist and prepares her for advancement in one company or a move to another, for which an evolution is prerequisite. This career development, although it may be urged by a boss acting as a mentor or by the research administration, is largely her own responsibility. Although someone else can point out the path, she must herself learn to follow it visibly, confidently, and successfully.

A working knowledge of synthetic organic chemistry remains crucial to making series of analogous target compounds. It is less helpful, however, in making the best choices of potential clinical candidates from a set of several potent substances. Such crucial choices advance a project, so they require collaboration with researchers and administrators from sciences other than organic chemistry. Understanding their points of view demands familiarity with their disciplines, which include biochemistry, pharmacology, and physiology. So synthetic organic chemists who develop into medicinal chemists learn something about mammalian physiology and the workings of the cardiovascular, gastrointestinal, immune, and central nervous systems. To acquire a knowledge of the biological fates of drugs—absorption, distribution, metabolism, excretion, and storage—is essential. Chemists also familiarize themselves with pharmacokinetics and modes of drug administration.

> *A knowledge of biology is no guide to synthesis, but is a medicinal chemist's turnstile to advancement.*

SOURCES AND SUPPLEMENTS

Barton, D. H. R.; private communication to the author and coworkers, Bloomfield, NJ, about 1975.

Clarke, F. H. (ed.); "How Modern Medicines Are Discovered," Futura, Mount Kisco, NY, 1973.

Davenport-Hines, R. P., and Slinn, J.; "Glaxo: A History to Nineteen Sixty-Two," Cambridge University Press, 1993.

Denkewalter, R. G., and Tishler, M.; "Drug Research—Whence and Whither," *Progress in Drug Research* **10**, 11–31 (1966).

Doyle, A. C.; "The valley of fear," in "The Complete Sherlock Holmes," p. 779, Doubleday, New York (undated; ISBN 0-385-00689-6).

Engel, L.; "Medicine Makers of Kalamazoo," McGraw–Hill, New York, 1961.

Goldberg, J.; "Anatomy of a Scientific Discovery," Bantam Books, New York, 1988.

Hailey, A.; "Strong Medicine," Doubleday & Co., Inc., Garden City, 1984.

Hammond, G. S.; *Chem. Technol.* **1**, 24–26 (1971).

Hofmann, A.; "LSD: My Problem Child," McGraw–Hill, New York, 1980.

Kogan, H.; "The Long White Line: The Story of Abbott Laboratories," Random House, New York, 1963.

Maxwell, R. A., and Eckhardt, S. B.; "Drug Discovery: A Casebook and Analysis," The Humana Press, Clifton, NJ, 1990.

Robinson, R.; "Memoirs of a Minor Prophet: 70 Years of Organic Chemistry," Elsevier, Amsterdam, 1976.

Savage, Mildred; "In Vivo," Simon and Schuster, New York, 1964.

Todd, A. R., and Cornforth, J. W.; "Robert Robinson, 1886–1975," *Biographical Memoirs of the Royal Society* **22**, 415–527 (1976).

Werth, Barry; "The Billion-Dollar Molecule," Simon & Schuster, New York, 1994.

Williams, M., and Malick, J. B. (eds.); "Drug Discovery and Development," The Humana Press, Clifton, NJ, 1987.

Williams, T. I.; "Robert Robinson: Chemist Extraordinary," Clarendon, Oxford, 1990.

Woodward, R. B.; "Art and science in the synthesis of organic compounds: Retrospect and prospect," in "Pointers and Pathways in Research," p. 28, CIBA of India, Ltd., Bombay, 1963.

Woodward, R. B.; Proc. Robert A. Welch Foundation Conference on Chemical Research, XII: Organic Synthesis, 1969, p. 3.

chapter

6

CHEMICAL DEVELOPMENT
Challenge in Organic Synthesis

"There is excitement, adventure, and challenge, and there can be great art, in organic synthesis."—R. B. Woodward

*"But at my back I always hear
Time's wingèd chariot hurrying near...."*—Andrew Marvell

INTRODUCTION

To appreciate the challenges of chemical development, as well as the chasms separating it from university or industrial discovery research, consider six distinguishing features.

1. Efforts to improve the synthesis of a medicine can continue for decades after a drug enters medical practice. By contrast, a successful industrial discovery group begins another project shortly after its drug begins clinical trials. A multistep synthesis completed in academia rarely if ever is applied to making hundredweights of the final product.

2. Urgency impels the chemical developers synthesizing the bulk drug that permits clinical trials. Not until the medicine passes a Phase II trial does the sponsor learn the drug works in humans. Eagerly and anxiously awaited, this outcome marks the end of the beginning. It raises hope even among veteran observers that several years hence the drug may justify its costs.

3. Some chemical developers assume grave financial responsibilities to their employer. Risking many tens of thousands of dollars on each occasion, they demonstrate in a pilot plant the feasibility, safety, and environmental soundness of one or more steps in a drug synthesis. The amounts of money at risk may represent not only the value of the starting material, reagents, and solvents, but also the costs of labor, overhead, and lost opportunity.
4. Governmental regulatory agencies play a larger role in chemical development than in chemical discovery research. They include (in the United States) the Food and Drug Administration, the Environmental Protection Agency, and the Occupational Safety and Hazard Administration. These agencies review the relevant research, records, buildings, and equipment of chemical developers, who work casting one eye on regulatory compliance.
5. Process chemists carry out chemical reactions on large scales in big reactors. Deemed typical by its Pharmacia and Upjohn developers, for example, is a certain run involving 528 kg of one reactant, 260 kg of another, and 325 kg of solvent (*Perrault et al.,* 1997). To work with massive quantities like these requires novice process chemists to reform any bad habits they learned as students. "Most research chemists," writes a development consultant, "prefer chromatography to crystallization..." (*Laird,* 1996). An academic chemist with industrial experience says that organic chemists choose the wrong solvents and use too much of each (*Sheldon,* 1994).
6. The arithmetic demon hampers developers' initial efforts to meet demand for the drug, disdaining the urgency of clinical trials. It limits the overall efficiency of the multistep syntheses needed to make many medicines. The difficulty is greater if the discovery researchers never maximized the yields afforded by their synthesis, which served purposes other than making one final product with great purity and efficiency. Nevertheless, the pressing need for high-quality bulk supplies can cause developers to use the research synthesis despite its inadequacies.

CHEMICAL DEVELOPMENT

In summary, chemical development in a FIPCO can entail long-lasting, tightly scheduled, and highly visible work that is costly, challenging, and subject to external regulation. It demands special skills and knowledge from its practitioners, some of whom operate plant equipment unfamiliar to discovery researchers (Figure 6-1). Developers work under pressures of time and money from which their counterparts in discovery research are exempt. Depending

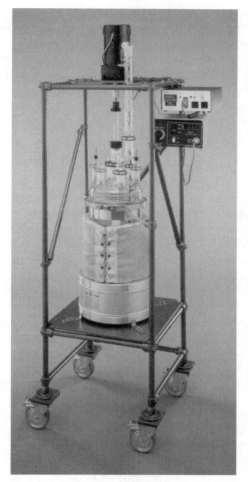

Figure 6-1 Ten-liter reactor (photo courtesy of Chem Glass)

on their responsibilities, chemical developers can expect to collaborate with various employees representing chemical discovery research and other disciplines as well. Examples of the latter coworkers include patent attorneys, pharmacists, pharmacologists, toxicologists, regulatory affairs specialists, chemical engineers, safety personnel, and marketers.

> *Within the pharmaceutical industry, chemical development is an essential specialty. An understanding of the work can help one decide whether to seek or accept a discovery post or a job in development research.*

It can also prevent awkward moments during site visits, when some interviewers ask candidates what interests they have in development research. Unaware of the distinction that FIPCOs make between discovery and development research, even some newly graduated doctoral chemists cannot express any preference. Familiarity with chemical development can assist a job hunter in evolving a preference. It can lead him to consider jobs not only with FIPCOs but also with other companies engaging exclusively in process research and development.

Information Resources

Chemical development and especially process chemistry, which it includes, form a subdiscipline of organic chemistry. Little taught in graduate schools, they are learned instead through experience on the job, in short intramural courses sponsored by chemical development groups, and in conferences attended and organized by practitioners. They are the subject of two recent books (*Lee and Robinson*, 1995; *Repic*, 1998), a journal (*Organic Process Research and Development*), and a chapter (*Laird*, 1990). Articles concerned with process research and development lie scattered in a variety of other chemical journals, particularly *Chemistry and Industry (London)* and the *Journal of Heterocyclic Chemistry*.

Scope

A compound worthy of chemical development offers commercial value as a medicine. Developmental work therefore involves only a drug candidate with desirable and selective pharmacological activities, which are specified as early as possible in the corresponding discovery effort. The chosen compound optimizes potency and selectivity in one or more animal models of the human disease. It shows high activity in a mechanism-based assay for antagonism, agonism, or enzyme inhibition and displays suitable pharmacokinetics. It lacks toxicity in short-term assays carried out with three to four animal species. The substance as well as its active isomers and metabolites are already patented, as are their therapeutic uses. Before passing the project to chemical developers, discovery managers and researchers rejected analogous but inferior substances from the same structural class.

This transition to development therefore calls only for applied chemical research from developers. It can mark a shutdown of the exploratory efforts that led to finding the clinical candidate. Chemical developers advance the chosen compound without seeking a pharmacologically superior substance deliberately or finding one serendipitously. Clinical physicians or discovery pharmacologists are more likely to make such a commercially valuable, serendipitous discovery than developmental chemists.

Driven by sales potential, each development campaign therefore concentrates on a single outstanding compound. Its biological activities and patented chemical structure define an unmistakable objective. The certainty associated with the fixed structure and specified activities lends to development work an appeal absent from discovery research. From outset of their project, chemical developers have clear goals important to their employers. Discovery research, by contrast, is more uncertain. Its scientists labor to identify their best compound until shortly before it enters clinical trials.

Timing

A chemical development campaign begins after research management recommends that a pharmacologically active compound of known structure advance to clinical studies. Such a decision implies that discovery efforts are successful in several respects. Chemically, at least small amounts of the clinical candidate can be made through the series of reactions used by the discovery scientists and called the research synthesis. The structure of the candidate drug is known with great certainty, often by single-crystal X-ray analysis. Comparison samples are at hand, as are various spectra and other data characterizing intermediate substances and the final product. Located in chemists' notebooks or a patent, experimental procedures for making these compounds are available. Written reports disclose helpful observations concerning the research synthesis of the clinical candidate.

Developers may initially adopt the research synthesis to make clinical supplies and subsequently adapt it to become the manufacturing synthesis. Frequently they improve it dramatically but sometimes abandon it utterly. Certainly every aspect of the research route receives scrutiny before chemical developers discard the synthesis or implement it on a manufacturing scale. All this takes time, so that efforts to prepare clinical supplies overlap with trials of new or improved processes. Different groups of scientists, comprising chemists at all degree levels, may work simultaneously on the two different tasks. A chemical development project typically lasts for 1 to 3 years and ceases when a suitable production synthesis results.

There are two noteworthy exceptions to this chronology. First, some chemical research to improve the manufacturing synthesis continues indefinitely when a successful medicine remains in a product line for decades. The second exception concerns the early involvement of development chemists in what hitherto had been exclusively a discovery project.

Development chemists often support a discovery project. Before a clinical candidate emerges, they make large batches of one or more late common intermediate substances. Such quantities let medicinal chemists devote their efforts to making new target compounds. Large batches free these chemists from the logistical work of making supplies and bringing them to the front. This cooperation is helpful or essential, so discovery and development efforts often coincide. When a project finally passes from discovery to development research, the discovery chemists provide samples or lore concerning the synthesis, reactions, and properties of the clinical candidate and related compounds. Members of either group can expect to work closely with one another at some times in their careers, and exchanges such as these benefit both parties' efforts. Close collaboration between discovery and development chemists is especially important when the chemical developers adopt an unimproved research synthesis to make urgently needed supplies for clinical trials.

Purposes of Chemical Development

Bulk Supplies

A chemical development campaign serves two purposes, both of them important and one urgent. Met within the first year of work, the latter goal calls for preparing a large supply of the drug for formulators, toxicologists, and clinicians. A relatively small quantity serves toxicologists carrying out a battery of medium- and long-term assays *in vivo*. Animal tests establish drug safety as much as possible in a nonhuman species, thereby satisfying a prerequisite to clinical trials. All these tests require quantities of the drug that can be calculated.

To complete a single assay, toxicologists may dose dozens of experimental animals for several weeks or months. They look for side effects in these animals, conducting rising-dose experiments in which some subjects receive, say, 3 milligrams of the drug per kilogram of body weight. Other groups receive doses of 30 and 300 milligrams per kilogram in a prolonged daily regimen. A single 90-day study using six 20-kg animals for each dose therefore consumes 3.6 kg of the drug. Similar rising-dose experiments with lighter animals from, say, two rodent species call for lesser additional amounts. Making as little as 3.6 kg, however, can tax the research synthesis, so preparing bulk supplies need not be trivial work.

A potent oral drug can act effectively and safely at a dose of a few milligrams per kilogram. This relatively low dose notwithstanding, clinical trials in adult humans may require greater quantities of drug than a few kilos. Adult body weights are 3 to 5 times greater than 20 kg, and many trials are needed. To establish a proper dosing regimen for humans, for example, calls for extended rising-dose studies with people, not animals. As many as 15,000 volunteers participate in certain trials of a single drug for one therapeutic indication. One trial can last a decade. The weight of drug needed for human studies—and for the requisite toxicological, formulation, and packaging experiments—can consequently reach or exceed a metric ton, depending on circumstances (*Robinson*, 1983). In Phase II studies alone, clinical trials of ateviridine mesylate, an AIDS drug, required multiton amounts (*Perrault et al.*, 1997). Chemical developers provide these quantities in batches of increasing sizes and according to a schedule.

Meeting material demands and keeping to schedule are crucial because so many complex arrangements depend on doing so. For example, one clinical trial may be carried out in far-flung treatment centers at different times. While the organizers are seeking and enrolling some volunteers in the Midwest, say, clinicians begin the study with other people in the East. To order supplies, reserve time in the pilot plant, arrange for formulation, and dispatch the finished drug, developers need to know how much of the medicine clinicians need and when and where they require it.

Of lesser importance at this stage of development, however, are the expense and efficiency of synthesis. Many experimental medicines fail during long-term animal toxicology studies or in their initial clinical trials. So, it is wise to expend development resources only on drugs succeeding in all of animal toxicology, Phase I, and Phase II studies. (Beginning a Phase II study to determine effectiveness in humans is contingent on success in Phase I, which shows safety in people. Not without regulatory approval and a showing of safety in animals does a drug begin its first Phase I study in humans.) Consequently, on a weight basis, it matters little if the drug given in clinical trials costs more than the same medicine made later by an optimized production process. Increases in overall yield and decreases in the per-kilo drug cost represent the goals and achievements of extensive process chemistry rather than the prerequisites of chemical development.

A Manufacturing Synthesis

The second purpose of chemical development is to devise and demonstrate a manufacturing synthesis that nonchemists can operate. Occurring on a plant scale, the synthesis must be safe, nonpolluting, and economic. This facet of chemical development is process chemistry, and its significance to a pharmaceutical company is evident in the following comment:

> *An important requirement for the derivation of a production process is speed, because the sooner a product can be introduced on the market, the sooner revenue can be earned to pay for all of the development work.* (Wilson, *1984;* italics added)

The work needed to derive a manufacturing process from a research synthesis begins about the time clinical trials start. It continues for months or years if necessary, but aims to reach completion before the sponsoring pharmaceutical company launches the medicine. Launching requires not only that the bulk drug be manufactured and registered, but also that it be formulated, packaged, labeled, and distributed.

Once a drug enters clinical trials or medical practice, its manufacturer is obliged to notify regulatory agencies of any changes to its synthesis. Such changes can affect the purity or bioavailability of new batches used to treat people. Contemplated alterations to an approved manufacturing synthesis therefore warrant concern, and developers make haste slowly. Any changes to the synthesis are conservative, lest drastic ones substantially alter the bioavailability or impurity profile of the medicine, requiring new animal studies or an extra clinical trial. Such trials are not only costly in their own right, but can delay launching or temporarily halt sales.

In accepting a new drug for clinical trials, a regulatory agency approves not only the safety of the medicine itself but also the safety of the impurities it contains, which depends on their concentrations and identities. Determined largely by the last reaction in the synthesis, the impurity profile refers to the number, structures, and proportions of inevitable minor drug contaminants. Any changes of starting material for this last reaction, or of solvent, reagent, conditions, work-up, or purification, can alter the profile, even though the impurity level remains at a total of 0.5%. Merely buying structurally identical starting materials, solvents, or reagents from a different supplier can introduce new impurities to the bulk drug. There is no guaranteeing that supplies procured from a new vendor arise from the same reactions or were isolated from identical natural sources. Different origins or changed procedures can vary impurities. A change in profile therefore raises the specter of more clinical trials or toxicological studies.

Further studies start after the patent clock begins ticking. This is so because FIPCOs make chemical, pharmaceutical, and clinical development contingent on the strong but time-limited protection that a composition-of-matter patent offers. Indeed, a granted patent, or a patent-office notice that certain claims are allowable, is prerequisite to advancing the medicine, unless it is an orphan drug. Toxicological or clinical studies necessitated by a change in impurity profile, however, effectively shorten the patent term. Worse still, a launching delay steals time from the end of the term, when drug income and earnings peak.

Some chemical research on the manufacturing synthesis does take place after production begins. Like the developmental efforts that precede it, it aims to reduce per-kilo cost of the drug, perhaps by showing that a newly available but less expensive or more advanced starting material is suitable. Alternatively, continued research can lower costs without changing the substrates or reactions of the synthesis. It improves yields, simplifies reaction work-ups, or achieves two chemical steps in one operation.

Continuing research work can be defensive and is undertaken to protect the patents covering the medicine and its production synthesis. A competing generic manufacturer of the same medicine infringes the patent holder's rights by selling the drug or operating the protected synthesis without seeking permission or paying royalties. To help stop the latter violation, researchers employed by the process patent owner determine the profiles of minor impurities in the competitor's drug and its own. Matching profiles suggest that the competitor is using the same synthesis as the owner and lend support to legal actions for patent infringement.

ORGANIZATION

By Discipline

A qualified new graduate seeking employment in the chemical development or discovery department of a FIPCO can expect other chemists to evaluate his credentials and to extend or withhold a job offer. Although personnel staffers may screen applicants to select qualified candidates and reject others, much of the hiring responsibility falls to experienced scientists. Once a job hunter accepts an offer and begins work, his bosses are likely to be fellow chemists; they, not nonscientific staff, will judge his performance.

By Assignment

For an understanding of how FIPCOs organize chemical development work, consider the single task of making a simple drug through a brief synthesis that the discovery chemists devised. Imagine that the chemists employed in the development department form groups A through, say, Z, and that the members of groups A, B, and C are free to take on new work. We suppose each group is composed of three to five members. There are then three logical ways to assign to these groups the three jobs that must be done (Table 6-1). These tasks include the preparation of clinical supplies, the development of a manufacturing process on a laboratory scale, and the demonstration of the synthesis in a pilot plant. In the first of the three assignments,

Table 6-1 Assignments of Chemical Development Work to Groups A–C of Process Chemists: Variations I–III for a Brief Synthesis of One Drug

Responsibility	Assignment no.		
	I	II	III
Preparation of clinical supplies	A	A	A
Chemical development			
Laboratory	A	B	B
Pilot plant	A	B	C

group A takes on all three jobs, beginning with the preparation of clinical supplies and ending with a large-scale demonstration of the production process. This arrangement (Assignment I in Table 6-1) benefits from the familiarity with the synthesis that the chemists acquire as they make clinical supplies and perfect the process. However, the work can suffer from linearity, if this arrangement defers efforts to improve the synthesis until clinical supplies become available. It thereby risks delay in demonstrating the production process on a pilot-plant scale. This arrangement requires good communications with discovery researchers, whose knowledge of the synthesis will be soon surpassed, and with chemical engineers.

In another arrangement, two different development groups, A and B, carry out the jobs of making the initial supplies for clinical trials (Assignment II in Table 6-1). This organizational scheme permits development to begin simultaneously and necessitates no delay. It does require open and frequent communication between two development groups so that each alerts the other about improvements, obstacles, and any hazards. Initially, the work may benefit through discussions with the discovery chemists who first made the drug candidate. Collaboration with chemical engineers also aids in making the final choice of a production process.

Finally, three development groups, A, B, and C, may participate in the project (Assignment III in Table 6-1). This arrangement means that laboratory-scale improvements to the original research synthesis translate as quickly as possible to pilot-plant-scale demonstrations of the production process. It lets some researchers demonstrate on a large scale improvements to one synthetic step, while others working on a small scale change another step. Such an arrangement also requires good communications, this time among all three development groups, the discovery researchers, and the chemical engineers.

If the research synthesis comprises many steps, then more groups join the effort. Each group of process chemists takes responsibility for perfecting one or several steps of the research synthesis rather than the entire sequence. A need for greater staffing also arises when other routes are investigated, and several dozen process chemists may work on one project. For example, to obtain broad process patent protection of the drug calls for simultaneously trying new syntheses. Most important is to invent and develop one or more new syntheses from which the final manufacturing process is ultimately chosen. These routes can bear little resemblance to the original research synthesis.

By Developmental Project Teams

The structure of Table 6-1 can inform another organizational scheme. Known as the team matrix, it is commonplace in large pharmaceutical companies. A large FIPCO contemporaneously carries out several developmental projects, say $a, b, c, \ldots n$, in each of which it advances a different drug candidate (Table 6-2). Depending on corporate interests as well as scientific and medical opportunities, these drug candidates represent different or identical therapeutic areas. Each project requires the services of specialists from various disciplines who form project teams. Representatives from a given team meet regularly to learn one another's results, assess progress, and plan and schedule further work. Combining several such teams with numerous projects creates the matrix illustrated. Its columns represent the different projects going on, its rows present the varying disciplines contributing, and its elements

Table 6-2 Matrix Structure in the Composition and Staffing of Developmental Project Teams[a]

Discipline	Staffing in projects a ... n				
	a	b	c	...	n
Biocatalysis	1	0	2	...	1
Chemical development	4	7	5	...	10
Chemical engineering	2	2	1	...	0
Clinical research	3	2	2	...	1
Hazards evaluation	1	1	2	...	1
Patent law	1	1	1		2
Pharmaceutical development	4	3	2	...	1
Regulatory affairs	1	2	1	...	0
Toxicology	3	2	4	...	5

[a]Adapted from C. G. Smith.

show the number and specialties of employees assigned to each project. The complexity, importance, and urgency of the projects determine these numbers, which rise and fall as the project advances. (The numerical values in Table 6-2 are representative, while the list of disciplines is selective.)

Process Research and Development

To devise and demonstrate a suitable manufacturing synthesis of a commercially valuable medicine defines the primary goal of process chemistry. This objective elicits the art of organic synthesis, offering a challenge worthy of its champions. Attaining it entails choosing as the manufacturing synthesis one route from several reaction sequences that discovery or process chemists invent and that the latter demonstrate on suitable scales. By looking into other syntheses, they seize another objective, which is to support a patent application claiming several distinct routes to the same drug. A patent limited to newly developed syntheses gives process protection, which often extends beyond the term covered by the composition-of-matter patent. Issued some years ago to the discovery chemists, the latter patent may contain no claims to chemical processes and so afford them no protection. Although the composition-of-matter patent discloses the best method of synthesis known to the inventors, this synthesis need not be claimed.

So refined is a production process that supervised operators who are not chemists or even scientists can carry it out. Their workplace may be so far from the development laboratories that frequent oversight by the researchers is impossible. Process chemists therefore compose clear, comprehensive instructions covering each reaction, work-up, and purification.

Process Research versus Process Development (by Dr. Thomas C. Nugent, Catalytica, Inc.)

Depending on the company, process chemistry may or may not be divided into the two branches of process research (PR) and process development (PD). Those companies that do not make this distinction generally have one site where one or several chemists follow a process from start (invention) to finish (manufacture). Both systems have advantages and disadvantages that will not be elaborated here.

Although clear distinctions exist for the responsibilities of PR and PD chemists (Table 6-3), each requires a general knowledge of both jobs. This is true more for PR chemists who "hand" their processes to development

Table 6-3 Distinctions between Process Research and Process Development

	Functions	Individual activities	Team interactions
Process research	Basic chemical research	Literature surveys Route selection Bench research 1– to 10-mmol reactions	Analytical services Information services

	Optimization
	  Yield
	  Price
	  Quality
	  Operability
		50– to 100-mmol
		reactions
			QA,[a] QC[b]
			Engineering
			Purchasing

	Demonstration and documentation	1– to 30-liter reactors	QA, QC Engineering Safety

Process development	Validation	Qualify starting materials
		Investigate any engineering problems:
		  Agitation
		  Exotherms
		  Filtrations
		  Equipment compatibility,
		  Basic operability
		Deal with QC issues
			QA, QC
			Engineering
			Purchasing

	Pilot plant support	Oversee pilot-plant process Troubleshoot Modify process if needed Scale: about 400 liters	QA, QC EH & S[c] Production technicians Engineering
	Manufacturing support	Trouble shooting Scale: about 4000 liters	As above

[a] Quality assurance: consistency in protocol.
[b] Quality control: quality of product.
[c] Environmental health and safety: hazard evaluation.

chemists. Also, because PR chemists are involved from the beginning of a new process, their work tends to be more creative and encompassing. The PD chemists' overall objectives are more practical; they are implementation oriented.

Briefly, process researchers invent syntheses that meet the criteria of safety and cost, while maintaining an acceptable purity for the target intermediate or drug. To do so PR chemists must design processes restricted to a few crucial and general concepts. These principles include the following factors, but are not limited to them: (1) inexpensive starting materials, reagents, and solvents; (2) lack of patent conflicts; (3) proven purification methods; (4) robust reactions; and (5) manufacturing capabilities matched to the hazards of the proposed chemistry.

Process development chemists, on the other hand, examine the proposed synthesis from PR and attempt to validate it using the principles of chemical engineering. Thus, development researchers must explore in much greater detail than process chemists the parameters for each chemical reaction in a synthesis. Knowledge of large-scale crystallization, distillation, heat transfer, agitation, polymorphs, and so on, enables them to reliably produce batches of consistently high-quality product time and again. It is the PD chemist's ultimate goal to manufacture the desired intermediate or drug target.

Organic chemists employed in process chemistry do some of the same basic work as discovery chemists. They run and monitor reactions; isolate, purify, and characterize products; and determine purities, identify samples, and elucidate structures. However, process chemists' specific goals, focus, apparatus, and skills distinguish them from their counterparts. As much as or more than discovery research, process research and development demand that chemists apply their knowledge of reaction mechanisms. Successful process chemistry requires scrutiny of chemical reactions, knowledge of physical organic chemistry, and a talent for devising experiments based on observations and understanding. In elaborating a manufacturing synthesis, moreover, process chemists consider factors relatively unimportant in small-scale work. Examples include the large-scale feasibility of otherwise standard work-ups, thermochemical hazards, the expense and risk of large amounts of waste solvent, and community safety around a plant.

To appreciate the painstaking research that is process chemistry, consider an example furnished by chemists at G. D. Searle, now part of Monsanto. In making $LiCH_2Cl$, they ascertained that the molar ratio of the reagents $BrCH_2Cl$ to n-BuLi ideally lay between 1.25 and 1.33 (*Liu et al.,* 1997).

Merck process chemists showed that an acid-regulated, one-pot transformation of a cephalosporin derivative to the antibiotic cefoxitin was complex (*Weinstock*, 1986). Not only did it entail eight discrete chemical reactions at least, but some of them had to occur synchronously to maximize the antibiotic yield. Moreover, the concentration of hydrochloric acid reached an optimum at 0.004 molar. The rates of some of the eight reactions were proportional to this concentration and others to its square.

Because of intense development efforts like these and others, the production process usually outshines the research synthesis in several respects. A Glaxo researcher provides a concrete example drawn from development of certain antiasthmatic imidazotriazinones (*Marshall*, 1983). Thanks to optimizations, the second synthetic route to be developed proved superior to the first in rate of production, costs of labor and materials, and in overall yield. The rate increased sevenfold, the costs fell by three-quarters, and the yield nearly doubled.

From a developmental chemist's viewpoint, an acceptable manufacturing synthesis comprises more than isolated chemical reactions, however high yielding they are. Rather, it represents a safe, sound, and economic process that integrates disparate elements. Some elements of a production process include procuring starting materials and assuring that supplies of them and other chemicals are reliable. They also include charging reactors; recycling solvents; minimizing and disposing other wastes; and crystallizing products—as certain polymorphs or as salts or solvates. A manufacturing synthesis therefore resembles a scheduled freight train in which the linked boxcars represent the individual chemical reactions and the couplings stand for the other essential process steps like drying. Without couplings or cars, the train neither departs nor arrives.

Equipment

To understand some of the work that process research and development entails, consider the apparatus used. Initially the flasks employed by researchers occupy as small a volume as a few milliliters, but soon reach beyond 22 liters as development progresses. Careful researchers, to avoid various troubles, gradually scale organic reactions upward. In a pilot plant, chemical changes occur in vessels as large as 1000 liters. Reactions from a fully developed manufacturing syntheses can take place in reactors occupying 10,000 liters. From research laboratory to manufacturing plant, vessel sizes span five to six powers of 10.

A volume as small as 22 liters represents the practicable upper limit of an unsupported spherical flask made from glass. Reactions occurring on larger scales than this demand stronger steel reactors or glass-lined steel vessels. From the viewpoint of a chemist with only academic training or discovery-group experience, such a vessel is special in several respects.

A reactor is opaque, so its contents are dark. Consequently, it thwarts simple efforts to monitor reactions occurring within. Viewing the contents is nearly impossible even with the help of a flashlight; the few ports on a steel reactor afford only narrow sight lines. Estimating the volumes of solvents added or removed therefore requires special measurements, not visual inspections.

Glass and steel expand on heating and contract on cooling, but the different materials do so at unequal rates. The materials composing a glass-lined steel reactor therefore limit the temperatures attainable within the vessel. The temperature sensitivity of the reactor demands caution in heating or cooling the vessel, lest the liner break. This sensitivity can influence reaction times and determine the kind of reaction for which the vessel is useful. Process chemists' reactions therefore occur under greater constraints than their counterparts' experiments in discovery research.

Pilot and production plants do not use the spherical flasks familiar to students of experimental organic chemistry. Instead, they employ cylindrical reactors because their shapes offer better control of heat transfer, which is important throughout chemical development and manufacturing. External cooling, for example, moderates an exothermic process only if heat transfer is fast, while slow transfer can lead to dangerous runaway reactions. Cylindrical reactors maximize heat transfer, which occurs at a rate proportional to their lateral surface areas. Their tops or bottoms are not heated or cooled. The lateral surface area of a right circular cylinder exceeds the surface area of the sphere when their volumes equal one another, if the cylinder height is more than three times greater than the sphere radius. The difference between these areas increases markedly as the volumes rise (Figure 6-2).

Changed Techniques and Conditions

A 1000-liter reactor is large, heavy, and stationary. Its situation and construction demand techniques little practiced in discovery or university research, call for modifications of methods or conditions used there, and forbid altogether some commonplace operations. For example, (1) the chemist using such a reactor cannot tilt it to pour its contents into a container more suitable for the next operation. One such operation, which entails washing with water a solution prepared with an organic solvent, must occur within the reactor, not in a separate funnel devoted to the procedure. As many successive operations as possible take place in the same vessel. Emptying the reactor requires that the contents flow freely, which lets them be pumped by suction, drained by gravity, or forced out by pressure. In any

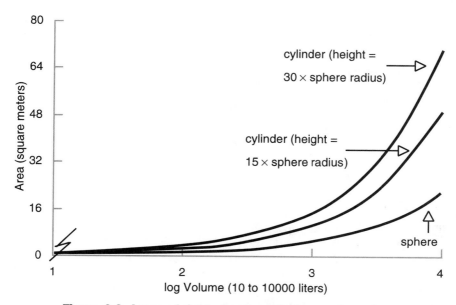

Figure 6-2 Areas of right circular cylinders and a sphere of equal volume. The area excludes the circular areas of the cylinder ends, where little heat transfer occurs in a reactor

case, the contents pass through a pipe leading to a receptacle located on a lower floor. Throughout such a transfer, the reactor contents remain advantageously confined to the apparatus.

(2) Modified conditions make some common laboratory operations feasible. Filtration, for example, is carried out under increased pressure applied to the suspension of solid. To filter under reduced pressure represents a laboratory technique that is undesirable on a plant scale for two reasons. It causes solvent to evaporate, which poses an easily avoided environmental problem, and it deposits by evaporation some of the originally dissolved solid within the pores of the filter. Such a deposit reduces yields of a costly substance and creates difficulties in cleaning the filter. In manufacturing plant operations, many filters, which are composed of cloth rather than paper, are large and costly enough to be saved and used again. Process chemists can combine filtering with centrifuging for better containment and greater convenience.

In a pilot plant, to take another instance, distillation in vacuum or hydrogenation under pressure remain practicable. However, they occur at respectively higher and lower pressures than in an ordinary laboratory of organic chemistry. Reactors do not attain or hold the same pressures as other apparatus.

(3) Some operations indispensable to discovery research, like concentrating a solution to dryness, are taboo in a pilot plant. To recover afterward any remaining solid or viscous resin from the reactor is prohibitively difficult. Consequently, a researcher working on a reactor scale cannot monitor reactions by weighing crude products obtained by completely evaporating solvents. This technique helps only discovery researchers, who can easily carry small flasks from bench to balance. Indeed, one of the jobs that a process chemist does in developing a manufacturing synthesis is to reduce the need to monitor reactions by weighing or other methods.

Reforms

There exist other commonplace but inappropriate practices that newly employed process chemists rethink. The laboratory habits that motivate reforms pervade small-scale chemical research, appearing whenever chemists set up reactions and isolate or purify their products. Changes save time and money and lend safety and environmental soundness to chemical development. Consider the following five examples.

1. To charge a flask with solvent and all the needed reactants and then to heat the contents invites an exotherm and risks a runaway reaction in large-scale work. For two reasons, such an exotherm escapes notice in small-scale work. Small spherical flasks rapidly transfer to the surrounding atmosphere all the heat evolved by the reaction. Second, the quantity of heat produced by a reaction occurring on a millimolar scale is often too small to threaten danger or even attract notice. Neither reason applies, however, when the same reaction utilizes several thousand moles of starting material on a scale greater by a million times.

2. Using one solvent for a reaction and another for the work-up entails a costly exchange that prolongs the whole procedure. On a small scale, such an exchange is affordable partly because it is quick to effect, even though it entails distilling one solvent before replacing it with a more suitable one. But even simple operations like distillation take long times on a plant scale because heat does not transfer instantaneously. Hours pass before, say, 5000 liters of toluene absorb enough heat merely to boil without distilling.

3. To dilute a reaction mixture with many times its volume of water helps isolate a water-insoluble product from a synthetically useful, water-miscible solvent like *N,N*-dimethylacetamide. It prepares for extracting only the desired product with a water-immiscible solvent. But, on a large scale, the volume of available reactors limits the quantity of starting material convertible in one batch to the desired product. This constraint makes volume-inefficient any work-up involving such a dilution, which is better avoided.

4. In a laboratory, adding a solid drying agent to remove traces of water from an organic solution remains a helpful and acceptable practice. In a manufacturing or pilot plant, however, this technique creates a string of avoidable problems. It creates a need for a crude filtration at least and for washing of the collected solid. The solvent that invariably clings to the spent dessicant must not escape to the atmosphere, and the solvent-wet dessicant demands proper disposal or economic recycling. So, process chemists seek to dry solutions and solutes by other means; for example, azeotropic distillation of a suitable cosolvent.

5. Although elution chromatography is not only feasible but also practiced on a large scale, it similarly brings the need to discard or recycle the used adsorbent. Not to speak of the huge volume of solvent that elutes the desired product from the adsorbent in a successful separation. Process chemists learn to prefer crystallization to chromatography, especially because a single crystallization often removes all the impurities accumulated through several reactions.

ELEMENTS OF A SUITABLE MANUFACTURING SYNTHESIS

Not only does a successful project make a passage from discovery to development research, but the criteria defining a good synthesis also change. Process chemists adopt criteria that medicinal chemists need not meet, including product specifications, safety, soundness, throughput (see text below), sourcing, cost, and overall efficiency. They abandon others that like versatility are important to discovery research. They fulfill some requirements common to research and manufacturing syntheses, namely robustness, simplicity, brevity, convergency, and adaptability to scale-up. To understand the challenges that process research and development presents to its practitioners, let us consider some of these needs, beginning with criteria peculiar to process chemistry.

Specifications

For human use, the final product of a successful manufacturing synthesis consistently meets certain specifications. It undergoes a battery of tests showing that it does so. The tests pertain to its identity, purity, enantiomeric excess if appropriate, and its appearance. The bulk drug contains no more than a total of about 0.5% of impurities, with no one contaminant amounting to more than 0.1% by weight. Its color is also specified since highly colored contaminants present at levels far less than 0.1% can determine the appearance of the bulk drug. By contrast, the final product of a discovery synthesis has to pass fewer tests because it is not subject to rigid specifications.

Safety

So numerous, pervasive, and important are safety considerations in process research and chemical development that FIPCOs employ specialists to deal properly with them. To appreciate this importance, let us consider only one topic in any detail, namely runaway reactions. Carried out on a large scale, such reactions can destroy property and injure or kill people; they thus threaten the plant, surrounding homes, employees, and community residents. Runaways arise from two causes, poor understanding of kinetics or chemistry and inadequate operators' training or procedures (*Etchells,* 1997). Some large-scale reactions begin with explosives or produce them, which, in a detonation, can liberate large amounts of heat and gas in microseconds. Other runaways can unexpectedly form gases like dimethyl ether, which result from certain methylations occurring under basic aqueous conditions and using methyl iodide (*Lee and Robinson,* 1995).

Yet other chemical changes are exothermic, releasing heat quickly. Reactions that are docile on a laboratory scale can, unless they are controlled, run away on a plant scale, for reasons basically concerning heat evolution, transfer, and dissipation. For example, in a unit of time the heat lost by a standard reactor can be three to six times less than by laboratory glass equipment (*Etchells,* 1997). Heat transfer and dissipation depend on various factors including the cooling capacity of coolant surrounding the reactor, the heat transfer coefficient of the reactor wall, and the reactor surface area. In a cylindrical reactor, this area is proportional to the square of the radius. The heat evolved, however, depends indirectly on the volume of reactants and solvent. Therefore, it effectively increases as does the cube of the reactor radius.

To prevent runaway reactions may require nothing more complicated than slow addition of liquid reactants or solutions of solid ones. Slowly adding solids offers greater difficulty since even screw mechanisms designed especially for a particular solid may jam. Alternatively, passing a suitable coolant through a jacket surrounding the reactor offers the possibility of good control, as does operating at a temperature significantly lower than that used on a laboratory scale. A rule of thumb calls for process chemists to reduce by about 10°C the temperature at which an exothermic reaction occurs. This change allows scale-up by 10-fold (*Laird,* 1990) and reduces the reaction rate to about one-half to one-third of its original value (*Daniels and Alberty,* 1961). Larger scale factors demand greater temperature reductions.

Certain development researchers specialize in investigating the thermochemical properties of each reaction in a production process. Forewarning requires this knowledge. Before a process chemist can decide how to control an exothermic reaction, he must expect to encounter one. The work begins at the inception of chemical development and entails adiabatic and differential scanning calorimetry. It detects exotherms and their onset temperatures and so offers guidance to researchers responsible for gradually scaling up the steps of a proposed manufacturing synthesis. The thermochemical investigation begins before the change from laboratory to pilot plant scale. This passage can involve a 100-fold increase in scale, while the increase from pilot to manufacturing plant entails only a factor of 10 (*Laird,* 1990).

Soundness

Drugs represent highly profitable substances made in low annual tonnages by relatively wasteful manufacturing processes. In one estimate, the ratio of by-products to drug reaches 25–100 to 1 (*Sheldon,* 1994). The costs of waste disposal, high as they were, were once acceptable if paying them helped launch drugs quickly (*Cusumano,* 1997). Oil refineries, by contrast to FIPCOs, produce less profitable petroleum products in higher yearly tonnages but with 250–1000 times less waste per kilogram of product. This situation prevails partly because so many medicines in the developed countries arise as the products of multistep organic syntheses. Such sequences of reactions suffer not only from overall yield inefficiency but from low atom utilization; now they draw attention because of cost and environmental concerns. More than ever, these pressures direct process chemists to improvements that come from waste minimization and therefore from environmentally sound reactions.

Individual reactions wastefully form by-products in several ways (*Lester,* 1996). Consequently, the economy of reactions proposed for a manufacturing process influences the final choice of a synthetic route. Stoichiometric changes (A + B → C + D) create a mole of waste product D for every mole of the desired product C. Bimolecular nucleophilic substitutions of benzyl halides exemplify these stoichiometric changes. Some conversions (C ← A + B → D) lack chemospecificity, forming by-product D in addition to the desired C. The product C desired from another reaction may undergo a further change (A + B → C → D). Catalytic hydrogenations of aromatic nitro compounds to reactive hydroxyl amines exemplify such changes; uncontrolled further hydrogenation of desirable hydroxyl amines forms unwanted anilines.

All these reactions challenge process chemists to minimize the number and quantity of waste products by exercising control or devising other conversions than form the same intermediate or final product C.

To quantitate the wastefulness or economy of a reaction, researchers consider its characteristic atom utilization. This number is a fraction reckoned by dividing the molecular weight of the desired product by the sum of the molecular weights of all the products and by-products. The closer this fraction approaches unity the better. In principle, it equals 1 in the case of addition reactions, for example the bromination of an olefin. Stoichiometric reactions, because they form more products than only the desired one, are less efficient.

To scrutinize in isolation the reactions of a proposed synthesis is necessary but not sufficient to realize a suitable production process. A synthesis entails more than the discrete sequential reactions that compose it, for it includes work-ups and purifications. Consequently, the entire manufacturing process demands a quantitative evaluation against environmental standards, which makes different proposed processes comparable (*Sheldon*, 1994). Process chemists therefore calculate a single number characterizing an integrated process by reckoning environmental load factors (ELF) for all the steps and then summing. Each factor measures the amount of materials wasted for a unit amount of intermediate or final product formed. It therefore applies to inorganic and organic by-products, for examples, and to solvents, catalysts, and resolving agents. It covers the usage of solvents and reagents that work-ups and purifications demand as well as what goes into a discrete reaction of the synthesis. For a single step of the process, the following equation gives a value of the environmental load factor when the appropriate weights are substituted as needed:

$$\text{ELF} = \frac{\text{waste materials} - \text{product}}{\text{product}}.$$

A refined analysis of the environmental costs considers energy requirements and the relative harmlessness of waste products (*Sheldon*, 1994).

A Glaxo researcher, *Mark Owen*, gives an example in which process research and development reduced an ELF by a factor of 2.7 (*Lester*, 1996). In the initial pilot plant runs, removal of a tertiary-butyldimethylsilyl protecting group produced 160 kilograms of waste for each kilogram of deprotected product. The ELF value fell to 59 after several improvements had been realized. Although the original process was obviously wasteful, calculating ELF values is nonetheless helpful; they objectively measure the success of improvements.

An environmental load factor may be negative, which means that the weight of product formed exceeds the weight of materials wasted. An example comes from the Hoechst company in Germany, of which the FIPCO

Hoechst Marion Roussell is part (*Christ,* 1996). Noble-metal catalysis instead of chemical reduction by iron reduced the waste to 1.5 kilograms per 100 kilograms of product. This value corresponds to an ELF of −0.98. Originally, the chemical reduction of an aromatic nitro compound to an amine produced 2.37 kg of waste for each kilo of product.

Catalysis, whether by noble metals or enzymes, offers an opportunity to reduce environmental loads. Replacing a chemical process with an enzymatic one, Hoechst research lowered the ELF value for making the valuable antibiotic intermediate 7-aminocephalosporin. The value fell from 30 to −0.7, a fourfold change (*Christ,* 1996). Despite an increase in waste water, the fall in ELF reduced environmental costs of this step by 90%.

To characterize an entire manufacturing process, a usage factor multiplies the ELF for every reaction before any sum is taken. It represents the relative contribution that each step makes to the whole. The usage factor equals that weight of an intermediate product needed to furnish a unit weight of the final product, so it effectively accounts for chemical yields. A sum of so-modified ELF values helps process chemists compare one synthesis to another and select a manufacturing route to the drug.

Sourcing and Cost of Raw Materials

Other important factors in choosing synthetic routes to develop are the structure, reactions, sources, and costs of the starting material. Its structure and chemistry influence the remainder of the synthesis in several respects. For example, to operate the manufacturing process that results from development research may require capital investment in special chemical equipment or even in plant construction. Consequently, the commercial source of the starting material must be reliable, providing the requisite quantity and quality of goods to serve the manufacturing schedule adopted. A Glaxo developer notes that to retrofit a plant designed and built for another process would be costly (*Wilson,* 1984). High-quality starting material may be less crucial than cost because of opportunities to purify that occur partway through the synthesis. The final step of a reaction sequence largely determines the quality of the bulk drug and its impurity profile.

Finding reliable sources of starting materials is itself a specialty occupation. Within the process research and development groups of some FIPCOs, a few experienced chemists make careers of it. Independent brokers exist to locate large quantities of starting materials for sale, serving the pharmaceutical and other chemical industries. Their knowledge of the chemical industry indirectly helps reduce per-kilo costs of finished drugs, so their assistance can be valuable. Affordable starting materials are not necessarily available from the fine-chemical suppliers that sell small quantities to discovery and development chemists and can be hard to find elsewhere. They can be expensive if they are available from fine-chemical suppliers.

A Boeringer-Ingelheim project exemplifies the prohibitive and affordable costs of a starting material available from different sources (*Roth et al.*, 1997). One fine-chemicals supplier asked $10 per gram for an enantiomerically enriched substance offered by another vendor at $0.0055 per gram. The former supplier could provide no more than one kilogram, however. The latter sold one metric ton of the substance with an enantiomeric excess (ee) of 93.5%, which served to make the antiasthmatic drug ontazolast. Purification taking place later in the synthesis raised the ee of the drug to over 99%.

Costing a Synthesis

Lasting pharmaceutical companies heed their operating expenses. Consequently, the per-kilo cost of drug manufacturing does not escape their notice, even though it accounts for only part of the prescription price. This cost becomes more important when patents expire and generic drug makers begin to compete; cost estimates are therefore essential in choosing a manufacturing synthesis from several tried routes. As they gain experience, process chemists learn to estimate the per-kilo cost to make a medicine, which depends on various factors. Two of them—efficiency and throughput—receive separate treatment below, while others are surveyed here, after a caveat:

> *Costing is a notoriously ambiguous [endeavor] in which the inexperienced chemist can get into difficulties. . . . costs . . .based on . . . a laboratory synthesis may mislead senior personnel . . . which may be detrimental.*
> (Laird, *1990*; italics added)

Obvious factors affecting the overall cost include, for each step, the prices of starting material, reagent, auxiliary if any, and solvent. Less obvious costs comprise the expense of recycling or discarding used solvents and other wastes. Also significant are overhead expenses including labor costs, which vary geographically, and plant occupancy costs. The latter expenses arise not only from time spent running reactions but also from time passed in working them up, which increases with scale. It is advantageous therefore to defer isolation and purification from one reaction to the next because doing so reduces processing time. This expedient, known as *telescoping,* effectively decreases the number of separate steps in a manufacturing synthesis.

In one sense, process chemists concerned with cost reduction and synthesis choice have some flexibility. The overall cost of a manufacturing synthesis can accommodate long plant occupancies if starting materials are inexpensive. Costly starting materials, however, demand short residence times. The overall cost depends on both expenses, so one may be high if the other is low.

Process chemists estimate the relative costs of several proposed syntheses as a basis for comparisons that identify the best one. In making estimates it helps initially to make certain simplifying assumptions and temporarily to neglect some of the foregoing factors. Therefore, Parke-Davis researchers assume 100% yields and low bulk prices for starting materials (Hoekstra et al., 1997). They neglect overhead, labor, and disposal costs to prepare idealized cost estimates. Costing a proposed production synthesis draws attention to those process elements most needing improvement.

Some countries like Ireland bestow favorable tax treatment on companies that establish themselves within their boundaries. Consequently, the site of manufacture can help reduce the overall cost of drug synthesis. Pharmaceutical companies that export manufacturing encounter fluctuating exchange rates that can abolish (or augment) these savings.

Efficiency, Brevity, and the Arithmetic Fiend

The per-kilo cost of a marketed drug depends partly upon the overall chemical yield of the synthesis used to manufacture the compound. Process chemists, to make their manufacturing synthesis as efficient as possible, maximize this yield by various means, making many improvements throughout development. However, the infamous demon resists their efforts. It wields against them the number of independent chemical steps in the synthesis. (The number of steps in a synthesis is important to count, although a step is hard to define. In one definition, a 1:1 ratio exists between steps and workups including purifications and sometimes sterilizations.) Unless each step is efficient, the synthesis produces a small quantity of the end-product but consumes a large amount of the starting materials. A multistage sequence imposes low upper limits on overall yields, making long syntheses uneconomic. However, high individual yields tend to raise overall efficiencies:

> *. . . the arithmetic fiend, which besets all multi-stage synthetic activity, can be conquered by careful and intensive developmental work. Every organic chemist knows how very rapidly fractions diminish when multiplied by one another. It is perhaps less widely appreciated, that as a fraction nears one it diminishes very much less rapidly on multiplication, and how*

> *many reactions can be coaxed into giving yields in the range of 90 to 100 per cent.* (Woodward, *1956*; italics added)

The number of steps in a synthesis and the average yield per step are conflicting influences. To see how they determine the overall yield, look at Figures 6-3 and 6-4. In viewing the figures, consider three-, six-, and nine-step syntheses that bring overall yields as much (or as little) as 10% (Figure 6-3). The average yield per step must equal 46, 65, and 77%, respectively. If an overall yield of 2% is affordable, then a synthesis averaging 75% per step could go on as long as 13 steps (Figure 6-4). Suppose this sequence includes the two chemical steps that introduce and remove that albatross of synthesis, a protecting group. If so, it sacrifices half the overall yield and doubles the cost of a shorter but otherwise comparable 11-reaction sequence. The longer a synthesis, moreover, the worse the crisis if something goes wrong at the end. A Glaxo process chemist notes that the final product embodies all the cumulative costs of manufacture. He writes, "Therefore, success or failure is more dramatic here than elsewhere" (*Wilson,* 1984).

Each stage of a long synthesis adds value to the final product. Consequently, a low yield sustained later in the process brings a greater loss than

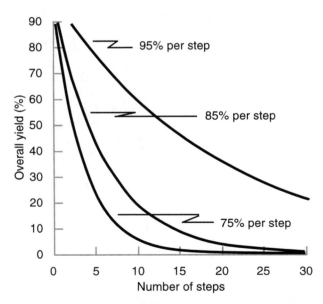

Figure 6-3 Average yields per step influencing the overall yield of a synthesis

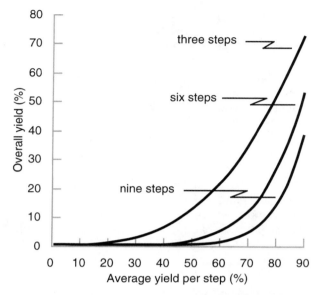

Figure 6-4 The number of steps influencing the overall yield of a synthesis

it inflicts earlier. If the starting material is cheap and readily available, a low yield that comes early is acceptable without being desirable.

Not all the stages of a manufacturing process entail purifications, workups, or simple chemical reactions. Medicines given by injection must be free of pathogens and require sterilization that was unnecessary in the research synthesis. Sterilizing bulk samples adds expense and an operational step to drug manufacture. To sterilize a solid drug might require dissolution of the sample followed by sterile filtration to remove infectious contaminants. Completing the job calls for crystallizing the drug, collecting it on a filter, and drying, analyzing, formulating, and packaging it. Trained personnel carry out all these operations under sterile conditions. Nor surprisingly, orally active compounds make especially valuable drugs because they need not be sterile.

Throughput

The numbers and efficiencies of stages in a production process are important but not alone in determining the drug cost per kilo. Throughput, which refers to the quantity of product emerging in unit time from one reaction, is also critical. Meager throughputs reflect the low concentrations of reactants typical in small-scale discovery research. They mean that on a large scale a single reaction occurring at the same concentration occupies expensive plant apparatus and personnel for a comparatively long time. So, the opportunity to take on other work disappears, and costs soar.

A rise in throughput and a fall in cost come from either of two increases in volume efficiency. Process researchers can decrease the solvent volume taken to work up the product mixture or reduce the amount of solvent used to run the reaction. For example, to obtain a certain phenylacetonitrile from 1 kilogram of the corresponding benzyl chloride, Parke-Davis process chemists dissolve the chloride in only 1 liter of toluene (*Dozeman et al.*, 1997). This volume represents an estimated 10–20 times less than what discovery chemists in any company might have used. On this basis and the 40-kg scale actually employed, the hypothetical change would have saved distilling some 360–760 liters of toluene to isolate the product. To reduce the solvent volume, however, although it can save time by increasing the reaction rate, can also bring a disadvantage. For the same quantity of product formed at a higher concentration, an equal amount of heat evolves from an exothermic reaction. This heat disperses, however, through a smaller solvent volume. To secure an advantage in throughput is therefore to pay a price in vigilance.

Robustness

A robust synthesis, which creates few problems that require troubleshooting, is desirable but not obligatory in discovery research. There it suffices if the synthesis produces the desired target compounds, so discovery chemists need not deliberately test its robustness. By contrast, this quality is essential to process research and development for economy, timing, and safety. These features are especially important if the synthesis is to be operated by nonchemists far away from the researchers who devised it. The yield and quality of the drug, as well as its impurity profile, must tolerate inadvertent, small changes to reaction or work-up conditions, including pH, temperature, time, and concentration. Insensitivity to these and other changes becomes the subject of separate experiments that test robustness. Parke-Davis researchers, for example, in concluding development of a cardiovascular drug, tested the effects of several variables in their production synthesis (*Dozeman et al.*, 1997). They used 10% excesses of two starting materials and reduced solvent volumes by 25%, all in independent trials. Also, they determined the result of allowing a 41° exotherm to occur. Passing these nontrivial tests meant that the Parke-Davis synthesis was robust.

Regulation

Government agencies minutely regulate the manufacture and quality of finished pharmaceuticals. In the United States, for example, the FDA makes more than 100 aspects of drug manufacturing subject to its evaluation. Consequently, pharmaceutical companies maintain quality control groups, with responsibilities clearly linked to individual employees. The analytical procedures devised to control purity are subject to validation before they can be adopted or changed. Control extends to buildings and equipment. Their design, maintenance, and use meet FDA standards, and pharmaceutical companies are subject to surprise inspections as well as seizures and shutdowns. Their obligations to keep records cover raw material, bulk active, and finished drug, and they extend to details of storage, distribution, returns, and salvage. Starting materials, bulk actives, and finished pharmaceuticals all conform to certain physical and chemical specifications, while finished pharmaceuticals show consistency from batch to batch.

Chemical developers comply with current good manufacturing practices (cGMP) and good laboratory practices (GLP) when they prepare supplies of a drug for clinical trials. Process chemists, in their search for a suitable manufacturing synthesis, work more freely, unless a sample of theirs is to treat people. After a production process becomes established and approved sales begin, the responsibility for regulatory compliance passes from process to manufacturing chemists.

Investigational New Drug and New Drug Approval (NDA) filings completely disclose the processes used to make bulk samples of the medicine. Included for all the reactions are the substrates, reagents, solvents, and products. Filings reveal the nature and level of impurities in the bulk drug and, for a solid medicine, link the polymorph prepared to the bioavailability expected. Data identifying the polymorph and, if necessary, a method for making it reproducibly form part of the filings. It falls to process chemists to furnish this information in the section of an NDA that is headed "Chemistry, Manufacturing and Control."

Making changes to the filings is possible if proposed changes are submitted in writing and approved before they are implemented. Although small changes to a synthesis are quick to effect, new toxicity or bioavailability studies lag. Approval is not subject to acceleration, so these precautionary studies protect patients from any toxicity of new impurities or from the ineffectiveness of a new polymorphic form that is not bioavailable.

The developed countries control their pharmaceutical industries in other ways. Environmental regulations, for example, cover transport of finished drugs and intermediate and starting chemicals. Bulk chemicals transported from one site to another are subject to toxicity testing that utilizes animal models of human diseases. Although the use of animals makes the work expensive and time consuming, it is essential because *in vitro* assays offer few substitutes.

The industrialized countries also oversee emissions and effluents that issue from manufacturing sites and research buildings. Pharmaceutical companies control effluent content to eliminate or minimize the release of harmful or potentially harmful chemicals. Regulations prohibit the discharge of, for example, heavy metals into sewage or storm run-off waters. FIPCOs therefore monitor their aqueous effluents, sometimes continuously.

Emissions to the air even of relatively nontoxic materials must be minimized. This necessity affects the choice of solvents. Preferable solvents like water or supercritical carbon dioxide are dischargeable. Other solvents are advantageous if they are recoverable through distillation. Volatile chlorinated solvents like dichloromethane are problematic unless they are recovered. Freely discharging them is prohibited, while incinerating them forms hydrogen chloride, which creates its own hazards. Proper disposal of solvents therefore influences the choice of a manufacturing process because the disposal of wastes and the synthesis of drugs are interdependent.

SATISFACTIONS AND SUCCESSES

Process and medicinal chemists alike share many of the same personal satisfactions that accrue to successful researchers, which are professional as well as financial. The former chemists, however, enjoy two distinct chances to advance projects of known importance and great urgency. In the first stages of a development campaign, a chemist's contributions are welcome if they clearly help maintain or accelerate the schedule for making drug supplies. The relatively large quantities needed permit the extensive toxicological studies that are prerequisite to Phase I clinical trials, which, as we saw, also demand drug supplies. The later stages of a development project bring recognition to any process chemist whose efforts reduce the per-kilo cost of making the drug. Reductions in this cost allow for quantitative evaluations of different researchers' contributions. To reduce the number and quantity of waste products coming from a synthesis demands creativity from a process chemist and brings gratification. As this chapter shows, the opportunity to devise a manufacturing synthesis embraces more possibilities than merely raising the average chemical yield per step.

SOURCES AND SUPPLEMENTS

Brandeis, M. A.; "Chemical engineering in support of development chemistry," *Chemistry and Industry (London)*, 90–94 (1986).

Christ, C.; *Chem. Technol. Eur.*, May–June, 19–25 (1996).
Coates, C. F., and Riddell, W.; "Assessment of thermal hazards in batch processing," *Chemistry and Industry (London)*, 84–88 (1981).
Cusumano, J. A.; "Synthetic success with catalysts," SCRIP *Magazine*, September, 42–45 (1997).
Daniels, F., and Alberty, R. A.; "Physical Chemistry," 2nd ed., Wiley, New York, 1961.
Dozeman, G. J., Fiore, P. J., Puls, T. P., and Walker, J. C.; "Chemical development of a pilot scale process for the ACAT inhibitor 2,6-diisopropylpheny[(2,4,6-triisopropylphenyl)acetyl]sulfamate," *Organic Process Research & Development* **1**, 137–148 (1997).
Etchells, J. C.; "Why reactions run away," *Organic Process Research & Development* **1**, 435–437 (1997).
Gadamasetti, K. (ed.); "Process Chemistry in the Pharmaceutical Industry," Dekker, New York/Basel, 1999.
Hoekstra, M. S., Sobieray, D. M., Schwindt, M. S., Mulhern, T. A., Grote, T. M., Huckabee, B. K., Hendrickson, V. S., Franklin, L. C., Granger, E. J., and Karrick, G. L.; "Chemical development of CI-1008, an enantiomerically pure anticonvulsant," *Organic Process Research & Development* **1**, 26–38 (1997).
Kappe, T., and Stadlbauer, W.; "DSC [Differential Scanning Calorimetry]— A valuable tool in heterocyclic synthesis," *Molecules* **1**, 255–263 (1996).
Laird, T.; "Development and scale-up of processes for the manufacture of new pharmaceuticals," Chapter 5 in Vol. 1 of "Comprehensive Medicinal Chemistry," Pergamon Press, Oxford/New York, 1990.
Laird, T.; "Working up to scratch," *Chemistry in Britain*, August (1996).
Lee, S., and Robinson, G.; "Process Development: Fine Chemicals from Grams to Kilograms," Oxford University Press, London/New York, 1995.
Lester, T.; "Cleaner synthesis," *Chemistry in Britain*, December, 45–48 (1996).
Liu, C., Ng, J. S., Behling, J. R., Yen, C. H., Campbell, A. L., Fuzail, K. S., Yonan, E. E., and Mehrotra, D. V.; "Development of a large-scale process for an HIV protease inhibitor," *Organic Process Research & Development* **1**, 45–54 (1997).
Marshall, D. R.; "Process investigations in the development of imidazotriazinones," *Chemistry & Industry (London)*, 331–335 (1983).
Marvell, A.; "To His Coy Mistress," in "Understanding Poetry," 3rd ed., pp. 308–309, C. Brooks and R. P. Warren (eds.), Holt, Rinehart and Winston, New York, 1960.
Peck, J.; "Overview of the Pharmaceutical Business," Alternative Futures Associates, Alexandria, VA, undated.
Perrault, W. R., Shephard, K. P., LaPean, L. A., Krook, M. A., Dobrowolski, P. J., Lyster, M. A., McMillan, M. W., Knoechel, D. J., Evenson, G. N., Watt, W.,

and Pearlman, B. A.; "Production scale synthesis of the non-nucleoside reverse transcriptase inhibitor atevirdine mesylate (U-87, 201E)," *Organic Process Research & Development* **1**, 106–116 (1997).

Repic, O.; "Principles of Process Research and Chemical Development in the Pharmaceutical Industry," Wiley, New York, 1998.

Robinson, G. E.; "Development of a manufacturing route for tiotidine," *Chemistry & Industry (London)*, 349–352 (1983).

Roth, G. P., Landi, Jr., J. J., Salvagno, A. M., and Müller-Bötticher, H.; "Optimization and scale-up of an asymmetric route to the LTB$_4$ inhibitor ontazolast," *Organic Process Research & Development* **1**, 331–338 (1997).

Sheldon, R. A.; "Consider the environmental quotient," CHEMTECH, March, 38–47 (1994).

Stinson, S. C.; "Wanted: Process chemists to tackle challenges," *Chemical and Engineering News*, June 21, 45 (1999).

Weinstock, L. M.; "Evolution of the cefoxitin process," *Chemistry and Industry (London)*, 86–90 (1986).

Wilson, E. M.; "A tale of two cephalosporins," *Chemistry and Industry (London)*, 217–221 (1984).

Woodward, R. B.; "Synthesis," in "Perspectives in Organic Chemistry," pp. 155–184, A. R. Todd (ed.), Wiley, Interscience, New York, 1956.

chapter

7

QUALIFYING AND SEARCHING FOR JOBS IN THE DRUG INDUSTRY

"The drug industry... cannot seem to find enough synthetic organic and process chemists."
—Madeleine Jacobs, *Editor-in-Chief*, Chemical and Engineering News

INTRODUCTION

You're qualified to do a chemist's job in the drug industry if you earned at least a baccalaureate degree in chemistry. But before you get a job or even an invitation to interview, you need to show your credentials to suitable employers and to know what they seek. What you must know and show is the subject of Chapter 7, which presents the common prerequisites that employed chemists meet and the various qualifications that B.S., M.S., and Ph.D. scientists offer. The chapter also discusses permanent residency and working visas as well as the important role played by the American Chemical Society in helping chemists find jobs. It concludes with methods for discovering the open positions that need chemists.

What You Need to Know and Show

Common Prerequisites

To secure interviews, job hunters need to show that they meet the common prerequisites of employment in a drug house or a satellite company. Perhaps the most important prerequisite is completing an appropriate education, so aspiring researchers generally need baccalaureate or graduate degrees.

Employers insist that B.S. candidates study chemistry as undergraduates, complete courses offering a sufficient and suitable background for the job sought, and receive acceptable grades. Personnel departments verify these facts, confirming that applicants earned the appropriate degrees. Chemists educated abroad may be asked to provide written translations of diplomas issued in languages other than English. Few Americans read Latin, for example, so European diplomas using that language are prime candidates for translation.

During an interview, a chemist with a graduate degree can expect to hear questions concerning his dissertation. Is the writing underway, for example, or is the thesis complete and has it received the necessary faculty approval? Answers to questions like these establish that the candidate's graduate education is advancing toward the appropriate diploma. Making progress is important for another reason.

Employers sometimes make job offers contingent on future acceptance of a finished thesis, allowing a new employee to begin work before his dissertation receives approval. If necessary, a candidate should actively seek an arrangement to take an offered, desirable job and to finish the thesis by an agreed, reasonable deadline. Such an arrangement—which an employer may not initiate or mention—calls for approval not only by the recruiter but also by the job hunter's faculty adviser. Getting approvals can be well worth the effort. However, job hunters must ultimately present acceptable theses and receive their degrees, even though recruiters agree to a postponement. For candidates to meet the requirements of a graduate degree is a matter that gravely concerns employers. For one thing, more is at stake than only the job seeker's credentials; a recruiter's own performance rating depends partly on finding serious candidates.

Another significant prerequisite sought from candidates at all degree levels is an ability to communicate science. It usually calls for intelligible spoken English, however accented or ungrammatical. Meeting this requirement can be critical to receiving job offers and is essential to reporting work planned or accomplished. Practiced interviewers listen carefully to ascertain whether job seekers' spoken English is understandable and whether they utter appropriate replies rather than non sequiturs. Some supervisors prefer having linguistic mix-ups clarified immediately, expecting their coworkers to acknowledge misunderstandings when they occur. Prompt clarification prevents prolonged confusion, so avoiding any inimical consequences. Such an interviewer may never become your supervisor, yet in conversing with him you must impress him favorably.

Personal Qualities

An offer to work in chemical research within the pharmaceutical industry rewards one or more successful, day-long interviews in a single company. All your conversations there must evince many avidly sought personal qualities and do so to interviewers of disparate backgrounds and even of different cultures.

> *Foremost among the traits crucial to pharmaceutical company researchers are enthusiasm, industriousness, a sense of urgency, and cooperativeness.*

Veteran recruiters expect enthusiasm from young chemists, so they are hardly surprised by it. A lack of enthusiasm is conspicuous, while a knowledgeable interest in a particular job opening is distinctive in an applicant and encouraging to a recruiter.

Industriousness is an obvious but essential virtue. It concerns every recruiter who, as a fellow scientist, might supervise your work. Your productivity affects not only your own career but his as well. So, no one receives an offer who does not persuade an interviewer that he is a hard worker who can sustain his output.

You will do well during interviews if you project a sense of urgency. It conveys your resolve to achieve significant results as soon as you begin work and demonstrates the seriousness of your interest in the position available. In doing drug research, a sense of urgency helps chemists shun distractions, assign priorities, and reevaluate the latter periodically.

Exposure to exotic languages brings one of the pleasures of work in a FIPCO. So, when you visit such a firm, pause to listen and look about you. You can expect to hear in the corridors a host of languages beside English: French, Kanada, Spanish, Mandarin, and Australian, to cite only a few. Many scientists come from distant lands and ancient civilizations, so progress in drug discovery and development depends not only upon cooperation but also on mutual respect. When you look around, notice the other employees who are to become your coworkers. They are maintenance personnel, purchasing agents, cleaners, research financiers, and guards, among others. Appreciate that every employed chemist necessarily establishes effective working relationships with a great variety of nonscientists as well as other scientists. An ability to get along with people from all walks of life is therefore indispensable, although it is not an intellectual quality.

Other important personal qualities are reliability, flexibility, initiative, and objectivity. A reliable coworker does the task he promises to do and completes it on time. Flexibility inspires researchers to learn and practice new technical developments, abandon inefficient working habits, and forsake failed projects. Every boss looks for an employee who begins a task on his own and calls for any needed help; initiative like this is the source of the autonomy that comes with job experience. To adopt an objective viewpoint, it can help to ask what a knowledgeable outsider might think, for example, of one's progress or regress in realizing an unexpectedly difficult synthesis.

Finally, a certain restraint is advisable.

> *During interviews, a wise but inexperienced chemist refrains from belittling as menial the research position sought.*

Progress in drug research demands hands-on efforts, and success can require several pairs of hands. Moreover, many interviewers are also accomplished chemists who do benchwork, or did so for decades, despite their graduate degrees, corporate ranks, and scientific accomplishments.

Qualifications

Inexperienced new chemistry graduates' qualifications depend on their academic degrees, with more credentials asked of doctoral chemists than B.S. or M.S. scientists.

B.S. Chemists

Because requirements for baccalaureate chemists are the least demanding, these scientists enjoy the widest choices of job types. For positions in chemical discovery or development research, baccalaureate chemists show basic knowledge of organic chemistry and acquaintance with experimental techniques as well as instrumental analyses. However, because their exposure to experimental organic chemistry is limited, they have much to learn in any job that demands organic synthesis. Their situation makes two important implications.

1. Despite the continuing demand for chemists, an inexperienced B.S. scientist may need more time to find a position than, say, an equally inexperienced M.S. chemist. Many companies advertise for experienced B.S.

chemists, who require little or no on-the-job instruction. The number of employers offering individual in-house training to B.S. chemists is few. Large established pharmaceutical companies are willing employers of B.S. chemists; more so perhaps than small start-ups, if generalization about such a diverse industry has any meaning.

2. An inexperienced B.S. chemist needs a boss who is willing to tutor, who has enough time for the task, and who can accomplish it. Nevertheless, a B.S. scientist's study of chemistry represents adequate and relevant vocational training, which is ample qualification for chemical research in a drug house.

Baccalaureate chemists who took part in undergraduate chemical research, or in cooperative programs alternating college studies with industrial research, enjoy a solid advantage over competitors lacking these experiences. Similarly, to have carried out a summer's chemical research as an intern within a pharmaceutical or other chemical company carries weight. (Students completing the standard, two-semester introduction to organic chemistry qualify for paid summer internships, some of which appear in Appendix B.) Recounted on résumés, these accomplishments hallmark a candidate with a strong research interest, considerable know-how, and a safe work record. Indeed, research experience is one of the strongest qualifications that a B.S. chemist can present to a prospective pharmaceutical company employer. No B.S. chemist's résumé is adequate that fails to mention such experience, for it brings job interviews.

The competitive advantage that arrives with research experience travels with obligation and opportunity. During interviews, B.S. and M.S. chemists can expect to be asked to describe any undergraduate research they did. Typical questions, not all of them technical, might include the following ones. What was the problem and why did it merit efforts to find a solution? What methods did you employ and why were they chosen? What progress did you make, what remains to be done, and how soon can it be accomplished? Do the results attained represent failure or success, or is the outcome still unknown? What did you learn from your efforts? A good oral account including a final judgment answers questions like these. It also speaks volumes about an applicant's willingness to prepare for an interview and to appreciate science as business.

Other qualifications include foreign languages and computer literacy. A reading knowledge of French, German, Italian, Japanese, or Russian represents a bonus in some recruiters' eyes. The number of native-English-speaking chemists who know other languages in which chemical research is published seems scant. Yet the need continues to understand, say, German-language articles in *Justus Liebigs Annalen der Chemie* (now the *European Journal of Organic Chemistry*). These days computer literacy is essential to employed chemists, but is not a decisive hiring qualification. Yet such a skill is commonplace among new B.S. graduates and, unlike languages, can be acquired on the job.

M.S. Chemists

Employers expect M.S. chemists to present more than baccalaureate scientists' background. Chemists with master's degrees offer the broader knowledge that is gained from the many organic chemistry courses that they take in graduate school. These chemists need to show proficiency in common experimental and analytical methods, which entails understanding and practice. Experience in carrying out multistep syntheses is exemplary. So is familiarity with combinatorial chemistry or semiautomated synthetic methods. A knowledge of how to prepare samples and choose solvents for NMR measurements is essential, as is an understanding of how to record and adequately interpret the resulting spectra.

A broader knowledge of organic chemistry includes the ability to analyze a simple organic structure and suggest practical routes for making it. Suggesting practical routes demands much foresight. For example, the reagents necessary to remove one protecting group early in a synthesis may also eliminate another that must be retained until later; such a situation calls for rethinking the proposed route. A rudimentary understanding of reaction mechanisms is desirable, as is a basic knowledge of organic stereochemistry. Practice in isolating, purifying, characterizing, and identifying new organic compounds is expected of M.S. candidates.

Chemists with master's degrees can offer two kinds of bonuses to prospective employers. (1) One premium is experience. All or most of these scientists are experienced researchers who complete enough benchwork to have written a coherent thesis. Furthermore, some chemists' research is successful and original enough to warrant one or more articles in chemistry journals. Master's-level chemists with research-based dissertations or published articles lay strong claims to positions in discovery or development groups. These writings help demonstrate chemists' prowess and thereby facilitate the comparisons and contrasts of candidates that recruiters make.

> *Perhaps the most sought-after inexperienced nondoctoral chemists are M.S. scientists who carried out graduate research in organic chemistry.*

(2) The other bonus is technical know-how, which is augmented by research done for a master's degree. As graduate students of organic chemistry, some scientists master techniques and equipment for which there was no time or money when they learned basic laboratory practices. As job seekers, therefore, these chemists must not fail to inform their interviewers that,

fect on the wages and working conditions of U.S. workers similarly employed. Determination of labor availability in the United States is made at the time of a visa application and at the location where the applicant wishes to work. (www.ins.usdoj.gov; italics added*)*

Labor certification basically rewards an employer's effort to demonstrate that no suitable citizen or permanent resident can be found to do the job offered the alien worker. To procure certification, the employer begins by formally notifying the DOL that it offered work (DOL Form ETA-750A) to a qualified alien (DOL Form ETA-750B). The DOL assigns a priority date on receipt of the application, transmitting the latter to the employment agency of the state where the alien works at the offered job. (A nonimmigrant H1B visa allows the worker to begin before any labor certification issues.) Next the employer places a detailed "help-wanted" advertisement in an appropriate periodical. Beginning the "help-wanted" section of *C&EN,* such notices typically include a highly technical job description running for several hundred words printed in small type. Any interested citizens or permanent residents apply for the position by sending their résumés to the state DOL rather than the employer, which the ad does not identify. Making note of these responses, the DOL passes the suitable ones to the employer, which obligatorily interviews all the promising applicants, sometimes by telephone only. The employer informs the DOL of the interview results. If none of the applicants qualify, the employer's submission states the reasons, which the DOL reviews. Finding the reasons for rejection acceptable, a Certifying Officer of the federal DOL issues a Notice of Approval of Application for Alien Employment Certification. This notice lets the alien continue working at the approved job; without it no alien may continue her employment in that post.

That his position may be abolished without labor certification is the first of two risks that a nonimmigrant worker faces. The second concerns timing: despite renewal, an H1B visa can expire before the DOL grants labor certification. Some of the contributory factors in this case include (but are not limited to) filing of the labor certification application long after the H1B visa begins, understaffing of the DOL, and backlogs in the state employment agency. When the number of cases is large, as it is for Indian and Chinese chemists, the delay can reach 2 years or more before any correspondence regarding a particular case reaches the employer. In 1998, for example, the delay in southern California equaled 4 years (*Newman,* 1998).

Immigrant Visas

Employment-based permanent residency requires labor certification, a showing that the immigrant's education and experience qualify him for the

Table 7-1 Selected Routes to Permanent Residency with Preference Category in Parentheses

Petition Type (on INS Form I-140)	Labor Certification	Offer of Permanent Job
(1st) Alien of extraordinary ability	Not needed	Not needed
(1st) Outstanding professor or researcher	Not needed	Not needed
(2nd) Member of the professions holding an advanced degree or an alien of exceptional ability	Needed but waiveable in the national interest	Needed
(3rd) Skilled worker (with at least 2 years' specialized training or experience) or professional	Needed but waiveable in the national interest	Needed

job, and filing of DOJ Form I-140 accompanied by the fee that in 1999 equaled $115. To complete the form, an applicant or employer must choose a preference category. For chemists, four of these categories are most relevant (Table 7-1); not all the categories are discussed here. Why the illustrated four categories represent preferences is evident. Aliens of outstanding ability, who as chemists might be Nobel laureates, and outstanding professors or researchers are admissible as permanent residents without job offers or labor certification. This himalayan first-preference category lies above the reach of inexperienced and newly graduated chemists, regardless of their nationalities. Most of these immigrant chemists do fit either of the remaining two categories, which require labor certification and a job offer. However, the DOL may waive these requirements if doing so serves the national interest.

National Interest Waivers

To obtain a national interest waiver serves the immigrant as well as the nation. For the immigrant, the waiver represents a means to shorten the time needed to receive employment-based permanent residency. It bypasses labor certification, which for many chemists can be lengthy.

Simultaneously to seek a national interest waiver and labor certification represents a sensible tactic.

Either way a successful petitioner gains permission to reside permanently in the United States.

However, the unwritten criteria that an immigrant must meet to receive a national interest waiver of labor certification are ill-defined. The INS staffers decide on a case-by-case basis, keeping an eye on precedents. For immigrant chemists working in a pharmaceutical company, one tactic is to argue that discovery and development of, say, an antiviral drug serve the national interest. The importance of seeking such a drug might be demonstrated by a collection of articles from the popular press and other sources; these articles would show how American public health benefits from such drugs. Petitions for national interest waivers also include letters of recommendation from various sources and lists of the chemist's patents, scientific publications, lectures, and awards.

How to Obtain INS Forms

Single copies of INS forms cost nothing to individuals who request them. The INS Web site allows anyone to download some of them or to order any of them online. Its forms are also available through a toll-free telephone number (1-800-870-3676). The service mails forms ordered by individuals only to U.S. addresses.

Foreigners planning to attend a U.S. university apply for the needed F1 student visa using INS Form I-20. The name of this form is "Certificate of Eligibility for Nonimmigrant (F1) Student Status—for Academic and Language Students." Ordinarily the university admissions office mails a copy of this form to the students' home outside the United States.

Do You Need a Lawyer?

No. A foreign-born worker seeking permanent residency has no legal obligation to employ an immigration lawyer. Indeed, the high cost of legal services argues against using such an attorney or a law firm. The fee can exceed $7500. This money can be saved if the chemist gathers the needed documents and files the necessary petitions herself.

To hire one's own lawyer may be unnecessary. A foreign-born chemist offered a job should ask if her employer sponsors petitions for immigrant visas. Some large FIPCOs do so and, on behalf of newly hired scientists, use immigration attorneys or paralegal professionals. Alternatively, they outsource the work to specialized independent law firms. This practice, however, is one dependent on the corporate financial health; it can be stopped short when budgetary constraints loom.

Hiring one's own lawyer, however, is likely to prove desirable, even if one assumes debt to do so. Immigration regulations and law are complex, while some requirements of employment-based visas are open ended. The tactics and knowledge of an experienced immigration attorney or even a paralegal assistant employed in a law firm can help or perhaps snatch victory from the jaws of defeat.

ROLE OF THE AMERICAN CHEMICAL SOCIETY IN JOB SEARCHES

Among a job hunter's strongest allies, the American Chemical Society provides an array of employment services and publications valuable to chemists seeking their first professional positions. (Author's note: The writer belongs to the ACS.) These services and publications, many of them free and some at nominal cost, give new graduates and students a strong incentive to join the society. Members and affiliates also derive other benefits like an easy camaraderie between new acquaintances and opportunities to establish or expand a network of colleagues.

The ACS employment services and publications primarily flow from two sources, *Chemical & Engineering News* and the Department of Career Services. Other sources comprise the national, regional, and local chemists' meetings that the society organizes as well as local sections and national divisions.

Chemical & Engineering News (C&EN)

Published by the ACS, the glossy weekly *C&EN* comes free to society members and provides important information to job seekers. University and certain public and departmental libraries receive it, as do many faculty members, graduate students, and postdoctoral researchers. The magazine covers among other news items chemical industry events and employment trends. Timely articles show which kinds of jobs are hot where, and many chemists secure interviews or find jobs by responding to advertisements appearing in this weekly. In the variety and print area of chemist-wanted advertisements placed by companies, *C&EN* leads all other U.S. publications. No other medium surpasses or even matches this coverage, not the big-city dailies and not the Internet job banks. (The only exception is trivial: the *ACS Job Bank* stores the advertisements that the magazine prints.)

- *Chemists seeking jobs read C&EN.*

FIPCOs and satellite companies from the pharmaceutical industry seek help in the pages of *C&EN*. One typical kind of help-wanted advertisement describes the particular job to be done and specifies an identifying job code.

Several such advertisements from different firms appear weekly. At longer intervals, the magazine devotes a feature article to the pharmaceutical industry, and many FIPCOs and other firms advertise in the corresponding issue. The magazine regularly names the forthcoming issues that are to contain such feature articles. Often the accompanying advertisements occupy full pages but lack job codes, describe the company instead of particular jobs, and invite chemists to apply for future positions. Many employers anticipate forthcoming openings by advance hiring.

Career Services

The ACS maintains a vigorous career services department run by helpful and intensely committed staffers. This department publishes an extensive library of brochures, booklets, and videotapes presenting the fundamentals of finding a job (Table 7-2). One of these booklets offers clear and concise

Table 7–2 Selected Publications of the ACS Department of Career Services

Brochures

Planning for a Career in Industry (1997)

Professional Employment Guidelines (1998)

What a B.S./B.A. Chemist Should Consider Before Accepting an Industrial Position (1999)

What an M.S. Chemist Should Consider Before Accepting an Industrial Position (1999)

What a Ph.D. Chemist Should Consider Before Accepting an Industrial Position (1999)

Booklets

Current Trends in Chemical Technology, Business, and Employment (1998)

Employment Guide for Foreign-Born Chemists in the United States (1997)

Starting Salaries of Chemists and Chemical Engineers (annual)

Targeting the Job Market (1995)

The Interview Handbook (1998)

Tips on Resume Preparation (1998)

Videos

Developing the Right Picture: Resume Preparation

Formula for Success: Turning Job Leads into Gold

The Essence of a Winning Interview

Your Career in Chemistry: Measuring Your Skills, Weighing Your Options guidance to immigrant chemists seeking permanent U.S. residency. This booklet is the *Employment Guide for Foreign-Born Chemists in the United States*. It covers not only visa issues but other topics ranging from culture and language through workplace etiquette to the art of dining with prospective employers. Even without social advice ("Drinks must be sipped and not slurped") the booklet commends itself to all chemists tracking their first permanent jobs.

At each of the twice-yearly ACS national meetings, the department operates several career-related workshops meant for new chemistry graduates looking for jobs. Veterans of the chemical industry lead discussions, offer guidance, distribute job-related literature, and answer questions. These workshops also take place at regional and some local gatherings; sometimes the speakers travel by arrangement to colleges and university campuses. Invited by student clubs or faculty departments, they advise student audiences. At recent national meetings, the ACS career services group offered the following workshops and programs: "How to Jump-Start Your Career," "How to Use ACS Career Services to Your Advantage," "Effective Job Searching for Graduate Students and Younger Chemists," "Employment Trends and Job Security in the Chemical Industry," "Electronic Job Searching," and "Network or Not Work."

The department offers to student affiliates, and to student and other members, the opportunity to consult volunteer career consultants. Interested job hunters and employed chemists telephone (1-800-227-5558) throughout the year to seek these consultants' services.

Career Resource Center

At national and other ACS meetings, career-department staffers privately review job seekers' résumés and hold mock interviews. Résumé reviews last 30 minutes and may include free-ranging discussions of other aspects of career establishment and development. Mock interview sessions entail videotaping of a 10-minute interview, followed by a 30-minute critique. These sessions give practice in asking and answering the incisive questions that recruiters pose. Perhaps most importantly, they help allay the anxiety that many new graduates feel in presenting qualifications, impressing interviewers, and negotiating terms.

The mock interviews, résumé reviews, and career workshops take place within a designated part of a large hall. Called the Career Resource Library, this separate area displays scores of current, commercially available employment books, annual corporate reports, and free packages of corporate recruiting literature. Wall posters present salary information compiled during the latest annual ACS survey of chemists' salaries.

The Center, which distributes free Career Services booklets and brochures, also includes several personal computers. Job seekers wishing to follow advice concerning their résumés use these computers. They revise their résumés on the spot, before taking part in the real interviews that the Career Service Department helps arrange through its job exchanges. These popular exchanges, as well as the Career Resource Center and Library are well attended. Ten minutes before opening on the first day of a 1999 national ACS meeting, 150 young chemists formed a line at the entrance.

National and Regional Employment Clearing Houses
(NECH AND RECH)

About 40 years ago, the ACS began to organize job exchanges, now attended by hundreds of employers and thousands of job seekers and by dozens of society staffers and volunteers eager to help. The largest FIPCOs take part, as do many other companies from the rest of the pharmaceutical industry. The job exchanges occur during national society congresses, take place within the Career Resource Center, and occupy the greater portion of its floor space. They introduce mutually interested recruiters and candidates to one another through résumés, candidates' summary forms, employers' position-available forms, and on-the-spot screening interviews.

Registering for an employment clearing house calls for reading the guidelines published in the *C&EN* issue that carries the preliminary meeting program. These guidelines also appear on the ACS Web site (Table 7-3). Some job seekers register by mail because preregistration promptly brings them to the attention of the greatest number of employers. Others register in advance through the on-line professional data bank. Yet other chemists register onsite, and all registrants receive admission badges giving their names and affiliations.

To participate in the clearing house, applicants complete a candidate summary form, which is a questionnaire. It calls for some 50 items of information—personal data, educational history, skills, and fields of specialization. Candidates state the kind of job sought, their own dates of availability, and their willingness to relocate. They also note their citizenship and professional objectives. Registrants present résumés not exceeding two pages in length. To aid in scheduling interviews, candidates also fill out a time form. It indicates the hours of each meeting day when the applicant will not be available to meet a potential employer.

At a 1998 national meeting, participants completing the questionnaires sat at half-a-dozen banquet tables, each accommodating as many as 60 candidates. Altogether the tables filled only part of a convention-center hall seemingly vast enough to hold an airplane, where columns of glittering dust

Table 7-3 Job Banks and Internet Addresses (URLs)

For Chemists

ACS Job Bank	www.acsinfo.acs.org
CHEM395: Cybermania	www.vni.net/~murtaza/chem395/
ChemWeb	www.ChemWeb.com
Jobs	www.5z.com/divinfo/forum/jobs.html
Search the Jobs Database	ci.mond.org/jobs/jobsearch.html
WebChemistry Jobs On-Line	www-chem.harvard.edu/webchemistryjobs/positions/industrial.html

For Scientists

Bio Online Career Center	www.bio.com
BioSpace Career Center	www.biospace.com
A Career Planning Center for Beginning Scientists	www2.nas.edu/cpc/
Drug Discovery Online	www.drugdiscovery.com
Nature	www.america.nature.com
SCIENCE On-Line	www.sciencemag.org
Science Global Career Network	www.edoc.com
Search for Jobs	www.pharmaceuticalonline.com

For Anyone

America's Job Bank	www.ajb.dni.us
Career Builder	http://careerbuilder.com
Career City	www.careercity.com
Career Consulting Committee	omni.cc.purdue.edu/~mohamad/kaar.html
Career Magazine Database	www.careermag.com/cgi-bin/searchbanner.cgi
Career Mosaic	www.careermosaic.com
CareerPath	www.careerpath.com
E-Span	www.espan.com
Federal Jobs Digest	www.fedworld.gov/jobs/jobsearch.html
FedWorld Federal Jobs Search	http://www.fedworld.gov/jobs/jobsearch.html
First Steps in the Hunt	www.interbiznet.com/hunt/companies
Help Wanted USA	iccweb.com
Minorities' Job Bank	www.minorities-jb.com
NationJob Network	www.nationjob.com
Online Career Center	www.occ.com
The Education & Career Center	www.petersons.com
The Monster Board	www.monster.com
Yahoo!	www.yahoo.com

motes, lit by far-away lamps, rose into the darkness overhead. Shuffled paper, distant footsteps, and muffled conversations made the only sounds 1½ hours after the exchange opened, when participants occupied about half the places. Baskets full of numbered job descriptions covered another three tables, while 15 staffers typed at NECHworks computer-search stations. Seated in a reserved section, about 40 recruiters spoke into cell phones or pored over thick loose-leaf binders containing résumés. Dozens of other binders—full but momentarily unused—lay nearby. Looking like a restaurant half-set for lunch, another section of the hall contained some 150 unoccupied white-cloth-covered tables saved for two-person interviews. Around the periphery stood rows of curtained booths ready for team interviews, each booth just large enough for three to four people.

Registration with NECH entitles candidates to onsite computer searches for suitable jobs. Applicants who find such openings send to onsite recruiters brief messages expressing interest, which serve as the cover letters that traditionally accompany mailed résumés. Each registered job hunter is also assigned a mailbox to which interested employers send replies to make interview appointments. Candidates need not visit their boxes to look for messages, but instead may call a special telephone number to learn whether messages await them. Registration also brings relevant literature: a written guide that orients candidates taking part in the exchange, a sample of the terse message-to-employers form, and instructions for formulating computer searches. In one-line entries, lists of numbered openings state any specialization that the available positions demand, the employer's name, the requisite academic degree, and the job title. A candidate interested in one of these openings uses its number to find the corresponding job description in packets of job descriptions, each one containing 20 such position-available forms. Hundreds of positions are available at a single national ACS meeting.

All participating employers present their open jobs in position-available forms. Each form provides the employer's name, mailing address, job location and type, and the number of chemists needed. It states the job title and the academic degree needed to qualify an applicant and describes the job, usually in one page. By checking a box, an employer may specify salary ranges like the following: below 20K, 20–30K, 31–40K, 41–50K, 51–60K, 61–70K, 71–80K, and over 80K. (K represents a factor of 1000, so 80K means an annual salary of $80,000.) Some employers check consecutive boxes to indicate a broader range than the salaries mentioned by any one box; other employers mark none of the boxes but beside them write the word *competitive*. Employers state the days and times when their recruiters plan to interview candidates during the meeting. Recruiters with positions to fill use keyword searches to find suitable candidates, much as the latter seek the former.

However, not all industrial employers participating in a NECH or a regional employment clearing house send recruiters to the congress. If none attend, then employers mail position-available forms to the clearing house, indicating on them where and how job seekers should apply. Temporarily doubling as recruiters, the participating interviewers often are full-time scientists seeking help for their own laboratories.

The Career Services Department organizes, operates, and supervises the clearing houses. It staffs job seekers' sign-up and position-available booths, employers' sign-up stations, and a data-entry and computer-search group. Over a dozen onsite temporary workers transfer information from summary questionnaires and position-available forms to a computer database. As many as 60 temporary workers assist at a single NECH, where they and other staffers place the forms as well as the résumés into three-ring binders made accessible to participants. At a 1999 NECH, the candidate summary forms occupied 72 binders while the résumés filled more than 50 of them.

All these services ease the task of getting a job, imparting value to ACS membership or affiliation.

FINDING JOB OPENINGS

Expect to feel a certain satisfaction mixed perhaps with relief when you receive your first offer of a permanent position. Although such an offer represents a destination reached, the journey on which you are to embark—from student to professional chemist—demands a ticket. You pay for the trip with your own labors, for getting a job is a job in itself. To find permanent employment, even a highly qualified new graduate must do more than look for a position, yet learning where to look forms a crucial part of the search.

This section shows specifically where and how to look for a chemist's job in the North American pharmaceutical industry. It covers print advertisements and job banks specifically for chemists, alerts of chemists' job openings, and outlines job banks and hotlines meant for all kinds of pharmaceutical company positions. The section points out that many entry-level job openings are created and filled with little or no advertisement. It concludes advising how undergraduate and graduate students of chemistry establish employment networks comprising employed, in-the-know industrial chemists.

Before you begin to search for openings, you'll need to choose the kind of company—large or small—where you would like to work. Perhaps you'll fit in both kinds but, if not, a decision to look only among one or the other can be momentous. It's smart to consider working in a small company, which the accompanying example shows.

fine mode (fax), white or light-colored paper (surface mail), and the word *résumé* in the subject line (e-mail) exemplify commonplace preferences. Several FIPCOs use optical scanning technology to produce résumé databases searchable by keywords. Indeed, this or other electronic methods of creating, storing, and drawing from candidate databases is sure to prevail throughout the industry. Ascertaining and following submission guidelines are therefore essential to making acceptable applications.

Corporate Job Banks

One of the remarkable innovations made possible by the Internet is the job bank. This feature of many corporate Web sites is a list of positions that recruiters need to fill. Listings change as newly hired employees take some positions, while other jobs are posted when they fall vacant. Several large pharmaceutical companies maintain job banks with a branch reserved for drug discovery and development posts. These companies require scientists of many kinds, so their banks are not restricted to chemists' jobs. Appendix B notes almost 200 job banks maintained by pharmaceutical companies and other institutions and gives Web addresses.

Other Job Banks

In addition to pharmaceutical companies, scientific societies and other organizations also maintain job banks (Table 7-3). The American Chemical Society and the Society for Chemical Industry (ci.mond.org in Table 7-3) list chemists' positions exclusively, as does ChemWeb. Although the last two organizations are British in origin, their job banks include positions in North America. In the ChemWeb and Society for Chemical Industry job banks' listings are not restricted to pharmaceutical or even industrial jobs; academic and other positions are also included. Other job banks specializing in scientists' positions are also tabulated.

Scientists who want to discover drugs can view the offerings from a job bank specific to their interests. This is the Online Job Center found at the Drug Discovery Online site (www.drugdiscoveryonline.com) and managed by an editor trained in organic chemistry. Among other features (embryonic at this mid-1999 writing), the start-up center offers a free job search routine. Making selections from drop-down menus, the user specifies the country, region, and province or state where he would like to work. Other choices include the preferred kind of duty and job type; for example, a pharmaceutical company and a research scientist, respectively. Carrying out the search is to yield a list of firms offering the position sought.

Table 7-4 Job Lines

Cancer Therapy & Research Center	210-616-5801
Cell Therapeutics, Inc.	800-656-2355
Desmos Inc.	619-455-3708
Glaxo Wellcome, Inc.	919-483-2565
Hauser, Inc.	303-441-5841, ext. 3000
Ibis Therapeutics	760-603-3858
Isis Pharmaceuticals	760-603-3858
Eli Lilly & Co.	800-892-9121
PathoGenesis Corp.	206-505-6958
Scripps Research Institute	619-784-WORK
Sloan-Kettering Institute for Cancer Research	212-639-5627

Corporate Job Lines

A few companies as well as a research institute maintain special telephone numbers for job seekers (Table 7-4). Dialing these numbers—one of them toll free—lets the caller hear recorded voices describing openings. To hear any description, the caller usually must listen to all previous ones; in some cases, the first description begins after a statement concerning equal employment opportunities. Nearly all the recordings say how the company prefers to receive applications. They specify the documents to submit, the means for transmitting them, and the simple format to use. Some of the recordings advise callers to consult the firm's Web site. As we saw, corporate Internet sites often provide application guidelines that entail no long-distance toll charges.

Job Alerts by E-mail

ChemWeb, the Society for Chemical Industry, and the American Association for the Advancement of Science, which publishes *Science* and maintains a Web site called SCIENCE Online (Table 7-3), offer free alerting services. Subscribers furnish their e-mail addresses and employment preferences. They can expect to receive periodic e-mailed notices of suitable vacancies.

Networking

It is important to know that not all desirable jobs are advertised and essential to act on the knowledge. Recruiters can fill some jobs without the expenses and deadlines associated with advertising, so no help-wanted notices are necessary. Consequently, no job hunter can neglect, or rely on, responding to printed or online advertisements. Even outstanding candidates' responses may, through no fault of theirs, arrive after the job has been filled. Alternatively, when an avalanche of documents descends on a hiring scientist, the latecomers can bury the early arrivals.

To create a network is essential because using it helps a job hunter and because another avenue is closed to newcomers. Fee-paid recruiting firms, which serve the North American pharmaceutical industry, help veterans more than beginners. They recruit industry-experienced chemists often with demonstrated supervisory skills, earning corporate fees by matching candidates with open positions. Headhunters do not make a business of placing in permanent jobs new chemistry graduates at any degree level or even postdoctorate researchers. [However, one of these agencies, which is Lybrook Associates, does offer advice on résumés and interviews that any chemist seeking work can procure (www.chemistry-jobs.com).]

Network Members

As newcomers to the pharmaceutical industry, chemists seeking their first permanent jobs create and cultivate networks of helpful acquaintances. Network members can comprise fellow students and recent graduates, faculty members, friends, relatives, friends of relatives, or even complete strangers. Recent graduates are especially appropriate if they take pharmaceutical industry jobs after commencement. Professors of chemistry or biology keep in touch with their own friends and former students employed in drug firms. Your parents or relatives may work in the industry or know someone who does and who would willingly advise. Good sources of up-to-date employment lore are the sales representatives who purvey scientific equipment and services to pharmaceutical companies. Some of these gregarious people receive unrestricted access to corporate laboratories, which they visit regularly. They also sell their goods in university chemistry departments and attend as exhibitors the national meetings of the ACS as well as regional conferences like the Eastern Analytical Symposium (EAS) and the Pittsburgh Conference (PITTCON). Talk to them.

Above all, a job seeker needs acquaintances who can assist by furnishing current information about hiring trends or about attractive and suitable positions in their own or their competitors' companies. These network members may also furnish the names and addresses of scientists empowered to hire and privy to corporate plans for doing so. To expand a drug discovery or development effort represents a complex, costly, and important enterprise, and FIPCOs formulate their plans to do so a year or more in advance. Which is not too early to apply for work.

Informational Interviews

Knowledgeable informants can be well disposed even to cold callers, feeling flattered to be asked to advise young chemists. Job seekers therefore plan to meet and consult industrial researchers who are strangers. The ensuing informational interviews, however, need an organizational structure that a responsible aspirant gives them. Thoughtful questions conceived in advance elicit the most useful information in the shortest possible time. No industrial scientist who grants such an interview thereby offers to engage in open-ended social small talk or to rehearse his own remarks. Nor does he imply that, by agreeing to speak to a job hunter, he has a position to fill. Rarely if ever can such informants make on-the-spot job offers because hiring a particular individual ordinarily represents a decision reached by consensus. Therefore, to solicit an offer from a scientist who agreed only to pass on employment lore can lead merely to a dead-end referral to a personnel department.

Successful networking, however, leads to job interviews, but requires a source of suitable informants. So, a job hunter needs first to learn their names and companies and to make sure that they are chemists. This information is available to any BS chemist who works to discover it and invests in his future. Here's how to get it.

Names for Networks

Published and readily accessible sources give the names of industrial chemists employed in drug discovery and development. These primary sources number three, none of them necessarily as complete or sufficient as a systematic job hunter might like. They include (1) the current e-mailing list and printed directory issued to members of the ACS medicinal chemistry division. The mailing list includes members employed by pharmaceutical companies, while the directory gives many of their names, affiliations, and surface-mailing addresses. For example, an M.S. chemist's e-mailed job inquiry, which reached many pharmaceutical-company researchers, brought him several 1997 interviews. The medicinal chemistry section of the ACS issues the list to its members. Read the directory in a college or departmental library or borrow it from a member.

(2) Relevant authors of scientific articles can join your network. Their names appear in journals of medicinal chemistry, of drug discovery and development, and of process chemistry. The first page of each article states the writers' names, affiliations and surface-mailing addresses. Sometimes the article includes the senior author's e-mail address, often in a footnote on the first page. However, lists of article authors, which frequently include both chemists or biologists, do not indicate the name of the science they practice. But a telephone call to an author rather than a personnel department can elicit this information.

It may also be possible to establish or confirm which authors are chemists by consulting *Chemical Abstracts*. Some chemists also publish in journals devoted exclusively to organic chemistry (as opposed to medicinal chemistry). A search of the Author Indices of *Chemical Abstracts* discloses articles written in both kinds of journals; common authorship marks the chemists and excludes the biologists who do not publish in chemistry journals.

(3) Chemists employed in pharmaceutical companies attend meetings sponsored by the ACS and other organizations. At the meetings they deliver lectures and present posters concerning their research. Books of abstracts published by the ACS announce these poster presentations and lectures, giving the authors' contact information. The ACS sells the books to meeting participants, who order them during online or mail registration. A reader learns from the abstracts who currently works in which pharmaceutical companies and where the companies are located.

Meeting programs, which *C&EN* publishes, state the date, time, and hotel conference room where the lectures and poster sessions occur. The programs effectively teach how to identify the speakers and presenters. If you are willing to invest in your own future, attend at your own expense one or more conferences to meet chemists working in the pharmaceutical industry. View the posters and hear the lectures, offered in organic, medicinal, or process chemistry symposia. Then, by all means, introduce yourself to the authors, who will be pleased to talk to you. To recruit others to your network, frequent the social mixers. There you can expect to find more chemists working for the same employer and representing both discovery and development groups. A single large company may send a dozen chemists to the same large meeting, only one or two of whom give talks or show posters. Be aware that many of these scientists have an incentive to meet new chemistry graduates and postdoctorate researchers. Industry veterans can earn substantial finder's fees from their employers by introducing serious job candidates to corporate personnel recruiters.

SOURCES AND SUPPLEMENTS

Groban, R. S., Jr.; "The immigration act of 1990: An employer's primer of its new provisions," *Employee Relations L. J.* **17**(3), Winter, (1991–1992).

Jacobs, M.; "Personal tragedies," *Chemical and Engineering News*, June 21, 5 (1999).

Kennedy, J. L.; "Hook up, get hired! The internet job search revolution," Wiley, New York, 1995.

Kennedy, J. L., and Morrow, T. J.; "Electronic Job Search Revolution," 2nd ed., Wiley, New York, 1995.

Kimmel, B. B., and Lubiner, A. M.; "Immigration Made Simple: An Easy to Read Guide to the US Immigration Process," 4th ed., Next Decade, Chester, NJ, 1998.

Newman, B.; "Alien Notions: The 'National Interest' Causes INS to Wander Down Peculiar Paths," *The Wall Street Journal,* August 20, A1 (1998).

Newman, B.; "Foreigners seeking work visas often land in hell instead," *The Wall Street Journal,* April 23, C1 (1998).

Silva, A.; "Employment Guide for Foreign-Born Chemists in the United States," Career Services Department, the American Chemical Society, Washington, DC, 1997.

Wernick, A.; "U.S. Immigration & Citizenship," 2nd ed., Prima, Rocklin, CA, 1999.

Woodward, R. B.; private communication to the author and coworkers, Basel, about 1971.

chapter

8

EVALUATING COMPANIES AND JOB OFFERS

"You can be young without money, but you can't be old without it."
—Tennessee Williams

INTRODUCTION

Among inexperienced chemists who found full-time jobs in 1998, there was nearly a 60% chance that any one of them would receive more than one offer of employment (Figure 8-1). These successful job hunters chose between offers, and so will many of you. They and those chemists who received only one offer made a related choice of the companies to which they applied for work. Obviously, this latter choice faces every job seeker, so each must adopt standards for judging employers. Evaluating companies to find ones where you would like to work is as important as looking into colleges to attend, cars to buy, or investments to make.

Furthermore, when you have worked several years for one employer, you would be wise to reevaluate your situation before you grow too old to move elsewhere. For about 5 years after completing Ph.D. studies, it will be comparatively easy to find work at a different company. Doing so will depend largely on the strength of your education, recommendations, patents,

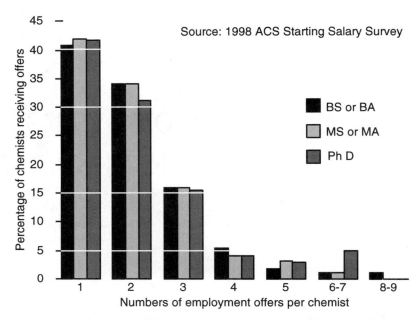

Figure 8-1 Offers of full-time employment to inexperienced chemists

and publications. Later, however, your employability at another drug company may depend more on managerial achievements and on contributions to finding and advancing clinical candidates than on a chemical researcher's skills. Be aware that luck plays an inevitable if not paramount role in the discovery of compounds that become clinical candidates.

Before you begin looking for an employer, ask what you want from the job you take. Answering this question is essential to finding and doing work that satisfies you throughout your career, which may last 25–40 years. Consider five important determinants of job satisfaction, which are challenge, autonomy, advancement, security, and balance, all of which are largely independent of salary and other forms of monetary compensation. Challenge and autonomy are influences significant to scientific researchers, especially to advanced ones. Every researcher who takes a job gives thought to the roles that advancement, balance, and security are to play in his or her career. Advancement may come more quickly in smaller companies than in larger ones, but whatever its source may entail unusually long workweeks, which detract from family life. Despite the possibilities of mergers and consequent layoffs, larger companies may offer greater job security than smaller ones.

Introduction **247**

A great many factors influence the desirability of a given job. So, it helps in making comparisons and ranking offers to tabulate the positions under consideration and the factors important to you. Suppose you use a 1-to-10 scale to rate each position with respect to every important factor. Then you can sum the ratings, characterizing each job by a single number reached in a semiquantitative way and rank the jobs in order of preference (*Fish*, 1999). Such an analysis helps define your preferences objectively. Of course the method can be extended to factors other than the foregoing ones.

To assist in evaluating companies, Chapter 8 presents criteria falling into several categories. It covers corporate status and future, living costs, home prices, forms of compensation other than salary, career prospects, bosses, patenting and publishing, and retirements. Appreciating the importance of these topics empowers job seekers to investigate openings, ask pertinent interview questions, and elicit answers helpful in evaluating companies. The chapter provides means to obtain information about companies.

One principle underlies much of Chapter 8.

> *In identifying desirable employers and evaluating job offers, prudent new graduates—despite their youth—must provide for their financial futures.*

Not surprisingly, encroaching age may diminish their capacity to earn their livings through gainful employment. Yet, for other reasons, looking ahead is more important now than at any other time since 1935. President Franklin Roosevelt's New Deal administration created the Social Security Act in that year and set the normal retirement age at 65 years. Congressional legislation enacted in 1983 advances this age by as much as 2 years. From 2003 to 2027, retirement age will increase gradually from 65 to 67 years. After 2003, no 65-year-old U.S. citizen who retires can expect to receive full Social Security payments. Receiving full benefits requires younger employees to work longer than their predecessors. Further, the age distribution of the North American population is shifting to reflect a greater number of older people. This increasing ratio of retired persons to workers will soon cause difficulty in sustaining the present levels of Social Security income in the United States. Retired people's Social Security payments depend upon immediately disbursing payroll taxes, not upon investing those levies to fund payments with interest or dividends. The effects of the demographic change on retirements 30 years hence may be profound. Finally, employers are dropping traditional pension schemes and adopting cash-balance plans. They are thereby transferring to their workers more responsibility for funding their own retirements through personal savings and investments. A great increase in personal responsibility for retirement demands more care than ever in selecting an employer.

CRITERIA FOR EVALUATING COMPANIES

Corporate Status and Future

Seeking or taking employment in a given company calls for critical judgment in making a series of choices. In looking for a job, one crucial choice lies in deciding whether you want to work for a large company or a small one. Small firms, especially start-ups, offer an innovative, entrepreneurial, and perhaps exciting climate well suited to some temperaments. Growing in many directions, they provide radiating opportunities for advancement. For larger companies, it may be difficult to sustain the exhilaration that research often brings. "... large firms, with their ponderous and bottom-line-oriented organizational structures, do not offer an ideal environment in which research can flourish," wrote *Jürgen Drews*. He was the U.S. President of Global Research at the Swiss pharmaceutical giant Hoffmann–La Roche.

In any case, it's smart to consider each interesting job as one that you hold for the next 30 years. Seen in this light, the problem of objectively evaluating an employer resembles the task of buying stock for the long run and, at least to some extent, can be approached similarly. By no means is this approach the only one to adopt.

Considerations

You can judge any publicly traded, established company by its stock price, by the ratio of this price to its earnings, by the ratio of profits to sales, and by its history of dividend payments. The share price reflects investors' confidence in the company, while a comparatively high price-to-earnings ratio (PE) characterizes drug companies with good prospects for fast earnings growth. In today's market, PE value exceeding 30 and reaching 50 is commonplace among growth stocks in the pharmaceutical sector. Robust sales are the consequence of making products.

- *How many drugs a company offers is therefore an important factor by which to judge the firm because its business is the riskier, the fewer medicines it makes.*

Patent expirations of best-selling drugs. Pharmaceutical companies in particular depend on patents to protect their marketed drugs from competition and on both licensing and their own research as sources of drugs new to their pipelines.

> A company that derives much of its revenue from sales of one drug risks a severe loss of income when its primary composition-of-matter patent expires.

Generic drug houses begin to manufacture and sell such a drug soon afterward. The originating FIPCO loses sales, reduces price and profit, and its share price falls. The patent expiration and, for the next several years, generic competition can adversely affect the fortunes of a pharmaceutical company that depends heavily on one drug. Therefore, in judging a prospective employer, you need to know how many crucial patent expirations it faces in the near future.

Drugs under development.

> The immediate future of a research-based pharmaceutical company, and therefore its employees' prospects, also hangs on the number of drugs it is registering and developing.

It depends especially on the number of blockbuster drugs undergoing clinical development when safety and effectiveness have already been established. This number is by no means an infallible predictor of forthcoming success; many drugs falter late in clinical studies or fail to win regulatory approval. Some are withdrawn from the market after gaining approval when a grave side effect undetectable even in large-scale trials turns up in medical practice. Nevertheless, think twice before taking or keeping a job with a FIPCO having few drugs in its pipeline or having many entering the pipeline rather than emerging from it. What drugs are undergoing development appears in a compendium called "The NDA Pipeline—1998" (*Carlson*, 1999), in the biweekly newsletter SCRIP, and in some corporate annual reports and Web sites. Visit www.labpuppy.com for a summary of certain sites offering this information.

In-licensing. The largest FIPCOs, although they are all research based, constantly search for other companies' candidate drugs that are worthy of in-licensing. They secure intellectual property rights to the drugs and develop, register, and sell what they did not discover themselves. Their pipelines therefore rely on others' efforts to discover new pharmaceuticals as well as on their own. In 1995, for example, the drug portfolios of all 50 highest ranking pharmaceutical companies contained licensed products. The ratio of their own products to licensed ones averaged 2.3, equaling 10 or more in four cases (*SCRIP Magazine*). Although this ratio changes with time and other circumstances, it is wise to look into it when you are considering a job offer. A low ratio of owned to licensed drugs casts bad light on the effectiveness of discovery research management. At the same time, it spotlights the successes of business, chemical, and clinical development and of the company's regulatory affairs department.

Survival and risk. In working for a start-up striving to become a FIPCO, an employee needs a certain ability to survive an abrupt corporate failure and to tolerate the risk of one. Often a nascent pharmaceutical company has no marketed drugs or any other products to sell, and its pipeline contains only one compound in Phase III development. Such a start-up can sink like a torpedoed ship. Adverse events during clinical trials can force it to suspend development of a drug at a late stage. Alternatively, a regulatory agency may reject an application to market the compound as a medicine. Investors, whose hopes concentrated on a drug approval that does not materialize, sell their stock at any price. Alack, the share price plummets (Table 8-1), and stock options lose all value. The company ceases to exist in some cases because its financial backing disappears along with its chances for marketing the one drug it had been progressing. Its erstwhile employees then face the prospect of finding work elsewhere, perhaps at an awkward time when their familial or financial responsibilities are heavy.

On the other hand, approval to market a drug can profit or even enrich any risk-tolerant employee or investor who holds a start-up's stock or options (see text below). Although an FDA panel initially rejected Integrilin for treatment of unstable angina (Table 8-1), the agency ultimately allowed marketing of the drug. According to *The Wall Street Journal* (April 3, 1998, p. B5), "FDA [is] to Clear COR Therapeutics Drug For Unstable Angina, Lifting Stock 79%."

Long before a start-up can bring a compound before a regulatory agency, it must discover the substance and its properties, which calls for millions of dollars in annual research investment. These funds come initially from venture capitalists and later from initial public offerings of stock. But they represent one-time infusions of cash rather than continuing income

Table 8-1 Headlines

"Liposome Shares Take 61% Nose Dive," (*New York Times*, June 26, 1997, p. D9)

"Stock Drops 70% as FDA Rejects Injection for HIV" (*The Wall Street Journal*, April 3, 1998, p. B7A)

"Zonagen Crushed by FDA Delay" (D. Wilkerson, CBS Market Watch, May 10, 1999)

"COR's Shares Plummet 53% on Drug Ruling" (*The Wall Street Journal*, January 30, 1998, p. B6)

"Stock Plunges 74% as Tests of Drug Show Poor Results" (*The Wall Street Journal*, September 2, 1999, p. B16)

from sales. Consequently, it is important to a job seeker to *ascertain how long the start-up can stay in business* when it draws on a fixed sum indefinitely to pay expenses. The available capital and the so-called burn rate of capital spending determine the time remaining until the company must close or derive income from drug sales or new investments.

Other kinds of start-ups are less risky than nascent pharmaceutical companies. For example, a newly founded custom-synthesis house or a contract manufacturer of intermediate compounds and bulk drugs may have orders booked long in advance. Depending for its revenues on several FIPCOs, such a firm is largely independent of the success or failure of any one substance undergoing drug development.

The fortunes of a large pharmaceutical company selling many medicines are not vulnerable to such abrupt failures, although there is job risk even in FIPCOS. Their employees are susceptible to the layoffs that sometimes follow friendly mergers, restructurings, or hostile takeovers. Although the likelihood of such a takeover grows smaller as the subject company's market capitalization becomes larger, no such dollar limit governs efforts to restructure. Recently the pharmaceutical industry witnessed several mergers. For example, the Swedish drug firm Astra joined the British firm Zeneca, while in the United States Warner-Lambert acquired Agouron. Note, however, that layoffs may not extend to scientific researchers, even when they fall heavily on other employees. Indeed, some growing FIPCOs may not have laid off scientists for decades.

Information Resources

Judging a company demands information concerning it. Alas, there is no one source of corporate or drug-portfolio information that is comprehensive, affordable to individuals, or readily available, for example, in public libraries. Pharmaceutical industry newsletters and magazines do contain drug-portfolio information, but they tend to be costly and narrowly circulated. You will probably need to supplement publicly available information with knowledge gleaned from networking and interviewing. Virtually all of this information on a given company originates with the firm itself; some of it the company makes public.

Annual corporate reports are freely available from the company's investor services department. This department's telephone number appears, for example, in the Yahoo!Finance pages devoted to the company; look for the number under *Profile* at www.yahoofinance.com. Another Web site makes these reports available: www.icbinc.com. You can also obtain annual reports through *The Wall Street Journal* (1-800-654-2582). Corporate filings with the Securities and Exchange Commission provide a wealth of information; find them at www.sec.gov. Yahoo!Finance, in addition to publishing pharmaceutical company profiles, makes fundamental financial research available at no charge. So does *ValueLine*, an investment newsletter offering 3- to 5-year forecasts of corporate prospects and available in public libraries.

Relatively new sources of corporate information, some of it contributed by employees, are www.VaultReports.com and www.drugdiscoveryonline.com. On their own Web sites, pharmaceutical companies publish much information concerning the contents of their pipelines, the therapeutic areas in which they concentrate their research, and their approaches to drug discovery. News reports and company press releases dealing with these and related topics are available over the Internet, for example, from Yahoo!Finance.

Geography

Where in North America you look for work obviously influences where you find employment and deserves consideration. You need to decide what you want from your surroundings, which might be mild Californian weather year-round, the changing seasons found elsewhere in North America, or low living costs. Proximity to family or friends may be decisive. Do you wish to pursue your education by taking classes at night? If so, you may find yourself working in or near a large city home to several universities and therefore offering choices in fields of study.

Wherever you seek work, you'll need to look into commuting costs and times. Although living far from where you work may be inexpensive, traveling increases expenses and decreases the time spent at home or in leisure pursuits. So, you have to set limits on how far and long you are willing to commute. Knowing what you want will help you sort possible locations and make rational choices. One such choice concerns the style and cost of the life you seek.

But Can You Afford to Live There?

In seeking or accepting an offer of employment, reflect on where you are to reside while you hold the job. You need to compare the salary offered to the costs incurred, either to decide where you are to apply for work or to choose between job offers. Living expenses vary across the North American continent and partly determine how much money you save from your earnings. The costs you bear as a new hire can therefore influence the comfort or discomfort of your life when you retire as a veteran employee. Because these expenses can differ drastically from one place to another, the buying power of salaries varies proportionately. Similar buying powers therefore correspond to equivalent but unequal salaries in different locales. Such salary figures effectively but indirectly represent variable living costs (Table 8-2).

to all their scientists; those that do include Wyeth-Ayerst, which forms part of American Home Products, and Bristol-Myers Squibb.

In offering options, the company guarantees to sell a given number of its stock shares to the recipient at a fixed price. This cost might reflect the current market price, a discount, or the offering price of a new stock issue. The offer has a time limit, which might be as long as 10 years. Sometimes the right to exercise the options is subject to a vesting requirement. Options ". . . typically vest, or become effective, 25 per cent at a time over four years" (*Ellin,* 1999). If the share price soon rises, the option holder may buy at the low, guaranteed price (called the *strike price*) and immediately sell at the higher market value. The difference is taxable as income, depending on how long the employee keeps the shares. Presently, it is taxable at a relatively low capital gains rate of 20% if the term of ownership equals or exceeds 1 year. The option recipient has no obligation to buy the stock at all, and, if he cares to buy, need not possess sufficient liquid cash to make the purchase. The options and stock can serve as security for a brief loan. Finding a buyer causes no difficulty, if the offered stock comes from a large, established drug firm. Daily trades of such shares number in the millions.

Alternatively, the optimistic owner can retain the stock for years, hoping that it will appreciate greatly. Suppose that you receive options allowing but not requiring you to buy 10 shares of company stock at $100. You hold the options for the longest possible term of 10 years, exercise them, and then sell the stock. Your before-tax profit would equal $967, if the stock price appreciated during those 10 years at an average annual real rate of 7% compounded yearly. This rate equals the return from U.S. stocks from 1802 to 1997 (*Siegel,* 1998):

$$\$967 = (1.07)^{10} (10 \times 100) - (10 \times 100).$$

Before-tax profit increases to $3707 if the average annual price appreciation doubles to 14%. Anyone holding this kind of option stands only to profit, not to lose. If the stock fails to rise beyond the strike price, the options expire unused.

Stock Ownership Plans

Some employers operate employee stock ownership plans (ESOPs). They guarantee to sell their own stock to their workers at a substantial discount, which can be as much as 15%. Purchases have a limit of 10% of annual salary. Appendix B notes 23 companies offering stock purchase plans.

Jobs for Spouses

Suppose your spouse or fiancé(e) is also a chemist looking or planning to look for a position at the same company where you are interviewing. Then, once you have an offer, be sure to ask if your employer permits a wife and husband to work in the same division, department, or group, as appropriate. If you and your spouse simultaneously apply for jobs with the same employer, then you might try to learn the company's policy through your network. In any case, be aware that some employers may forbid married couples to work closely together, while others allow it. You need to know which rule applies, lest one or both of you waste your efforts. However, if you regard the company as an especially desirable employer, apply despite a corporate policy not to employ both spouses in similar positions.

> *Solicit an exemption to a strict policy once you have a job offer in hand, when your bargaining position reaches its zenith of strength.*

Career Prospects

To judge an offer you receive or a company where you might like to work, it's important to estimate your prospects for advancement. How any given company cultivates its scientists' aspirations is not widely known if it is publicly available information at all. A few companies' Web sites do, however, describe their policies. For the most part, familiarity with their practices comes from networking or interviewing, with a notable exception. *A. M. Thayer's* 1998 essay "Dual Career Ladders" offers valuable insights into a corporate structure commonplace among research-based companies.

Scientific and Managerial Ladders

The system of dual career ladders provides different means by which scientific and managerial personnel can separately advance. It affords equivalent titles, privileges, salaries, and other forms of compensation to those scientists who do make personal progress through welcomed corporate contributions. Such a policy helps recruit and keep employees and gives them performance incentives. In part, its purpose is to keep talented scientists doing science and to discourage them from departing for alternative careers elsewhere in a corporation.

Although the details of dual-ladder arrangements differ from one company to another and, within one company, change from time to time, the system exhibits several common features.

1. Upward moves on the scientific ladder reward technical accomplishments, while merely assuming greater responsibility can suffice to bring an ascent on the managerial rungs.

2. Some companies broadly sketch their criteria for promotions, not specifying them in detail, but making the requirements known to employees.
3. The number of rungs on one ladder may not equal the count on another.
4. Often scientists serve a minimum time on each rung before rising; 6-month tenures for them are infrequent.
5. The number of positions available or populated on any one rung decreases as altitude increases.
6. Occupancy of the lowest rungs may be restricted on either ladder.

> *Interviewed for Thayer's C&EN article, a recruiting manager from the chemical company Rohm and Haas noted that graduate school prepares no one to take immediately after commencement a managerial job.*

This observation applies equally to aspiring scientists and fledgling research managers. To take another example, Ph.D. chemists find it easier to ascend the scientific ladder than chemists with lesser academic diplomas. Students of management recognized this impediment decades ago. One of them, *E. Roseman,* wrote about:

> *... chemists who ... earned a master's degree before entering ... industry, and ... had no opportunity to earn a doctorate. Their ... progress is blocked by lack of an advanced degree.* (italics added)

Hard questions. To devise equitable means for rewarding performance is not an easy task. For that reason it is essential to ask critical questions about the dual career ladder in any company that adopts this system. Here's a selection of such questions, which may find answers through networking or interviewing; tact in voicing them is obviously desirable. Perhaps the best advice you can follow is to ask these questions when you have an offer in hand.

Does each step upward bring a tangible, immediate reward that the recipients value? For example, do increases in salary accompany a promotion and exceed the inflation rate? Such raises are preferable to equal bonuses, which arrive only once.

To become eligible to mount the ladder, what requirements must a climber meet? Are the criteria for occupying each rung specific and objective and are they made known to aspiring scientists (a) in advance and (b) in writing? May an interested employee keep a copy of the document setting forth these necessities? If consistently performing at the next higher rank provides a standard for promotion, may an aspiring scientist read and retain the job description?

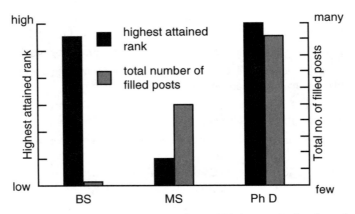

Figure 8-3 Hypothetical distribution of highest attained ranks and total numbers of filled positions on a scientific ladder

How are ranks distributed among holders of B.S., M.S., and Ph.D. degrees? For example, do the highest ranks and greatest number of positions on the ladder go largely or exclusively to doctorate scientists, as shown (Figure 8-3)? In the imaginary distribution depicted, doctorate scientists fill many high-ranking positions, while scientists with master's degrees occupy fewer posts that rank lower. One richly rewarded B.S. chemist invented either a marketed drug or the synthesis used to manufacture it, an extraordinary accomplishment for anyone. Of course, distributions other than this one are also possible. What is important about any of them are four factors. The factors include the numbers of rungs on the ladder and of scientists populating them, rewards for achievement rather than diploma or rank, and timeliness. Surely no chemist contents herself to serve time waiting for promotion while his worthy accomplishments go unrecognized.

Did any factor other than performance also influence the distribution in either dimension? One example of such an influence might be sex, although examples need not be limited to this factor. Anyway, depending on your sex, you may wish to ascertain women's prospects for advancement and to weigh answers to questions like the following ones. Within the past decade, how many women chemists from discovery or development research became vice-presidents, directors, research fellows, section leaders, or their equivalents? How many did so without moving to other companies? What fraction of all such occupied positions do the women represent in their own companies, and how large or small should such a fraction be? Were those women promoted from within the organization or were they imports?

Finally, do transfers of employees from the managerial ladder to the scientific ladder occur summarily? Such transfers, as well, perhaps, as simultaneous appointments to both ladders, devalue the dual ladder system, diminishing its credibility.

Lateral Moves

After a few years' experimental work, some chemists seek different challenges, taking on some responsibility for managerial work or moving to a similar research position in another company. But, there is sometimes an appealing third route to a new destination, namely a lateral move to another part of the same company in which the scientist originally accepted employment. (The specific jobs usually open to chemists inform Chapter 3). Here, suffice it to note that networking—but not interviewing—can be a politic source of information regarding recent career changes made by chemists within a given company. Ask your contact how his company makes open jobs known to its employees and what those jobs are. Familiarity with alternative career paths is hardly decisive in seeking work or accepting an employment offer; and during an interview the recruiter may not welcome discussion of such a topic. Nevertheless, knowledge that you can hedge your bet may become useful to you in a few years.

Publishing and Patenting Policies

To assure yourself of continued employment as a scientist, especially in a firm other than the first one that hires you, you have to keep yourself employable. Consequently, you need to prepare now to show verifiable corporate accomplishments several years hence. For a chemist working as an entry-level discovery or development researcher, a showing of achievements takes three forms. It demands publishing your work in chemical journals, delivering lectures or presenting posters at chemists' external meetings, and patenting your inventions. All these activities reduce the time devoted to doing your ordinary work, whereas patenting also entails considerable expenditure of a company's money and an attorney's efforts. Different firms therefore adopt varying policies concerning patenting and publishing. These policies admittedly change with time, but can still last for a decade or more; and even as a job seeker you need to know them. You needn't learn every detail, but stand to benefit by asking if a given company encourages publishing. Does the firm favor precautionary patenting of its scientists' many inventions? Some do even though few companies' patents can be expected to include a marketed drug or a manufacturing synthesis. Many pharmacologically active compounds prove worthier of patenting than of chemical, clinical, and pharmaceutical development. The need to patent them arises before a decision to advance or abandon them.

A pharmaceutical company favors publication if it meets several conditions. It promptly grants needed approvals, reimburses travel costs, and allows its scientists to compose articles, lectures, or posters during a portion of the workweek. It pays for artwork, at least by providing personal computers and suitable software, if not by offering an in-house graphics art department or paying for outside services. It merely tolerates publishing by insisting that the preparations be made during its researchers' spare time.

To establish that your prospective supervisor and coworkers do publish their research, ask for copies of their publication lists, which should cite articles they wrote and the patents granted them. These lists demonstrate how often those scientists publish, while their articles and patents show that the results they report were obtained at the company where they work. The Author Indices of *Chemical Abstracts* also furnish this information. If you visit the Web site of the U.S. Patent and Trade Office (www.uspto.gov), you can search the Bibliographic Database by an inventor's surname to find the patents granted her.

Choosing Your Boss

During a day of interviews, you can expect to meet several chemists who represent discovery or development research and who have vacant jobs to fill. This circumstance makes two implications worth considering. Of course, it suggests you face a choice between one endeavor or the other and need to express credible interest in both. Moreover it implies that in a large company you may be able to influence the assignment of your own supervisor. Lest no one offer you a choice, or even notify you of its availability, plan to ask to work with one person in particular. To prefer one prospective supervisor to another, you may desire to match what you need to learn and what your boss can teach. Virtually all new graduates who work as researchers in the pharmaceutical industry need to understand what practical drug discovery or development entail. Some chemists stand to learn how to do organochemical research in either area.

Generally, whether you need to learn to do chemical research depends on your academic training and degree. With a master's or doctor's degree, you already know more than enough to begin. If you have a bachelor's degree but lack research experience, you need a boss who not only serves as your supervisor but who also acts as your mentor. She must guide you in carrying out research in synthetic organic chemistry, just as she would do in a university if she served as your faculty research adviser. Consequently, your boss must be willing to teach you what you need to know and be patient enough for you to learn. Not all industrial chemists are willing and capable teachers; some have too many other responsibilities to pay much attention to a single newly hired researcher. Furthermore, a desirable boss is an experienced supervisor of inexperienced newcomers as well perhaps as a chemist competent in, for example, combinatorial chemistry or parallel synthesis. Such a person knows enough to teach you the newest technical methods that are widespread in discovery or development research. Her qualifications should be evident from her list of publications, and she should be glad to talk about them in an interview. Your chance to inquire comes when she says, "Do you have any questions for me?"

Much of drug discovery and development is usually learned through close daily contact with a knowledgeable supervisor. As a discovery researcher, a desirable boss should be experienced in advancing drug candidates through preclinical discovery research to clinical trials. A supervisor working in chemical development research should be successful in devising a manufacturing synthesis for a drug or in pragmatically making large quantities of a drug for clinical studies. A boss with no successful experience in these areas may be finding her own way through the labyrinth that drug discovery and development represent. Such a supervisor has little valuable know-how to impart.

What your boss can teach you is worth finding out before you take the job, provided that you win more than one employment offer or a choice of supervisors within one company. To ascertain your prospective supervisor's accomplishments, examine her medicinal chemistry articles that conclude saying one substance from a series of biologically active compounds advanced to a clinical trial. This information appears in the abstract, introduction, or conclusion, and sometimes in a footnote. For a chemical development researcher, check the chemical literature for papers detailing, and patents protecting, manufacturing syntheses. Tactfully ask your interviewer to give you a copy of her résumé and to recount her corporate accomplishments. Listen carefully for answers that bear directly on what you need to learn about practical chemical drug discovery and development.

Benefits beyond Salary and Other Compensation

Jobs in the drug and other industries offer yet other financial advantages additional to salary, namely group insurance and retirement benefits. A few companies detail their offerings on their Web sites, while other employers' sites briefly summarize them. Perhaps the most general method of learning about a particular employer's benefits is to discuss them with a personnel representative during an interview. To talk with such a person for half-an-hour or more is a standard feature of day-long interviews. You may want to prepare an organized list of questions and ask for any company benefits literature, including a description of pension plans. When you chat with other employees, find out whether they are satisfied with the benefits the company does offer and what desirable benefits it does *not* extend to its workers. Younger workers may evince little knowledge or interest in benefit plans, so speak to an older chemist facing retirement. To consider retirement schemes when you choose between offers, it helps to return from an interview with pension literature or an understanding of what the plans entail. Failing that, it is essential to speak again to the personnel representatives whom you met during your site visits. Phone them.

Group Insurance

Perhaps the most important financial benefit that large employers offer to new employees is a comprehensive insurance plan. Such a plan covers the worker, his spouse, and their dependents. It provides family members with health and dental care and insures against the employee's death or dismemberment. Comprehensive group insurance does not come free to employees, buy they pay for it at lower rates than they would meet elsewhere.

Recent developments in group insurance plans include health maintenance organizations and flexible spending accounts. These organizations offer health care at lower costs than traditional medical insurers but restrict coverage of treatment and choice of physician. Flexible spending accounts permit holders to pay certain medical costs with money not subject to federal income taxes. As a prerequisite to using such an account, the holders' income tax bracket must be high enough to make the untaxed payments advantageous. The annual deposit made to the account has an upper limit, and the Internal Revenue Service seizes unused funds left in the account at year's end. To make the best use of such an account requires a careful estimate of annual medical expenses so that deposits are small enough to be consumed.

Retirement

Young chemists must consider their pensions when they seek work for the first time, even though they will not receive them for decades. Therefore, to judge a job by its benefits, you have to ascertain that it does bring a pension, which is too important to your future to take for granted. Private U.S. companies grant pensions to fewer than 50% of all employed workers. However, pharmaceutical companies generally do offer pensions to retiring permanent workers, which represent a lasting advantage of regular employment.

In offering pensions, U.S. employers and the federal government strive to replace a retiring worker's salary with an equal retirement income. This income flows from three sources, which are Social Security payments, the employee's preretirement savings and investments, and pension benefits. Pensions strike a balance between lower present salaries and dependable future incomes when employees retire.

> *Differences between pension plans can give reasons to accept one offer of employment and reject another.*

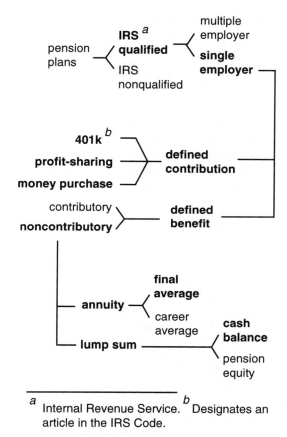

Figure 8-4 Features of selected pension plans

[a] Internal Revenue Service. [b] Designates an article in the IRS Code.

To give due consideration to differing pension plans, however, requires an understanding of their features (Figure 8-4).

Pension plans that meet the requirements of the U.S. Internal Revenue Service (IRS) are commonplace in the medicinal industry and elsewhere. Let us restrict our consideration to them, understanding that a given employer may offer different plans to certain groups of workers. The ensuing discussion covers the topics emboldened in Figure 8-4.

There are six defining features of IRS-qualified pension plans.

1. Assets of the plans are held in a separate trust.
2. By law the assets lie beyond corporate creditors' reach.
3. They exclusively benefit plan participants.

4. Before they are distributed, their earnings are not subject to taxes.
5. Employers' contributions to the plans are deductible from corporate income and so are not taxed.
6. Although distributions to employees are taxable, no tax or payout falls due until retirement begins.

In some pension plans several employers contribute to a single fund, usually to cover unionized workers. These plans are not relevant to North American chemists, who have no union. The pertinent and also commonplace arrangement calls only for a single company to contribute to one pension fund, which may include a *defined-benefit* or *defined-contribution* scheme. Some corporate pension plans include elements of both these schemes, so it's vital to understand them and the differences between them.

Defined-benefit pension plans. If, in retirement, you receive from your former employer a fixed monthly payment, your pension exemplifies the defined-benefit type. The structure of such a plan makes your employer legally responsible for implementing all of it. It requires the company but not the employee to make a minimum annual contribution, which comes from investments. The employer must make up any investment losses to maintain its annual contribution. It may not end the plan without sufficient assets to provide the promised benefits, and its obligation to continue payments to employees survives bankruptcy. The Pension Benefit Guarantee Corporation, which is wholly owned by the U.S. government, secures payments to employees who participate in defined-benefit schemes. Consequently, participants can never exhaust their benefits while they live, and their spouses may continue to receive pension payments after the participants die. Regular lifetime payments are important features of defined-benefit pension plans, about two-thirds of which disallow lump-sum distributions at retirement (*Kehrer,* 1995).

PARTICIPATION, VESTING, ELIGIBILITY, AND SERVICE. Three significant dates and an interval figure in defined-benefit pension plans.

1. Participation in the plan formally begins when the employee completes the minimum service requirement and when the next entry date arrives. The minimum service is 1 year's work or less, and in some companies participation begins when the employee meets only this requirement. Other firms impose entry dates, which might be the next January 1 or July 1 that arrives after the minimum requirement is met. Effectively, this policy adds as much as, but not more than, 6 months to the needed length of service. Before the employee becomes a participant in the

plan, her work earns no monetary pension entitlements; quitting then will bring no pension payments when retirement comes. Participation requirements of defined-benefit plans tend to prevent workers from moving from one employer to another. Make five moves, and you will have lost as much as 7½ years of benefits accrual.

2. Vesting follows participation in the plan, usually after 5 years of service. It entitles an eligible participant who leaves a given employer to receive a pension in retirement. Nonvested participants who quit their jobs forfeit their accrued benefits, which subsidize other employees' retirements. With only two job changes, you can lose as much as 8 years of accruals, unless you negotiate a restoration of benefits with your second employer.

3. Eligibility refers to the date on which a participant qualifies to retire, often her 55th birthday. Vested participants too young to meet the age requirement are ineligible to receive a pension if and when they retire early. Retiring early at age 55 drastically reduces the pension payments that a vested eligible participant receives. It also means that the worker is too young to receive Social Security benefits, which otherwise represent indispensable contributions to the incomes of many retired workers. In defined-benefit schemes, reduced payments represent the price of early retirement.

4. How long you work for one employer substantially influences how much of a pension you receive in retirement. Length of service is counted from the first date of formal participation in your employer's pension plan, *not* from the first date of your employment. *How* it is counted varies from company to company and may include an elapsed-time method. Some companies require that to earn pension credit for a year's work you must put in at least at 1000 hours within a 12-month period. In general, breaks in service reduce the working time credited to an employee.

To appreciate the effect of service years on pension entitlements, consider certain worst-case examples that are neither completely impossible nor highly likely. They pertain to a chemist who completes 3 years of postdoctoral research, earns a Ph.D. degree and an M.S. diploma in 7 years, all following her 22nd birthday. Suppose she then works for one employer until she reaches age 65, which at present is the earliest retirement age that can bring full pension. Deducting a 1½-year participation requirement, her counted years of service amount to 31.5. Then the sum of her age and service years equals 96.5, so her pension would reach only 96.5% of the maximum. Immediate participation, by contrast, would have raised this chemist's pension to 98% of the limit. Alternatively, she would have received the maximum benefit despite the minimum service requirement, if her postgraduate training had included only 5 years of Ph.D. studies. However, the higher salaries paid to researchers with postdoctoral training offer some compensation for the smaller pensions that shorter tenures sometimes earn.

ACCRUAL OF BENEFITS. Entitlements accrue slowly in defined-benefit pension plans, with most of them coming late in an employee's working lifetime. A representative plot of pension entitlements versus age appears in Figure 8-5. Such a rate tends to keep vested participants working for a single employer for a long time, although the steep rise in pension credits does make early retirements possible.

PENSION SIZES. Different companies use varying formulas to calculate pension amounts, which depend on years of counted service, a constant, and the final or career average pay. The constant equals as much as 1.5%, and the salary contribution typically represents the average of the final 5 years of earnings. This average can be calculated from starting wages if they equal the median 1998 salaries paid to inexperienced chemists working in the pharmaceutical industry. Suppose a 22-year-old B.S. chemist takes a job at an annual salary of $32,000 and receives inflation-adjusted raises that amount to 1% annually. Imagine that she retires at age 67 after 44 years of counted service. This chemist's final 5-year average salary and annual pension respectively equal $49,092 and $32,401. The pension can be reckoned according to the following formula. It assumes that the sum of the chemist's age and counted years of service equal 100, entitling her to the maximum 100% benefit.

$$\$32{,}401 = \$49{,}092 \times 0.015 \text{ years}^{-1} \times 44 \text{ years}$$

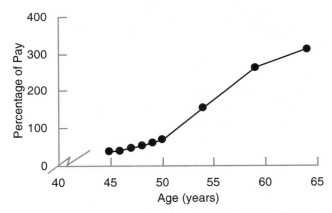

Figure 8-5 Present value of a typical defined-benefit pension plan (Source: I. G. Kastrinsky, *Employee Benefits Journal*, June, 1999, pp. 15–19)

An M.S. chemist who retired at 67 with 42 counted years of employment would enjoy an annual pension of $43,653. This figure assumes her starting salary at age 24 had been $47,000 and that raises brought an average of 1% more in salary each year. It takes the final 5-year average salary to have been $69,291.

Consider a Ph.D. chemist who, at age 27, began with a $62,000 salary. If she worked 40 counted years, received 1% yearly salary increases, and retired at age 67, her final average salary would amount to $91,404. Her pension would initially equal $54,843.

RAVAGES OF INFLATION. The pensions paid to these three scientists during their first year of retirement amount approximately to two-thirds of their final 5-year average salaries. However, these pensions are not only close in value to the scientists' starting salaries—despite 4 decades of employment—but may decline when measured in constant dollars.

> *Fewer than 7% of defined-benefit pension plans provide inflation-indexed increases (Kehrer, 1995), which are likely to result only if pensioners petition their employers.*

For this reason, retired people depend on other means to raise their fixed pension incomes to preretirement levels. A plot of pension values (in 1999 dollars) versus retirement duration indirectly shows the importance of other

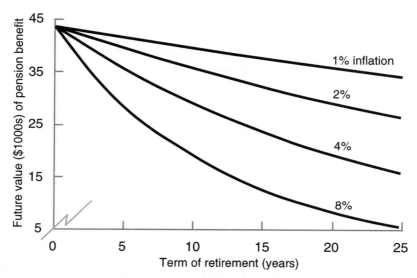

Figure 8-6 Inflation-ravaged pension benefits

income sources (Figure 8-6). It depicts the adverse effects of increasing inflation rates on a pension initially equal to $43,653. In 17 years, for example, a 4% annual inflation rate would halve this pension.

CASH-BALANCE PLANS AS PORTABLE PENSIONS. For chemists seeking jobs now, the immediate future promises a structural change in pensions that will affect their retirements years hence. In the United States and other countries, large private employers are switching from the traditional defined-benefit plan that brings a fixed monthly payment throughout retirement and that represents an annuity. They are instituting a variant entailing a one-time lump-sum distribution that occurs at retirement. This type is known as a cash-balance plan. Hundreds of U.S. companies have already made the change, including one FIPCO, SmithKline Beecham, which switched early in 1999.

Cash-balance pensions show four distinguishing features in comparison to annuity-based pension plans.

1. An employee who invests her lump-sum distribution when she retires receives the rewards and bears the risks of investment. Making appropriate investments allows her to increase her capital faster than inflation shrinks it. As we saw, other defined-benefit pensions make none but ad hoc adjustments for inflation, and retired employees do not choose the investments that underlie their pensions. With these annuity-based, guaranteed plans in which employers determine investments, retired persons cannot outlive their benefits. However, exhausting or losing one's capital is possible in cash-balance plans, after the lump-sum payment is received at retirement.

2. Cash-balance plans avert the complete forfeits that other defined-benefit schemes entail, bringing higher pension credits to a worker who quits one job to take another. Representing portable pensions, they facilitate the career development that working for another employer can bring. How quickly pension credits accrue in a representative cash-balance plan is evidenced by the plots in Figure 8-7. The plots assume that cash balances at retirement reflect the differences in 1998 median starting salaries paid inexperienced chemists with doctor's, master's and bachelor's degrees. (Other assumptions of the underlying calculations are the same as those related for Figure 8-8.)

   Baccalaureate chemists who work 20 or 30 years at one job spend respectively 50 and 75% of their working lifetimes with one employer. However, if they quit or retire early, they would depart with considerably less than the same percentages of the final cash balance that they could earn by staying on (Figure 8-7). Two decades of employment represents half a working lifetime but brings less than 30% of the final cash balance. Thirty years of a possibly 40-year-long career earns under 60%

Figure 8-7 Accrual of cash balances

of the final balance, not three-quarters of it. Almost any working time brings a lesser credit than the final one, for the plotted lines are curved as shown. Only at the two extremes of each plot do working times bring equivalent pension credits. These plans, like traditional annuity-based pensions, favor long-staying workers.

3. Cash-balance plans credit workers with higher pension entitlements earlier in their careers than do traditional schemes, but these plans also tend to reduce early-retirement subsidies. Traditional, annuity-based schemes accelerate the accrual of pension credits that begin when older workers become eligible to retire (Figure 8-5). The boost makes early retirements possible, even though it brings smaller pensions than longer-staying workers finally receive.

4. Employees know throughout their careers what to expect when they retire because cash-balance plans provide periodic statements of accounts. In ordinary plans, no detailed reckoning of pension benefits takes place until a worker becomes eligible at about age 55. None occurs then except at the employee's behest. The new plans help make what-if calculations early in a career, which can be important to anyone considering an early retirement. In a cash-balance plan, an employer creates a pension account in each worker's name and annually contributes funds on the employee's behalf. The contributed amount represents some percentage of the employee's salary and increases with pay raises. For the

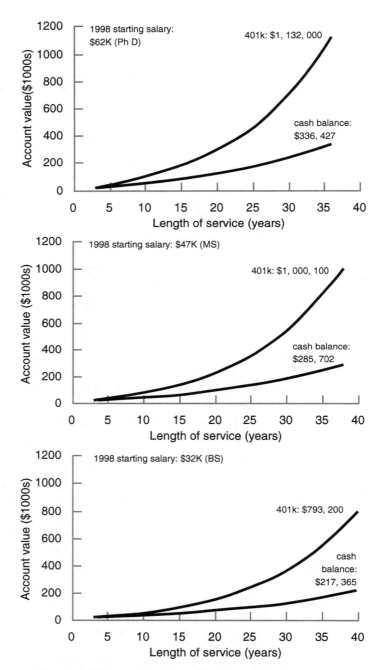

Figure 8-8 Accruals of tax-deferred, inflation-adjusted cash-balance and 401k pensions

worker's sake and its own, the employer pools all the deposits to invest them in stocks, government securities, high-yielding corporate bonds, or other investment types. Pooling allows the employer or its agent to choose riskier investments for the sake of higher gains. Individual employees do not determine or influence this choice during their working years. The investments bring an annual return, some portion of which the employer credits as interest to each worker's account. This credited interest equals some benchmark rate, like the fluctuating interest rate offered by the U.S. Treasury on its 30-year bonds. In the years 1919–1996, these yields ranged from 2 to 14%, according to Standard and Poor's.

An employer's yearly investment returns may exceed the interest credited to each employee's account. If so, the excess over the credited interest goes to reduce the employer's costs in funding the pension plan, not to increase the employees' gains. In this respect, cash-balance plans resemble traditional pensions.

In a representative cash-balance plan, the annually credited interest equals 3.96%, an employer's contribution reaches 5%, and annual salary increases average 2% (Figure 8-8, the lower curve of each pair). With this information and a knowledge of starting salaries and working lifetimes, the cash balance in any account can be calculated by a seeded iteration and then plotted. The graphs in Figure 8-8 (and Figure 8-7) assume inexperienced chemists' 1998 median starting salaries of $62,000 (Ph.D.), $47,000 (M.S.), and $32,000 (B.S.). They make several other assumptions: the credited interest represents the 30-year Treasury bond yield (5.96%) prevailing on July 22, 1999, minus an estimated inflation rate of 2%.

Defined-contribution pension plans. If you are familiar with cash-balance schemes, then you already know something about defined-contribution pension plans because of their similarities. In a defined-contribution plan, an employer periodically contributes a fixed dollar amount to a tax-deferred account established in each covered employee's name. Contributions begin as soon as the worker formally begins to participate in the plan and continue until she retires. Afterward, employers make no more contributions, but distribute the principal and earnings as a lump sum at retirement. Throughout retirement, the employee bears decision-making responsibility for investing the principal and reinvesting its earnings, which are both reported to her in quarterly statements. The Pension Benefit Guarantee Corporation does not protect the principal after the retiring employee receives it; before that, the employer may end the plan at its discretion. Such circumstances represent risks borne by plan participants.

Serious responsibility for financing much of one's own retirement begins earlier and lasts longer in a defined-contribution plan than it does in a cash-balance scheme. It starts when the employee becomes a plan participant and continues throughout the rest of her life. A chemist who takes employment at age 22 and lives to 85 years can carry out this duty for 6 decades. Choosing investments wisely and spending carefully in retirement, a chemist need not lose or outlive the money. Sixty years offers ample opportunity for a buy-and-hold investor's capital to appreciate. In the long run, only stock investments defeat inflation that vanquishes investments in cash, gold, or bonds of all kinds (*Siegel,* 1998).

Defined-contribution pension schemes take three main forms, which are 401k, profit-sharing, and money-purchase plans. Some employers supplement them with defined benefit plans, employee stock purchase plans, or stock options.

401K PLANS. The federal government encourages its citizens to save and invest through the 401k plans offered by certain employers. Indeed, the majority of employers with over 500 workers offer these plans. They allow any eligible wage earner annually to save a limited but substantial fraction of salary. Employers' contributions to 401k plans currently have an annual upper limit of 25% of salary or $10,000, whichever is less. The cap is periodically adjusted for inflation. The employer pays no tax on its monthly contributions, if it contributes at all. The recipient incurs no tax when the contributed money is deposited, and interest or dividends accumulate tax free. Only when the retiree draws the money is it taxed as income. This taxation can occur at a lower rate than it otherwise would do. However, taxes and penalties fall due if the owner withdraws any of the saved money prematurely.

Employers deduct the savings from paychecks, credit their employees' accounts, and help to manage the invested money on the employees' behalf. The saved money is invested, and the investment(s) serviced, by a financial services house that the pharmaceutical firm selects. The employee can choose the kinds of investment, allocating varying percentages of each year's savings to the different funds. Allowed investments exclude collections of valuables—objets d'art, for example. They include company stock, various mutual funds, and annuities; but are not limited to these vehicles. Different employers offer varying numbers and kinds of investment choices. To maximize the return on investment, the money can readily be switched from one fund to another, usually by a telephone call.

In a 401k plan, employees pay for the benefits, not employers. However, some employers may also make matching, but not necessarily equal, contributions. Such an employer typically contributes 3% of its employees' pay each year while the workers may save another 6% of salary. Matching contribution plans like these are advantageous because the money can grow faster than it increases in a cash-balance plan. Let us consider how this can happen with a representative 401k plan (Figure 8-8).

Inspection of the returns from 401k and cash-balance plans makes evident the greater value of participation in a 401k plan. The lower yields of cash-balance plans come partly from the low annual company contributions.

> *For a disciplined, conservative stock investor, the likelihood of a higher long-term return is reason enough to work for a company offering a 401k plan.*

However, both plans deserve a detailed evaluation, for either may differ from the corresponding returns shown here.

The 401k plan calculations underlying Figure 8-8 make the following assumptions. The principal and earnings grow at 7% a year, and compounding occurs annually. Each year, the employee contributes 6% of salary while his employer contributes 3% of that amount. Contributions increase because the worker's salary grows at an average of 2% yearly. Finally, the calculations begin with the median 1998 starting salaries of inexperienced chemists working in the pharmaceutical industry. [Individualized 401k computations are available through an online calculator that compounds 26 times annually (www.dismalscientist.com).]

Contributions to a 401k account incur no tax until the employee draws on the money. Consequently, the balance compounds faster that it would otherwise do. To appreciate the desirability of saving and investing tax-deferred money, inspect Figure 8-9. It shows how compounding increases the value of tax-deferred savings compared to the worth of tax-paid savings.

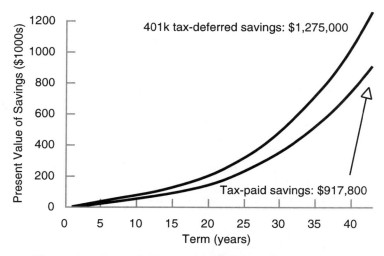

Figure 8-9 Growth of tax-deferred and tax-paid savings

In both cases, the underlying calculations presuppose a median 1998 starting salary of $42,000. They also include average annual salary increases of 1% over a 43-year working life and a yearly real rate of return equaling 7% with annual compounding. In the case of 401k savings, the computations assume that the investor's contributions equal 9% of salary. They assume in the other case that she pays income taxes of 28% on this contribution, which effectively reduces it to 6.48% of salary. After 43 years of saving and investing, the tax-deferred account value exceeds the other by nearly $360,000.

PROFIT SHARING. Some employers offer their workers limited, proportionate shares of the profits earned each year. The sizes of these shares, which represent defined contributions to a pension plan, are independent of the employees' annual performance. Added to their accounts once a year, the contributions—depending on how they are invested—may compound more slowly than a comparable amount credited through equal monthly deposits in cash-balance and 401k plans. Employers pay no tax on their contributions as long as the amounts do not exceed 15% of salaries, so a typical contribution equals this figure or a smaller one. Although the profits of FIPCOs often surpass 20% annually, sometimes doing so year after year, employee's shares may amount to no more than three-quarters of this percentage.

During lean times, which occur when the patents of best-selling drugs expire and generic manufacturers capture sales, employers' contributions to employees' accounts approach zero. Periods like these can last for years until new drug sales replace those lost to competition.

> *The value of a profit-sharing plan to employees depends upon sustained corporate earnings. In turn, these profits require distant patent expirations of profitable drugs and a wide, full pipeline of medicines flowing rapidly to the marketplace.*

Contributions to profit-sharing plans are not necessarily linked directly to corporate earnings. Some companies make them age dependent, distributing more to older employees than to younger ones. Other firms offer a percentage of salary that is determined by the ratio of the employee's wage to the total payroll. *Gerald Cole,* an attorney working as an employee benefits researcher, gives an example in which payroll equals $5,000,000 and salary amounts to $50,000. In this case the employee receives 1% of what the employer contributes to the plan. Employers have complete discretion in determining their profit-sharing contributions.

The responsibility, risks, and rewards of investing profit-sharing money devolve upon employees, as they do with 401k plans. They begin as soon as the first contribution is credited and last through retirement. Money received as profit shares is not taxable on receipt, but incurs income taxes only when an employee draws it. Early withdrawals are not only taxed but penalized. An employee who withdraws money from a 401k or profit-sharing plan before reaching age 59½ pays an immediate IRS penalty of 10%. She also loses the opportunity to compound the withdrawn sum through tax-deferred investments.

MONEY-PURCHASE PLANS. Defined-contribution pension plans include a variant in which the employer promises annually to contribute a certain percentage of salary to each employee's account. These plans, known as money-purchase plans, allow the employer to avoid taxes on as much as 25% of the recipient's salary. In this sense they are comparable to 401k plans. Money-purchase plans, unlike profit-sharing plans, forbid employees to withdraw funds before retirement.

SOURCES AND SUPPLEMENTS

Anonymous; "Top companies in R & D," SCRIP *Magazine,* January, 56 (1996).
Borchardt, J. K.; "Sizing up your future boss: Ask the right questions during an employment interview," *Today's Chemist at Work,* September, 25–30 (1998).
Borchardt, J. K.; "Hooray! You've got the job offer: Now sit back and figure," *Today's Chemist at Work,* July/August, 31–37 (1996).
Burnlingame, H. W., and Gulotta, M. J.; "Cash balance pension plan facilitates restructuring the workforce at AT & T," *Compensation & Benefits Review,* November/December, 25–31 (1998).
Carlson, M. (ed.); "The NDA Pipeline®—1998," F-D-C Reports, Chevy Chase, 1999.
Cole, G. E.; "An explanation of pension plans," *Employee Benefits Journal,* June, 3–13 (1999).
Drews, J.; "In Quest of Tomorrow's Medicines," Springer-Verlag, New York, 1999.
Gaudio, P. E., and Nicols, V. S.; "Your Retirement Benefits," Wiley, New York, 1992.
Ellin, A.;"When the glitter of stock options turns to dust," *The New York Times,* August 22, BU10 (1999).
Fish, R. A.; "Is That Your Best Offer?: How to Negotiate a Higher Salary and More Perks," WetFeet.com, Inc., San Francisco, 1999.
Heylin, M.; "Salaries up, but jobs still tight: Salary and employment survey," *Chemical and Engineering News,* August 2, 28–39 (1999).
Kastrinsky, I. G.; "Making a sound transition to a cash balance pension plan," *Employee Benefits Journal,* June, 15–20 (1999).

Kehrer, D.; "12 Steps to a Worry-Free Retirement," 2nd ed., The Kiplinger Washington Editors, Washington, DC, 1995.

Laabs, J. J.; "Learn what's hot in portable pensions," *Personnel Journal*, October, 34–42 (1995).

Roseman, E.; "Confronting Nonpromotability: How to Manage a Stalled Career," AMACON, New York, 1977.

Siegel, J. S.; "Stocks for the Long Run," 2nd ed., McGraw–Hill, New York, 1998.

Thayer, A. M.; "Dual career ladders," *Chemical and Engineering News*, November 2, 51–55 (1998).

Uchitelle, L.; "Signing bonus now a fixture farther down the job ladder," *New York Times*, June 10, A1 (1999).

appendix

B
∙∙∙∙∙∙∙∙∙∙

INDICES OF THE NORTH AMERICAN PHARMACEUTICAL INDUSTRY

".... most arguments in favor of not doing an experiment are too flimsily based."—R. B. Woodward

INTRODUCTION

Seize inspiration from the quoted words above and confidence as well. Thus equipped, you can defy convention in applying for work, even as you spurn contradictory advice given here or elsewhere. A brash experiment—importuning an eminent professor on the phone—brought me my first job in the pharmaceutical industry. As a crude product of graduate school, I telephoned the finest organic chemist of the 20th century, winning the Harvard interview that led to employment. The job lay in a storied foreign country, Switzerland; within a prosperous pharmaceutical and chemical company, CIBA-Geigy AG (now Novartis); and in a distinguished laboratory of organic chemistry, the *Woodward* Research Institute.

Here are two indices to help you discover, learn about, and approach other employers. The "Geographical Index of the North American Pharmaceutical Industry" presents the two countries, 41 states, and six provinces containing organizations that employ organic chemists as discovery or development researchers. The "Name Index of the North American Pharmaceutical

Industry" specifies more than 500 companies and other organizations. Matching any entry in the geographical index is another within the name index, which furnishes details. A job hunter who decides where to work and what kind of research to do can use the geographic index to find suitable organizations located in the preferred state or province. Then, using the name index, he can begin a search for desirable employers among those organizations and learn from the index where and how to apply for a position. In many entries, the name index relates the application method(s) that the employer prefers or discourages and specifies the documents that an applicant must submit. It also guides the job seeker to further sources of information about particular companies.

Kinds of Indexed Organizations

The name index presents start-up companies employing fewer than 2 dozen people. As well, it lists the largest FIPCOs, which are staffed by thousands of employees and located in Canada and the United States. They engage in all the pharmaceutical businesses, from preclinical synthesis and testing of experimental drugs to postmarketing surveillance for side effects. The indexed organizations do most of the drug discovery and development research that occurs in North America.

The name index is not limited to pharmaceutical companies. It extends to other nonacademic organizations that play a role in the drug industry and that employ organic chemists, mostly as researchers. This index also includes biotechnology and diagnostics companies, patent law firms employing chemists as scientific advisors, and a brokerage company that indirectly provides custom synthesis services to the pharmaceutical industry. Also noted are government and private institutes where chemical research takes place (Table B-1). So are drug discovery companies that carry out research in medicinal chemistry but do not engage in the other businesses of a FIPCO. Their work results, for example, in combinatorial libraries of compounds, which they license or sell to FIPCOs. Some companies devise and make semi-automated apparatus for synthesis or purification of combinatorial libraries, and the indices include them because they employ organic chemists.

The largest drug houses carry out developmental chemical research, so any organic chemist should consider them as possible employers. Developmental research—to the exclusion of discovery research—also occurs in smaller companies entered in the name index. Some small companies, for example, specialize in custom synthesis or radiosynthesis of isotopically labeled drugs or intermediate compounds. Others devise processes to make pharmaceutical intermediates as well as bulk actives regulated by current good manufacturing practices (cGMP). The name index specifies fine-

Table B-1 Selected Government and Private Institutes

Government	Private
Center for Advanced Materials	Cancer Therapy and Research Center
CHUL Research Center	Evanston Northwestern Healthcare Research Institute
Food and Drug Administration	
National Institutes of Health	Kettering Medical Center
Roswell Park Cancer Institute	Mayo Clinic Jacksonville
Steacie Institute for Molecular Sciences	Picower Institute for Medical Research
	Research Triangle Institute
U.S. Patent and Trade Office	Sloan-Kettering Institute for Cancer Research
Walter Reed Army Institute of Research	Southern Research Institute
	WCRI

chemical houses, chemical software developers, and the *Chemical Abstracts* Service. (Publishing houses serving chemists have not been included.) The indices name the Food and Drug Administration, the U.S. Patent and Trade Office, the National Institutes of Health, and the CHUL Research Center. All of them offer to organic chemists jobs that are permanent but mostly nontenured. Finally, the name index lists the scientific staffing agencies named in Chapter 3. Some of these temporary-employment agencies operate many North American offices or affiliations. For that reason the geographic index omits these agencies, while the name index gives no address but presents one or more telephone numbers instead.

Contents of Each Entry

The name index gives the names and mailing addresses of the institutions included (Table B-2). It provides fax numbers and e-mail addresses because many employers ask job seekers to submit applications and résumés electronically. These numbers and addresses generally belong to personnel departments or individual recruiters, but may, especially in the case of a small company, belong to the company president or another high official. Many employers invite applicants to explore their job banks through job lines, and these telephone numbers are included.

Large companies or institutions employ chemists at many sites but keep only one Web site or accept applications at only one location. In such cases, only the first entry for the organization contains the URL or the place to write.

Table B-2 A Typical Entry in the Name Index

Name of company or other institution
Street and suite
P.O. box
City, state or province, postal code and country, if Canada
Facsimile number
E-mail address
Job hotline
Address of Internet site with note of job bank
kind of company or institution if not a FIPCO
how to apply: by fax, surface or e-mail or by completing an online form; note of whether the orgnization prefers or discourages any means of application or inquiry
note of stock options or purchase program
notice of co-operative programs and summer jobs or internships

Excluded Information

Five other kinds of data make no appearance in the indices.

1. The name index generally omits telephone numbers because many employers explicitly discourage calls from job hunters. However, some companies including temporary staffing agencies encourage phone calls, so the name index includes their numbers.
2. It names few personnel recruiters for either of two reasons. Small companies employ no staffers dedicated to recruiting. Alternatively, rapid staff turnover or changing work assignments makes the recruiters' names irrelevant. Consequently, addressing them risks sending applications astray.
3. Both indices neglect companies devoted exclusively to generic drugs or contract manufacture. To produce bulk drugs and large quantities of pharmaceutical intermediate compounds, such businesses employ experienced manufacturing chemists or chemical operators. The latter personnel are neither baccalaureate chemists nor researchers, but are high school graduates with special on-the-job training.
4. Also omitted are firms striving to identify suitable biological targets for drug therapy. Such firms, also known as pharmaceutical companies, do no chemical development or discovery research. Consequently, they employ no organic chemists as researchers.
5. For the same reason, the indices generally exclude contract research organizations limited to clinical development of drug candidates. Contract

research organizations are known as pharmaceutical companies, but their businesses begin where preclinical discovery and chemical development end.

To maximize the numbers and kinds of potential employers presented here and in preceding chapters, this book uses a broad, elastic definition of the pharmaceutical industry. In one sense, however, the present index is not complete, for it makes no effort to specify the many other kinds of companies that employ organic chemists. These scientists and other kinds of chemists find work in perfumeries, for example, and in animal-health and agricultural chemical firms. Readers interested in seeking work there can use the same methods for finding openings that they bring to a search for jobs in the human pharmaceutical industry.

"...a tide...leads on to fortune" (Shakespeare)

Act quickly on the knowledge offered here, for its accuracy decreases with time. Growing companies shed smaller quarters and move to new addresses. Mergers and acquisitions soon make corporate names unrecognizable. Some start-up companies leave the pharmaceutical business overnight. Abrupt departures can occur when a regulatory agency denies approval to the only drug prospect that the company has. Telephone and fax numbers change, while new area codes arise to meet demand. Home pages acquire new addresses as old URLs become obsolete.

To update your knowledge of the job market, you'll need to read *Chemical and Engineering News* and *Science* weekly. Alternatively, you can consult the online job banks that the ACS, the American Association for the Advancement of Science, and other organizations maintain (Table 7-3). More than 190 indexed employers also keep online job banks, while a dozen offer job hotlines. Individual entries within the name appendix note both job banks and lines.

Sources

Most of the indexed information originates in advertisements to hire chemists or to sell goods or services. The latter ads make it clear that their sponsors employ organic chemists as discovery or development researchers. For example, a commercial offer to make ^{14}C-labeled drugs for metabolism studies requires synthetic chemists trained in handling radioactive compounds.

Readings of *Nature* and *The Scientist* provided some of the data related here, as did *Canadian Chemical News*. Two periodicals alone—*Chemical and Engineering News* and *Science*—furnished most of the indexed information.

More than 500 relevant chemical employers advertised for chemists during the researching and writing of this book (1997–1999), but not all employers of such scientists did so. To increase coverage, therefore, information concerning such employers but published elsewhere supplements data from want ads. Chemical journals and industry newsletters name some of the companies listed by the indices. Informative publications include *Arzneimittel Forschung, Anti-cancer Drug Design, Antiviral Chemistry & Chemotherapy, Bioorganic and Medicinal Chemistry, Bioorganic and Medicinal Chemistry Letters, Current Medicinal Chemistry, Current Pharmaceutical Design, Drug News and Perspectives, Emerging Pharmaceuticals*, the *European Journal of Medicinal Chemistry*, the *Journal of Medicinal Chemistry, Medicinal Chemistry Research, Organic Process Research and Development, SCRIP,* and *SCRIP Magazine*. *Yahoo!* offers at its Web site (www.yahoo.com) a sweeping list of corporate names and Internet addresses. Also contributing corporate names and addresses are the membership lists of the Pharmaceutical Research and Manufacturers Association (www.phrma.org), the Synthetic Organic Chemical Manufacturers Association (www.socma.com), and the Pharmaceutical Manufacturers Association of Canada (www.pmac-acim.org). Companies involved with molecular diversity, solid-phase organic synthesis, robotics, and automation can also be found online (www.5z.com). U.S. Patents furnish a few company particulars as well as the inventors' names; and the weekly *Patent Gazette* gives notice of which patents issue when. A Web site (http://rx-aid.com) lists the names, e-mail addresses, and URLs for Canadian pharmaceutical companies.

Cambridge NeuroScience, Inc.
CambridgeSoft Corp.
CarboMer, Inc.
Chemicus, Inc.
ChiRex
ChiRex Technology Center
Consensus Pharmaceuticals, Inc.
CPBD, Inc.
Cubist Pharmaceuticals
CytoMed Inc.
DuPont Pharmaceutical Co.
Eisai Research Institute
Enanta Pharmaceuticals
Epix Medical, Inc.
Eukarion, Inc.
GelTex Pharmaceuticals, Inc.
Genetics Institute, Inc.
Genzyme Corp.
GLSynthesis, Inc.
Hybridon, Inc.
Inotek Corp.
Lexigen Pharmaceuticals
LeukoSite, Inc.
Mass Trace
Metasyn, Inc.
Millennium Pharmaceuticals, Inc.
Mitotix, Inc.
Mosaic Technologies, Inc.
Myco Pharmaceuticals, Inc.
NitroMed
Organix, Inc.
Paratek Pharmaceuticals, Inc.
PerSeptive Biosystems, Inc.
Pharm-Eco Laboratories, Inc.
Pharmaceutical Peptides, Inc.
Phytera, Inc.
polyORGANIX Inc.
Praecis Pharmaceuticals Inc.
Procept, Inc.
ProScript, Inc.
Repligen Corp.
Research Biochemicals International
RSP Amino Acid Analogues, Inc.

SCRIPTGEN Pharmaceuticals, Inc.
Sepracor, Inc.
UCB Research, Inc.
Vertex Pharmaceuticals Inc.

Michigan

Ash Stevens, Inc.
Cayman Chemical Co.
Parke-Davis Pharmaceutical Research
Pharmacia & Upjohn
Upjohn Laboratories
Synthon Corp.
Zeeland Chemicals, Inc.

Minnesota

3M Pharmaceuticals
LKT Laboratories, Inc.
Mayo Clinic Rochester

Mississippi

Quality Chemicals, Inc.

Missouri

Gateway Chemical Technology
Monsanto Life Sciences Co.
Reliable Biopharmaceutical Corporation
Searle
Sigma Chemical Co.
Tripos, Inc.

Montana

Ribi ImmunoChem Research, Inc.

Nebraska

BioNebraska, Inc.

New Hampshire

Diatide, Inc.

New Jersey

Algroup Lonza
Austin Chemical Co., Inc.
Bristol-Myers Squibb
  Pharmaceutical Research
  Institute
Cambrex Corp.
CasChem, Inc.
Celgene Corp.
Chemical Design, Inc.
Chemo Dynamics, Inc.
ChemPacific Corp.
Ciba Pharmaceuticals Division
Coelacanth Corp.
Cosan Chemical Corp.
Creanova, Inc.
Cytogen Corp.
Davos Chemical Corp.
David Sarnoff Research Center and
  Orchid Biocomputers, Inc.
Degussa Corp.
DSM Fine Chemicals USA, Inc.
DuPont Pharmaceutical Co.
Enzon, Inc.
Hoechst Marion Roussel, Inc.
Hoffmann-La Roche
Indofine Chemical Company, Inc.
Innovative Scientific Services, Inc.
Intercardia Research Laboratories
Janssen Research Foundation
R. W. Johnson Pharmaceutical
  Research Institute
J-Star Research, Inc.
Merck Research Laboratories
Neurochemistry Research
Novartis Pharmaceuticals Corp.
Ohmeda Inc.
Palatin Technologies, Inc.
Pharmacopeia, Inc.
PharmaGenics, Inc.
Provid Research
Sandoz Pharmaceuticals Corp.
Schering-Plough Research Institute
Small Molecule Therapeutics
SST Corp.
Synaptic Pharmaceutical Corp.
Syntech Labs, Inc.
Trophix Pharmaceuticals, Inc.
Tyger Scientific, Inc.
Wyeth Ayerst Research
Xechem, Inc.
Yukong R & D Laboratories

New Mexico

Daylight Chemical Information
  Systems

New York

Albany Molecular Research
American Advanced Organics, Inc.
Barr Laboratories, Inc.
Binrad Chemicals
Bristol-Myers Squibb Co.
Cadus Pharmaceutical Corp.
College of Physicians and
  Surgeons, Columbia University
Coromed Inc.
Emisphere Technologies, Inc.
Fish and Neave
ImClone Systems Inc.
Innovir Laboratories, Inc.
L. Y. Research Corp.
Nepera, Inc.
Oncogene Science, Inc.
OSI Pharmaceuticals, Inc.
Paracelsian, Inc.
Pennie & Edmonds LLP
Picower Institute for Medical
  Research
Roswell Park Cancer Institute
Sloan-Kettering Institute for
  Cancer Research
Wyeth Ayerst Research

North Carolina

AAI
Becton Dickinson and Co.
Burroughs Wellcome Co.
Catalytica Pharmaceuticals, Inc.
FAR Research Inc.
GlaxoWellcome
*ICA*gen, Inc.
Magellan Laboratories, Inc.
Morflex, Inc.
MYCOsearch
Novalon Pharmaceutical Corp.
Pisgah Labs
Pharmaceutical Product
   Development, Inc.
Protein Delivery, Inc.
Research Triangle Institute
Sphinx Pharmaceuticals
Triangle Pharmaceuticals
Trimeris, Inc.
Wright Corp.

Ohio

Chemical Abstracts Service
Gliatech, Inc.
ISP Fine Chemicals
Isotec, Inc.
Kettering Medical Center
Procter & Gamble Co.
Quality Chemicals, Inc.
Research Organics
Ricerca, Inc.

Oklahoma

ZymeTx, Inc.

Oregon

Avi Biopharma
Chemica Technologies
Molecular Probes, Inc.

SyntheTech
TCI America

Pennsylvania

3-Dimensional Pharmaceuticals, Inc.
Adolor Corp.
Bearsden Bio, Inc.
BIOMOL Research Laboratories
Centocor
Cephalon, Inc.
Heico Chemicals, Inc.
Janssen Research Foundation
Luminide Pharmaceutical Corp.
Magainin Pharmaceuticals, Inc.
Merck Research Laboratories
Message Pharmaceuticals
Nycomed Amersham
Pressure Chemical Co.
PPG Industries, Inc.
Rhône-Poulenc Rorer Research
   and Development
Rütgers Organic Corp.
Sanofi Winthrop, Inc.
SmithKline Beecham
   Pharmaceuticals
Symphony Pharmaceuticals
Tripos, Inc.
Wyeth Ayerst Research

Rhode Island

Hoechst Celanese Corp.
Organomed Corp.

South Carolina

Eastman Chemical Co.
IRIX Pharmaceuticals, Inc.
Nipa Hardwicke, Inc.
Oakwood Products, Inc.
Ortec Inc.
Roche Carolina, Inc.
Schweizerhall Development Co.

Tennessee

Chattem Chemicals, Inc.
Eastman Chemical Co.
Zeneca Pharmaceuticals

Texas

ACCESS Pharmaceuticals
Alcon Laboratories, Inc.
Ambion, Inc.
Aronex Pharmaceuticals, Inc.
BioNumerik Pharmaceuticals, Inc.
Cancer Therapy & Research Center
Carbotek, Inc.
Chemicals Inc.
Diagnostics Systems Laboratories
Dixie Chemical Co., Inc.
Eastman Chemical Co.
Gene Medicine, Inc.
Ilex Oncology Inc.
Institute for Drug Development
Johnson & Johnson Medical, Inc.
PPG Industries, Inc.
Sachem
Texas Biotechnology Corp.
Zonagen, Inc.

Utah

Cognetix, Inc.
NPS Pharmaceuticals, Inc.

Virginia

Biotage
Insmed Pharmaceuticals, Inc.

Washington (State of)

Bristol-Myers Squibb Pharmaceutical Research Institute
Campbell and Flores LLP
CDR Therapeutics
Cell Therapeutics, Inc.
Darwin Molecular Corp.
ICOS Corp.
Immunex Corp.
Molecumetics
NeoRx Corp.
Oridigm Corp.
PathoGenesis Corp.
Prolinx, Inc.
ZymoGenetics Corp.

West Virginia

Rhône-Poulenc

Wisconsin

Aldrich Chemical Co., Inc.
Cambridge Chemical, Inc.
Clarion Pharmaceuticals, Inc.
Gilson, Inc.
PPD Pharmaco
Promega Corp.
Sigma-Aldrich
Specialty-Chem Products Corp.

Name Index of the North American Pharmaceutical Industry

AAI
1206 North 23rd Street
Wilmington, NC 28405
Fax: 910-251-6755
Internet site: www.pharmdev.com
chemical synthesis services

Abbott Laboratories
200 Abbott Park Road
Abbott Park, IL 60064-3500
Internet site with job bank:
 www.abbott.com
*solicits electronic résumés online
 for the pharmaceutical products
 division*
*offers summer jobs to undergraduate
 chemistry students; see Web site
 for details*

Abbott Laboratories
Human Resources, D39Y, A1
1401 Sheridan Rd
North Chicago, IL 60064-4000
chemical process development

Acadia Pharmaceuticals
3911 Sorrento Valley Boulevard
San Diego, CA 92121
Fax: 858-558-2872
E-mail: hr@acadia-pharm.com
Internet site: www.acadia-pharm.com
offers stock options

ACCESS Pharmaceuticals
2600 N. Stemmons Ferry, Suite 210
Dallas, TX 75252
Fax: 214-905-5101

Aceto Corp.
One Hollow Lane—Suite 201
Lake Success, New York 11042-1215
Fax: 516-627-6093
E-mail: aceto@compuserve.com
Internet site: www.aceto.com
*custom synthesis, FDA inspected
 facilities; chiral building blocks,
 carbohydrate intermediates,
 amino acid derivatives*

Adolor Corp.
371 Phoenixville Pike
Malvern, PA 19355
Fax: 610-889-2203
Internet site: www.adolor.com

Advanced Chem Tech, Inc.
5609 Fern Valley Road
Louisville, KY 40228
Fax: 502-968-1000
Internet site with job bank:
 www.peptide.com
*pharmaceutical intermediates, fine
 chemicals, custom radiochemical
 synthesis*

Advanced Medicine, Inc.
280 Utah Avenue
South San Francisco, CA 94080
Fax: 650-827-8690
E-mail: career@advmedicine.com

Advanced Science Professionals
Telephone: 800-555-8405
Fax: 800-886-7358
a temporary employment agency

Aerojet[1]
P.O. Box 13222
Sacramento, CA 95813
Fax: 800-700-9734
E-mail: jobs@aerojet.com
Internet site with job bank:
 www.aerojet.com
*develops, scales up, and
 manufactures pharmaceutical*

[1] A division of GenCorp.

intermediates and bulk actives; accepts e-mailed résumés in ASCII text

Affymax Research Institute[2]
3410 Central Expressway
Santa Clara, CA 95051
Fax: 408-730-1393
E-mail: ari_jobs@affymax.com
Internet site with job bank:
  www.affymax.com

Affymetrix[3]
3380 Central Expressway
Santa Clara, CA 95051
Fax: 408-481-0422
E-mail: hr@affymetrix.com
Internet site with job bank:
  www.affymetrix.com
accepts résumés by e-mail

Agouron Pharmaceuticals, Inc.[4]
3301 N. Torrey Pines Court
La Jolla, CA 92037
Fax: 858-622-3297
E-mail: jobs@agouron.com
Internet site with job bank:
  www.agouron.com
On-line open house: www.monster.com/agouron
offers stock options to permanent employees and summer jobs to undergraduate chemistry students; employs college students year-round

Alanex Corp.[5]
3550 General Atomics Court
San Diego, CA 92121-1194
E-mail: h-resources@alanex.com, and jobs@alanex.com
Internet site with job bank:
  www.alanex.com
a drug discovery company offers part-time employment to college students

Alanex Corp.
3301 North Torrey Pines Court
La Jolla, CA 92037-1022

Albany Molecular Research, Inc.
21 Corporate Circle
Albany, NY 12203
Fax: 518-452-5774
E-mail: hr@albmolecular.com
Internet site with job bank:
  www.albmolecular.com
contract chemical research in discovery and development; custom synthesis accepts résumés by fax, e-mail, or surface mail; offers guidelines for composing résumés and submitting them electronically; offers stock options.

Albemarle Corp.
451 Florida Street
Baton Rouge, LA 70801
Fax: 504-388-7848
Internet site: www.albemarle.com
advanced intermediates and pharmaceutical bulk actives, bromine fine chemicals

Alcon Laboratories, Inc.
P.O. Box 6600
Fort Worth, TX 76115
Fax: 817-551-4629
E-mail: ardis.kvare@alconlabs.com and careers@alcon.com
Internet site: www.alconlabs.com
researches, develops, manufactures, and markets ophthalmic prescription drugs; Web site summarizes benefits

Alcon Laboratories, Inc.
6201 South Freeway
Fort Worth, TX 76134

Algroup Lonza[6]
79 Highway 22 East
P.O. Box 993
Annandale, NJ 08801

---

[2]Part of GlaxoWellcome.
[3]Part of GlaxoWellcome.
[4]Part of Warner-Lambert.
[5]A subsidiary of Agouron Pharmaceuticals, Inc.

[6]Formerly Lonza, Inc.

Fax: 908-730-1546
Internet site: www.lonza.com
*pharmaceutical intermediates:
amino acids and peptides, chiral
compounds; fine chemicals, and
process development*

Aldrich Chemical Co., Inc.[7]
P.O. Box 355
Milwaukee, WI 53201-0355
Internet site with job bank:
 www.sial.com
fine chemicals

Allelix Biopharmaceutical, Inc.
6850 Goreway Drive
Mississauga, Ontario
L4V 1V7 Canada
Fax: 905-677-5344
E-mail: hr@allelix.com
Internet site: www.allelix.com

Allergan Inc.
2525 Dupont Drive
Irvine, CA 92715-1599
Fax: 714-246-4220
Internet site with job bank:
 www.allergan.com
*invites résumés by fax and surface
mail and accepts them online; Web
site details benefits*

Alpha-Beta Technology
One Innovation Drive
Worcester, MA 01605
Fax: 508-754-2579
Internet site: www.alpha-beta.com
complex carbohydrates as drugs

The Althexis Company, Inc.
1365 Main St.
Waltham, MA 02451-1624
a start-up pharmaceutical company

Altus Biologics, Inc.
40 Allston Street
Cambridge, MA 02139
Fax: 617-499-2480

Internet site: www.altus.com
*contract discovery and development
research in enzyme-catalyzed
reactions for pharmaceutical
manufacturing
names of three references should
accompany résumés*

Ambion, Inc.
2130 Woodward Street
Suite 200
Austin, TX 78744-1832
E-mail: moinfo@ambion.com
Internet site with job bank:
 www.ambion.com
*a biotechnology company employing
synthetic organic chemists
e-mails notices of changes in jobs list*

American Advanced Organics, Inc.
P.O. Box 11170
Syracuse, NY 14218
Fax: 315-477-5819
Internet site: www.aaorganics.com
E-mail: shultis@aaorganics
*a custom-synthesis house serving the
drug industry*

American Chemical Society
1155 16th Street N. W.
Washington, DC 20036
Fax: 202-872-4077

American Peptide Company, Inc.
777 East Evelyn Avenue
Sunnyvale, CA 94086
Fax: 408-733-7603
E-mail: apc@americanpeptide.com
Internet site with job bank:
 www.americanpeptide.com
*peptide custom synthesis
solicits résumés by fax or e-mail*

Amersham Pharmacia Biotech Inc.
303 East 17th Avenue, Suite 303
Denver, CO 80203
Fax: 303-861-8147
E-mail: recruit@hrsi.com
Internet site: www.apbiotech.com
*custom synthesis of oligonucleotides
and radiolabeled nucleotides.*

---

[7]Part of the Sigma-Aldrich Corporation. See also the Sigma Chemical Co.

relocating in Summer, 1999, to Piscataway, NJ

Amgen
3200 Walnut Street
Boulder, CO 80301
E-mail: jobs@amgen.com
Job hotline: 800-446-4007
Internet site: www.amgen.com
accepts e-mailed résumés every 6 months (ASCII text, with résumé as the subject), and solicits résumés online

Amgen
Amgen Center
1840 De Haviland Drive
Thousand Oaks, CA 91319-2569
Fax: 805-499-9981 or 805-447-1985

Amylin Pharmaceuticals, Inc.
9373 Towne Centre Drive
San Diego, CA 92121
Fax: 858-552-2212
Internet site with job bank: www.amylin.com
solicits e-mailed résumés (as text not as attachments)

Ancile Pharmaceuticals
10555 Science Center Drive, Suite B
San Diego, CA 92121
isolates and develops botanical drugs, specializing in traditional Chinese medicines

Ångstrom Pharmaceuticals, Inc.
11772 Sorrento Valley Road, #101
San Diego, CA 92121

AnorMED, Inc.
20641 Logan Avenue, Suite 206
Langley, British Columbia
V3A 7R3 Canada
E-mail: information@cmdf.com
Internet site with job bank: www.anormed.com
specializing in coordinated chemicals as drugs; e-mails notices of job postings

Antex Pharma, Inc.[8]
300 Professional Drive
Gaithersburg, MD 20879
Fax: 301-590-1251
E-mail: info@antexbiologics.com
Internet site: www.antexbiologics.com

Apotex Inc.
150 Signet Drive
Weston, Ontario
M9L 1T9 Canada
Fax: 416-401-3828
E-mail: mhunnise@apotex.com
Internet site with job bank: www.apotex.com
fine chemicals; research, development and manufacturing of generic pharmaceuticals

Apotex Fermentation, Inc.[9]
40 Scurfield Boulevard
Winnipeg, Manitoba
R3Y 1G4 Canada
Fax: 204-488-4063
E-mail: hr@apoferm.mb.ca
Internet site with job bank: www.apoferm.mb.ca
bulk pharmaceuticals—research, development, and manufacturing accepts résumés by e-mail, offers cooperative positions

Arena Pharmaceuticals, Inc.
6166 Nancy Ridge Drive
San Diego, CA 92121
Fax: 858-453-7210
E-mail: nheeley@arenapharm.com
an early-stage drug discovery company offering stock options to employees
asks that each applicant send a résumé and the names of three references by fax or e-mail

Ares Advanced Technologies, Inc.[10]
280 Pond Street

[8] A subsidiary of Antex Biologics, Inc.
[9] A subsidiary of Apotex, Inc.
[10] The research and development arm of the Ares Serono Group.

Randolph, MA 02368
Fax: 781-963-6381
Internet site: www.serono-usa.com

Argonaut Technologies, Inc.
887 Industrial Road, Suite G
San Carlos, CA 94070
Fax: 650-598-1359
E-mail: hrinfo@argotech.com
Internet site with job bank:
   www.argotech.com
*proprietary synthesizers,
   combinatorial chemistry*

Ariad Pharmaceuticals, Inc.
26 Landsdowne Street
Cambridge, MA 02139
Internet site: www.ariad.com

Aronex Pharmaceuticals, Inc.
8707 Technology Forest Place
The Woodlands, TX 77381
Fax: 281-364-4636
E-mail: jelliott@aronex-pharm.com
Internet site: www.aronex.com

ArQule, Inc.
200 Boston Avenue
Medford, MA 02155
Fax: 781-395-4438
Internet site with job bank:
   www.arqule.com
E-mail: resumes@arqule.com

Array BioPharma
1885 33rd St.
Boulder, CO 80301
Fax: 303-449-5376
E-mail: dkreider@arraybiopharma.com
Internet site with job bank:
   www.arraybiopharma.com
*a drug discovery company
offers summer jobs to undergraduate
   chemistry students*

Ash Stevens, Inc.
5861 John C. Lodge Freeway
Detroit, MI 48202-3398
E-mail: info@ashstevens.com
Internet site: www.ashstevens.com
cGMP process development

Ash Stevens, Inc.
18655 Krause Street
Riverview, MI 48192
*process development for
   pharmaceutical chemicals*

Astra Arcus USA, Inc.
3 Biotech
1 Innovation Drive
Worcester, MA 01605

Astra Research Center Boston[11]
128 Sidney Street
Cambridge, MA 02139-4239
Fax: 617-576-3030 or 617-576-4668
Internet site with job bank:
   www.AstraZeneca-us.com

Astra Research Centre Montréal
7171 Frédérick-Banting
Ville St-Laurent, Québec
H43 1Z9 Canada

Athena Neurosciences, Inc.[12]
800 Gateway Boulevard
South San Francisco, CA 94080
Fax: 415-877-8370
E-mail: eliebler@best.com
Internet site: www.athenaneuro.com
custom synthesis, process research

AntheroGenics, Inc.
Chemistry Department
3065 Northwoods Circle
Norcross, GA 30071

Aurora Biosciences Corp.
11010 Torreyana Road
San Diego, CA 92121
Fax: 858-404-6720
E-mail: hr@aurorabio.com
Internet site with job bank:
   www.aurorabio.com
*solicits e-mailed résumés and
   cover letters; sometimes offers
   summer jobs to chemistry
   students*

[11] Part of AstraZeneca.
[12] A subsidiary of Elan Corporation, plc.

Austin Chemical Company, Inc.
1565 Barclay Boulevard
Buffalo Grove, IL 60089
Fax: 847-520-9160
Internet site: www.austinchemical.com
markets custom synthesis to the pharmaceutical industry

Austin Chemical Company, Inc.
24 South Holmdel Road
Holmdel, NJ 07733
Fax: 908-946-8682

Avi Biopharma
4575 SW Research Way, Suite 200
Corvallis, OR 97333
Fax: 503-754-3545
E-mail: avi@avibio.com
Internet site: www.antivirals.com
custom synthesis with oligomers a specialty, solid-phase organic synthesis, combinatorial chemistry

Axcan Pharma Inc. R&D
3885 Industrial Blvd.
Laval, Québec
H7L 4S3 Canada
Fax: 514-669-5161
Internet site with job bank:
www.axcan.com

Axiom Biotechnologies, Inc.
3550 General Atomics Court
San Diego, CA 92121-1194
Fax: 858-455-4501
E-mail: mmcandrew@axiombio.com
Internet site: www.axiombio.com
offers proprietary high-throughput pharmacological screening; needs computational chemistry and chemometrics

AXYS Pharmaceutical Corp.[13]
180 Kimball Way
South San Francisco, CA 94080
Internet site with job bank:
www.axyspharm.com

E-mail: human_resources
@axyspharm.com
solicits e-mailed résumés and describes benefits on Web site

Bachem California, Inc.[14]
P.O. Box 3426
Torrance, CA 90510
Internet site: www.bachem.com
solution or solid-phase custom synthesis: amino acids, peptides, and other organic compounds; pharmaceutical intermediates; cGMP

Barr Laboratories, Inc.
2 Quaker Road
P.O. Box 2900
Pomona, NY 10970
Fax: 914-353-3317
E-mail: hr@barrlabs.com
Internet site with job bank:
www.barrlabs.com
a generic drug manufacturer solicits e-mailed résumés

BASF Bioresearch Corp.[15]
100 Research Drive
Worcester, MA 01605
Fax: 508-755-8511
E-mail: basf@ssihiringsolutions.com
Internet site with job bank:
www.basf.de
solicits résumés online and by e-mail or fax

Baxter Healthcare[16]
Cardiovascular Group
17221 Red Hill Avenue
Irvine, CA 92614
Fax: 714-250-3487
Internet site with job bank:
www.baxter.com
chemical modifications of surfaces holding bioactive substances; Web site lists universities where Baxter

---

[13]Formerly the Arris Pharmaceutical Corp.
[14]Part of Bachem AG in Switzerland.
[15]A subdivision of the BASF Group and part of BASF Pharma.
[16]Part of Baxter International, Inc.

*recruits; offers summer jobs to
undergraduate chemistry students*

Bayer Corp.
Pharmaceutical Recruiter
Pharmaceutical Division
400 Morgan Lane
West Haven, CT 06516
Fax: 203-812-3249
E-mail: pharma.hr@bayer.com
Internet sites: www.bayerpharmana.
  com (with job bank) and
  www.bayerus.com; see also
www.careermosaic.com/cm/bayer/
*solicits résumés by e-mail, fax, or
online; asks for salary history and
two copies of mailed résumés
offers summer jobs and cooperative
positions to undergraduate
chemistry students*

Bearsden Bio, Inc.
34 Mt. Pleasant Dr.
Aston, PA 19014

Becton Dickinson and Company
Becton Dickinson Research Center
21 Davis Drive
Research Triangle Park, NC 27709
Fax: 919-549-7572
Internet site with job bank:
  www.bdms.com
*diagnostic products; peptide
  chemistry
offers stock options; Web site details
  other benefits
offers summer jobs and cooperative
  positions*

Berlex Biosciences, Inc.[17]
15049 San Pablo Avenue
Richmond, CA 94804-0099
Internet site: www.betaseron.com/
  berlex.html

Beta Chem
10300 Howe Dr.
Leawood, Kansas 66206

Fax: 913-541-8808
E-mail: El-Sherif@worldnet.att.net
Internet site: www.betachem.com
custom synthesis of radiochemicals

Beta Chemicals, Inc.
25 Science Park
New Haven, CT 06511
Fax: 203-786-5437
E-mail: ctsynteck@aol.com
*custom synthesis, process
  development*

Binrad Chemicals
50 N. Harrison Avenue
Congers, NY 10920
*contract research and custom
  synthesis*

Biochem Therapeutic, Inc.[18]
275 Armand Frappier Boulevard
Laval, Québec
H7V 4A7 Canada
Fax: 514-978-7992
E-mail: cv@biochem-pharma.com
Internet site: www.biochem-
  pharma.com

BioCryst Pharmaceuticals Inc.
2190 Parkway Lake Drive
Birmingham, AL 35244
Fax: 205-444-4640
E-mail: HumanResources
  @biocryst.com
Internet site: www.biocryst.com
solicits online résumés

Biogen, Inc.
14 Cambridge Center
Cambridge, MA 02142
Fax: 617-679-2546
E-mail: resumes@biogen.com
Internet site with job bank:
  www.biogen.com
*solicits online résumés; subsidizes
  employee purchases of Biogen
  stock; details this and other
  benefits at Web site, which gives
  links to internship information*

---

[17]U.S. subsidiary of the German company Schering AG.

[18]A subsidiary of Biochem Pharma.

BioGenex Laboratories
4600 Norris Canyon Road
San Ramon, CA 94583
Fax: 925-275-0580
E-mail: bashkerb@biogenex.com
Internet site with job bank:
www.biogenex.com
histology and cytology diagnostic products

BIOMOL Research Laboratories
5100 Campus Drive
Plymouth Meeting, PA 19462-1132
Fax: 610-941-9252
E-mail: chemistry@biomol.com
Internet site: www.biomol.com
fine chemicals, pharmaceutical intermediates, custom synthesis sometimes offers summer jobs to undergraduate students of chemistry

BioNebraska, Inc.
3820 NW 46th Street
Lincoln, NE 68524-1637
Fax: 402-470-2345
E-mail: bionebraska@compuserve.com

BioNumerik Pharmaceuticals, Inc.
8122 Datapoint Drive
Suite 1250
San Antonio, TX 78229

Biosome, Inc.[19]
701-3 Cornell Business Park
Wilmington, DE 19801
E-mail: SASporn@ABSbioreagents.com
develops technology to detect enzyme inhibitors and receptor antagonists

Biotage[20]
1500 Avon Street Extended
Charlottesville, VA 22902
E-mail: valorz@cfw.com
Internet site: www.dyax.com

automated HPLC for purification of organic libraries

Bohdan Automation, Inc.
562 Bunker Court
Vernon Hills, IL 60061-1831
Fax: 847-680-1199
E-mail: hr@bohdan.com
Internet site with job bank:
www.bohdan.com
apparatus for automated organic synthesis; solicits applications by surface mail or e-mail; invites job seekers to submit a personnel profile online

Borregard Synthesis, Inc.[21]
9 Opportunity Way
Newburyport, MA 01950
Fax: 508-465-2057
pharmaceutical intermediates, fine chemicals

Borregard Synthesis
721 Route 202/206
Bridgewater, NJ 08807
Fax: 908-429-1112

Boehringer Ingelheim Pharmaceuticals, Inc.
900 Ridgebury Road
P.O. Box 368
Ridgefield, CT 06877-0368
Fax: 781-663-2431
E-mail: bipi@bi-careers.com
Internet site: www.boehringer-ingelheim.com
asks job seekers to send résumés to BI Staffing Center, P.O. Box 534, Waltham, MA 02454–0534

Boehringer Ingelheim (Canada) Ltd.
Bio-Méga Research Division
2100 Cunard Street
Laval, Québec
H7S 2G5 Canada
Fax: 514-682-8434

---

[19]An affiliate of Analytical Biological Services, Inc.
[20]Part of Dyax Corporation.

[21]See also polyORGANIX, Inc.

E-mail: hrch@bio-mega.boeringer-ingelheim.ca

Brinks Hofer Gilson & Lione
455 N. Cityfront Plaza Drive, Suite 3600
NBC Tower
Chicago, IL 60611
Internet site: www.brinkshofer.com
intellectual property law firm employing Ph.D. chemists as patent coordinators, and reimbursing employees for law school tuition
sometimes offers summer jobs to undergraduate chemistry students

**Bristol-Myers Squibb Co.
Pharmaceutical Research Institute**
P.O. Box 4000
Princeton, NJ 08543-4000
Fax: 609-581-8841
E-mail: mailto:hr@usccmail.bms.com
Internet site with job bank: www.bms.com
asks for scannable résumés and offers guidelines for composing them; accepts e-mailed and online applications
awards stock options to eligible employees
offer internships at some locations: inquire at college@usccmail.bms.com

Bristol-Myers Squibb Co.
P.O. Box 4755
Syracuse, NY 13221-4755

Bristol-Myers Squibb Co.
Pharmaceutical Research Institute
3005 First Avenue
Seattle, WA 98121
Fax: 206-727-3606

Bristol-Myers Squibb Co.
Pharmaceutical Research Institute
P.O. Box 5101
5 Research Parkway
Wallingford, CT 06492-7661
Fax: 203-677-7762

**Bristol-Myers Squibb
Pharmaceutical Group**
2365 Côte-de-Liesse Road
Ville Saint-Laurent, Québec
H4N 2M7 Canada
Fax: 514-333-7943
E-mail: HR_PG_CANADA @ccmail.bms.com
asks for attachments in WP or MSWord formats; prefers candidates bilingual in French and English

Burroughs Wellcome Co.[22]
Division of Organic Chemistry
Research Triangle Park, NC 27709

Cadus Pharmaceutical Corp.
777 Old Saw Mill River Road
Tarrytown, NY 10591-6705
E-mail: hr@cadus.com
Internet site with job bank: www.cadus.com
solicits applications by e-mail or surface mail

Calbiochem-Novabiochem Corp.
10394 Pacific Center Court
San Diego, CA 92121
Internet site: www.nova.ch
custom peptide synthesis, combinatorial chemistry

Calyx Therapeutics, Inc.
3525 Breakwater Avenue
Hayward, CA 94545
Fax: 510-780-1020
seeks drugs in plants

Cambrex Corp.
1 Meadowlands Plaza
East Rutherford, NJ 07063
Fax: 201-462-5945
Internet site: www.cambrex.com
pharmaceutical intermediates and bulk actives

Cambridge Chemical, Inc.
N115 WI9392 Edison Drive

[22]Part of GlaxoWellcome.

Germantown, WI 53022
Fax: 414-251-5577
custom synthesis

Cambridge Isotope Laboratories
50 Frontage Road
Andover, MA 01810
Fax: 978-749-2768
Internet site: www.isotope.com
radiosynthesis

Cambridge NeuroScience Corp.
1 Kendall Square
Cambridge, MA 02139
Fax: 617-225-2741
E-mail: hr@cambneuro.com
Internet site with job bank:
 www.cambneuro.com
invites applicants to send résumés by fax or surface or e-mail

CambridgeSoft Corp.
875 Massachusetts Avenue
Cambridge, MA 02139
Fax: 617-491-8208
E-mail: jobs@camsoft.com
Internet site with job bank:
 www.camsci.com
solicits résumés by e-mail; employs chemists to develop, test, support, and sell software

Campbell and Flores LLP
4370 La Jolla Drive, Suite 700
San Diego, CA 92122
Fax: 858-597-1585
E-mail: ccampbell@candf.com
biotechnology patent law firm employing Ph.D. chemists as scientific advisors

Campbell and Flores LLP
1111 Third Avenue
Seattle, WA 98101

Cancer Therapy & Research Center
8122 Datapoint Drive
Suite 250
San Antonio, TX 78229
Fax: 210-616-5875

Job line: 210-616-5801
Internet site with job bank:
 www.ctrc.saci.org

CanSyn Chem. Corp.
200 College Street
Toronto, Ontario
M5S 3E5 Canada
Fax: 416-978-8605
E-mail: mckague@chem-eng.utoronto.ca
organic synthesis and consulting

CarboMer, Inc.
P.O. Box 721
Westborough, MA 01581
Fax: 508-898-0432
E-mail: carbomers@aol.com
Internet site: www.carbomer.com
custom synthesis, cGMP; fine chemicals; sometimes offers summer jobs to chemistry students

Carbotek, Inc.
16223 ParkRow
Houston, TX 77084
Fax: 281-578-8265
Internet site: www.carbotek.com
pharmaceutical intermediates, process development, fine chemicals

CasChem, Inc.[23]
40 Avenue A
Bayonne, NJ 07002
Fax: 201-437-2728
E-mail: ahansen@M1.cambrex.com
Internet site: www.caschem.com

Catalytica, Inc.
430 Ferguson Drive, Bldg. #3
Mountain View, CA 94043
Fax: 650-968-7129
E-mail: ek@mv.catalytica-inc.com
Internet site with job bank:
 www.catalytica-inc.com
*pharmaceuticals and intermediates; process research
solicits résumés by e-mail*

---

[23]An operating company of Cambrex Corporation.

Catalytica Pharmaceuticals, Inc.[24]
P.O. Box 1887
Greenville, NC 27834
Fax: 252-707-7731
E-mail: pharmaceuticals@catalytica-inc.com
contractual process development and scale-up of pharmaceutical intermediates

Cayman Chemical Co.
1180 E. Ellsworth Road
Ann Arbor, MI 48108
Fax: 734-971-3640
E-mail: hr@caymanchem.com
Internet site with job bank:
www.caymanchemical.com
supplies biochemicals like inhibitors, antagonists, and agonists to pharmaceutical and other researchers; accepts faxed or e-mailed résumés and cover letters regularly offers summer jobs to one or two undergraduate chemistry students

CB Research and Development, Inc.
524 First State Boulevard
Wilmington, DE 19804-6038
Fax: 302-994-7941
custom synthesis

C/D/N Isotopes, Inc.
88 Leacock Street
Pointe-Claire, Québec
H9R 1H1 Canada
Fax: 514-697-6148
E-mail: info@cdniso.com
Internet site: www. cdniso.com
custom synthesis of stable-isotope-labeled compounds

CDR Therapeutics
P.O. Box 19497
Seattle, WA 98109

Celgene Corp.
7 Powder Horn Drive
Warren, NJ 07059

[24]Part of Catalytica, Inc.

Fax: 732-271-4184
E-mail: hr@celgene.com
Internet site: www.celgene.com
start-up pharmaceutical company hiring M.S. and Ph.D. chemists for medicinal and process chemistry

CellGate, Inc.
552 Del Rey Avenue
Sunnyvale, CA 94086
Internet site: www.cellgateinc.com
Fax: 408-774-4561
applies proprietary technology to drug discovery and delivery; offers stock options

Cell Therapeutics, Inc.
201 Elliot Avenue West, Suite 400
Seattle, WA 98119
Job line: 800-656-2355
Fax: 206-284-6206
E-mail: resume@ctiseattle.com
Internet sites with job banks:
www.bio.com and
www.cticseattle.com
*invites applicants to fax or e-mail résumés or CVs, and solicits telephone calls and online résumés
offers stock options and an employee stock purchase program*

Centaur Pharmaceuticals, Inc.
484 Oakmead Parkway
Sunnyvale, CA 94086
Fax: 408-822-4341
E-mail: hr@centpharm.com
Internet site with job bank:
www.centpharm.com
solicits faxed or e-mailed résumés

Centocor
200 Great Valley Parkway
Malvern, PA 19355
Fax: 610-651-6330
E-mail: jobbank@centocor.com
Internet site with job bank:
www.centocor.com
*accepts applications by e-mail
employs protein chemists*

Cephalon, Inc.
145 Brandywine Parkway
West Chester, PA 19380-4245
E-mail: pvandenb@cephalon.com
Internet site with job bank:
   www.cephalon.com

Charkit Chemical Corp.
330 Post Road
Darien, CT 06820
Fax: 203-655-8643
Internet site: www.charkit.com
pharmaceuticals

Chattem Chemicals, Inc.
3708 St. Elmo Ave.
Chattanooga, TN 37409
manufactures fine chemicals and active pharmaceutical ingredients

Chemica Technologies
20332 Empire Ave., Suite F-3
Bend, OR 97701
Fax: 541-385-0390
E-mail: mailroom@chemica.com
Internet site: www.chemica.com
custom synthesis of specialty organic chemicals for the biotech, pharma and medical device industries; chemical development
offers "generous vacation time"

Chemical Abstracts Service
P.O. Box 3012
Columbus, OH 43210
Fax: 614-447-3816
E-mail: jobs@cas.org
Internet site: http://info.cas.org
invites applicants to mail, e-mail, or fax cover letters and résumés
employs B.S. and M.S. chemists to abstract and index technical documents like patents; needs readers of Japanese, French, German, and Korean

Chemical Design, Inc.[25]
Suite 120
200 Route 17S
Mahwah, NJ
Fax: 201-529-2443
Internet site with job bank:
   www.chemdesign.com
software

Chemical Design (West Coast) Inc.[26]
8380 Miramar Mall Suite 224
San Diego, CA 92121-2550
Fax: 858-678-8713

Chem-Impex Intl., Inc.
935 Dillon Drive
Wood Dale, IL 60191
Internet site: www.chemimpex.com
fine chemicals, especially amino acids and peptides; laboratory to pilot plant

Chemicals Inc.
12321 Hatcherville Road
Baytown, TX 77520
Fax: 281-576-5712
E-mail: akm@chemicalsinc.com
Internet site: www.chemicalsinc.com
custom manufacturing of specialty chemicals including pharmaceuticals

Chembridge Corp.
16981 Via Tazon, Suite G
San Diego, CA 92127
Fax: 858-451-7401
E-mail: chem@chembridge.com
Internet site: www.chembridge.com
custom chemical synthesis and manufacturing, screening libraries

Chembridge Corp.
1858 Johns Drive
Glenview, IL 60025

Chemicus, Inc.
109 School Street
Watertown MA 02172
Fax: 617-924-4917
E-mail: chemicus@chemicus.com
Internet site: www.chemicus.com

---

[25]Part of the Oxford Molecular Group.

[26]Part of the Oxford Molecular Group.

Chemo Dynamics, Inc.
3 Crossman Road South
Sayreville, NJ 08872
Fax: 732-721-6835
E-mail: info@chemodynamics.com
Internet site:
www.chemodynamics.com
custom synthesis, process development, combinatorial chemistry, and analog preparation

ChemPacific Corp.
240 Dr. Martin Luther King Jr. Blvd.
Newark, NJ 07102-1982
Fax: 973-504-8686
E-mail: weidl@worldnet.att.net
Internet site: www.chempacific.com
custom synthesis, contract research, and bulk pharmaceutical ingredients

ChemSyn Laboratories[27]
13605 West 96th Terrace
Lenexa, KS 66215-1297
Fax: 913-888-3582
Internet site with job bank:
www.chemsyn.com
radiosynthesis, process development, preclinical and clinical drug supplies

ChiRex[28]
Wellesley, MA
Internet site: www.chirex.com
a contract manufacturing organization carrying on process development: chiral chemicals, bulk active, chemical intermediates; cGMP; fine chemicals

ChiRex Technology Center, Inc.[29]
56 Roland Street
Boston, MA 02129
Fax: 617-628-9207
E-mail: RPettman@chirex.com

---

[27]A subsidary of Eagle-Pitcher Industries, Inc.
[28]Affiliated with Sepracor.
[29]A subsidiary of ChiRex, Inc.

sometimes offers summer jobs to undergraduate chemistry students

Chiron Corp.
4560 Horton Street
Emeryville, CA 94608
Fax: 510-655-9910
E-mail: jobs@cc.chiron.com
Internet site with job bank:
www.chiron.com
invites applications for specific positions by surface and e-mail offers stock options and an employee stock purchase program

Chiroscience R & D, Inc.[30]
1631 220th Avenue SE, Suite 101
Bothell, WA 98021
Fax: 425-489-8019
E-mail: personnel@chiroscience.com
Internet site with job bank:
www.chiroscience.com
combinatorial chemistry, solid-phase organic chemistry, robotics, and automation

Chugai Biopharmaceuticals, Inc.
6275 Nancy Ridge Drive
San Diego, CA 92121
Fax: 858-535-5973
E-mail: jmustelin@chugaibio.com
Internet site with job bank:
www.chugaibio.com
Ph.D.s with Japanese language skills receive special consideration; accepts résumés online

CHUL Research Center, RC-709
Laboratory of Molecular Endocrinology
2705 Laurier Boulevard
Québec City, Québec
G1V 4G2 Canada
Internet site: www.crchul.ulaval.ca

Clarion Pharmaceuticals, Inc.
565 Science Drive
Madison, WI 52711
Internet site: www.clarionpharma.com

---

[30]Formerly Darwin Molecular Corp.; a division of Chiroscience plc.

Clinical Micro Sensors, Inc.
126 West Del Mar Boulevard
Pasadena, CA 91105
Fax: 626-584-0909
E-mail: hr@microsensor.com
Internet site: www.microsensor.com
develops DNA microchips as diagnostics

CoCensys, Inc.
201 Technology Drive
Irvine, CA 92618
Fax: 934-753-6135
E-mail: jobs@cocensys.com.
Internet site with job bank:
   www.cocensys.com
offers stock options and an employee stock purchase plan; solicits salary history or requirement and CVs or résumés by mail, e-mail or fax

Codon Pharmaceuticals, Inc.
200 Perry Parkway
Gaithersburg, MD 20877

Coelacanth Corp.
279 Princeton-Hightstown Rd.
East Windsor, NJ 08520
Fax: 609-448-8299
E-mail: resume@coelachem.com
Internet site with job bank:
   www.coelachem.com
start-up biopharmaceutical company

Cognetix, Inc.
421 Wakara Way, Suite 201
Salt Lake City, UT 84108
Fax: 801-581-9555
Internet site: www.cognetix.com

College of Physicians and Surgeons
Columbia University
630 W. 168th Street
New York, NY 10032
Fax: 212-305-9882
E-mail: srb40@columbia.edu
employs a synthetic radiochemist to discover and develop positron-emitting radiotracers

CombiChem, Inc.
9050 Camino Santa Fe
San Diego, CA 92121
Fax: 858-530-9998
E-mail: bbosley@combichem.com
Internet site with job bank:
   www.combichem.com

CombiChem, Inc.
11099 N. Torrey Pines Road, Suite 200
La Jolla, CA 92037
Fax: 858-453-3102

CombiMatrix Corp.
887 Mitten Road, Suite 200
Burlingame, CA 94010
biochip technology to speed drug discovery
employs synthetic organic and combinatorial chemists

ComGenex USA, Inc.
Princeton Corporate Center
5 Independence Way
Princeton, NJ 08540
Fax: 609-520-0897
E-mail: info@comgenex.com
Internet site: www.comgenex.com
analog synthesis, plant extracts, compound libraries

Consensus Pharmaceuticals, Inc.
200 Boston Avenue
Medford, MA 02155
E-mail: consensus@consensus-pharm.com
Internet site with job bank:
   www.consensus-pharm.com

COR Therapeutics, Inc.
256 East Grand Avenue
South San Francisco, CA 94080
Fax: 650-615-9639
E-mail: hr@corr.com
Internet site with job bank:
   www.cortherapeutics.com
offers stock options, stock purchase plan, and public transit assistance

Coromed Inc.
Rensselaer Technology Park
185 Jordan Road

Troy, NY 12180-8343
Fax: 518-283-1807
E-mail: info@coromed.com
Internet site: www.coromed.com
comprehensive contract research services in medicinal chemistry and chemical development; custom synthesis

Cortex Pharmaceuticals
15231 Barranca Parkway
Irvine, CA 92618
Fax: 949-727-3657
E-mail: jmello@cortexpharm.com
Internet site: www.cortexpharm.com
offers stock options

Corvas International, Inc.
3030 Science Park Road
San Diego, CA 92121
Fax: 858-455-0457
E-mail: jobs@corvas.com
Internet site with job bank:
 www.corvas.com
solicits mailed or faxed letters of interest accompanied by CVs

Cosan Chemical Corp.[31]
400, 13th Street
Carlstadt, NJ 07072
Fax: 201-460-7525
E-mail: ahansen@M1.cambrex.com

Coulter Pharmaceutical
550 California Avenue, Suite 200
Palo Alto, CA 94306-1440
Fax: 650-553-2028
E-mail: jobs@coulterpharm.com
Internet site with job bank:
 www.coulterpharm.com
solicits résumés by surface or e-mail or by fax
in 1999 this company will move to South San Francisco

CPBD, Inc.
300 Putnam Avenue
Cambridge, MA 02139
Fax: 617-868-2538
E-mail: hchu@cpbd.com

Creanova Inc.[32]
220 Davidson Avenue
P. O. Box 6821
Somerset, NJ 08873-6821
Fax: 732-560-6686
Job hotline: 1-888-485-7562
Internet site with job bank:
 www.Huls.com
mfgr pharmaceutical intermediates
accepts résumés online, by electronic or surface mail or by fax; Web site offers guidelines for submitting résumés by fax or e-mail and lists benefits

Cubist Pharmaceuticals
24 Emily Street
Cambridge, MA 02139
Fax: 617-576-0232
E-mail: emee@cubist.com
Internet site with job bank:
 www.cubist.com
combinatorial chemistry, solid-phase organic chemistry, robotics, and automation
offers an employee stock purchase plan
sometimes offers summer jobs to undergraduate students of chemistry

CuraGen Corp.
322 E. Main Street
Branford, CT 06405
Fax: 203-401-3331
E-mail: jobs@curagen.com
Internet site with job bank:
 www.curagen.com
accepts mailed, faxed, or e-mailed applications; include CV, a statement of career goals and research interests, and three letters of recommendation
offers stock options

CV Therapeutics, Inc.
3172 Porter Drive
Palo Alto, CA 94304
Fax: 650-858-0287

---

[31]An operating company of Cambrex Corporation.

[32]Formerly Hüls America, Inc.

E-mail: Diane_Larsen@cvt.com
Internet site with job bank: www.cvt.com
*offers stock options and a stock
purchase plan*

Cypros Pharmaceutical Corp.
2732 Loker Avenue West
Carlsbad, CA 92008
Fax: 760-929-8038
Internet site: www.cypros.com
*solicits applications by fax; include
salary history and résumé
offers stock options*

Cytel Corp.
3525 John Hopkins Court
San Diego, CA 92121-1121
Fax: 858-452-1064
E-mail: Teresa_Stephenson
@cytelcorp.com
Internet site with job bank:
www.cytelcorp.com

Cytogen Corp.
600 College Road E.
Princeton, NJ 08540-5308
Fax: 609-987-1157
Internet site with job bank:
www.cytogen.com
*offers stock options and an employee
stock purchase plan*

Cytokinetics, Inc.
280 East Grand Avenue, Suite 2
South San Francisco, CA 94080
E-mail: hr@cytokinetics.com
posts jobs at www.biospace.com

CytoMed Inc.
840 Memorial Drive
Cambridge, MA 02139
Fax: 617-661-7364
E-mail: hr@cytomed.com
Internet site: www.cytomed.com
*solicits résumés by surface or e-mail
or by fax*

Dade Chemistry Systems[33]
Glasgow Business Community

[33]Part of Dade Behring.

P.O. Box 6101, Mail Stop 517
Newark, DE 19714
Fax: 302-631-0348
E-mail: dadechem@dadebehring.com
Internet site with job bank:
www.dadebehring.com
medical diagnostic products

**David Sarnoff Research Center
and Orchid Biocomputers, Inc.**
201 Washington Road
Princeton, NJ 08540-6449
Fax: 609-750-2250
Internet site with job bank:
www.sarnoff.com
*combinatorial chemistry, solid-phase
organic chemistry, robotics, and
automation; invites job-seekers to
create a personnel file online; Web
site lists schools where Sarnoff
recruits*

**Daylight Chemical Information
Systems**
419 East Palace Avenue
Santa Fe, NM 87501
Fax: 505-989-1200
E-mail: info@daylight.com
Internet site: www.daylight.com
*software development: high-
performance chemical
information processing tools*

Davos Chemical Corp.
464 HudsonTerrace
Englewood Cliffs, NJ 07362
Fax: 201-569-2201
E-mail: info@davos.com
Internet site: www.davos.com

Degussa Corp.
65 Challenger Road
Ridgefield Park, NJ 07660
Fax: 201-807-3183
Internet site: www.degussa.com
*optically pure intermediates for the
pharmaceutical industry*

Delmar Chemicals
9321 Airlie Street
LaSalle, Québec

H8R 2B2 Canada
Fax: 514-366-5665
E-mail: delmar@delmarchem.com
Internet site: www.adicq.qc.ca/delmar
develops commercial chemical processes for the pharmaceutical industry; offers custom synthesis of contaminants

Desmos Inc.
3550 General Atomics Court, 02-503
San Diego, CA 92121
Fax: 858-455-3962
Job line: 858-455-3708
E-mail: ccolegrove@desmos.com
Internet site with job bank:
www.desmos.com/~desmos/pr.html
invites protein chemists to mail résumés

Dharmacon Research, Inc.
3200 Valmont Road #5
Boulder, CO 80301
Fax: 303-415-9879
E-mail: info@dharmacon.com
Internet site: www.dharmacon.com
custom synthesis of RNA oligonucleotides
asks for résumé and introductory cover letter

Diagnostics Systems Laboratories, Inc.
445 Medical Center Boulevard
Webster, TX 77598-4217
Fax: 281-554-4220
E-mail: cpatton@dslabs.com
Internet site with job bank:
www.dslabs.com
develops, makes, and sells endocrine test kits

Diatide, Inc.
9 Delta Drive
Londonderry, NH 03053
Fax: 603-437-8977
E-mail: j_fletcher@diatide.com
Internet site with job bank:
www.diatide.com
invites résumés by fax and employment inquiries by telephone (603-437-8970) or e-mail
sometimes offers summer jobs to undergraduate students of chemistry

3-Dimensional Pharmaceuticals, Inc.
Eagleview Corporate Center
665 Stockton Drive, Suite 104
Exton, PA 19341
Fax: 610-458-8325
E-mail: job-posting@3dp.com
Internet site with job bank:
www.3dp.com
solicits cover letters and résumés by fax, surface mail or e-mail; asks for the names of three references

Diversa Chemical Technologies, Inc.
100 Jersey Avenue, Bldg. A-103
New Brunswick, NJ 08901
Fax: 732-545-7227
start-up company seeking "to create and exploit commercially important new organic chemistry"

Diversa Corp.[34]
10665 Sorrento Valley Road
San Diego, CA 92121
Fax: 858-623-5170
E-mail: information@diversa.com
Internet site: www.diversa.com

Dixie Chemical Company, Inc.
P. O. Box 130410
Houston, TX 77219-0410
Fax: 713-863-8316
E-mail: joydixie@aol.com
manufactures specialty chemicals including pharmaceuticals
employs organic chemists in process development

Dixie Chemical Company, Inc.
10701 Bay Area Boulevard
Pasadena, TX 77507

---

[34]Formerly RecombinantBioCatalysis, Inc.

The Dow Chemical Company
Applicant Center
P.O. Box 1655
Midland, MI 48641-1655
Internet site: www.dow.com
process development and contract manufacturing services; bulk pharmaceuticals; Web site specifies campuses where the company recruits workers; invites applications by surface mail, asking for e-mail addresses

DSM Fine Chemicals USA, Inc.[35]
217 Route 46 West
Saddle Brook, NJ 07662
Fax: 201-845-4406
E-mail: Info@dsmna.com
Internet site: www.dsmna.com

The DuPont Pharmaceuticals Company[36]
P.O. Box 80400, Room 2413
Wilmington, DE 19880-0400
Internet site with job bank:
www.dupontpharma.com
asks each applicant to send résumé only to job contacts by surface or e-mail; posts e-mail addresses on Web site
Web site summarizes the benefits conferred on permanent employees and details the summer jobs offered to undergraduate chemistry students

The DuPont Pharmaceuticals Company
Chemical Process R & D
Deepwater, NJ 08023

The DuPont Pharmaceuticals Company
331 Treble Cove Road
N. Billerica, MA 01862

---

[35]Part of DSM, a Dutch company with headquarters in Heerlen.
[36]Formerly the Dupont Merck Pharmaceuticals Co.

This site houses the radiopharmaceutical and medical imaging divisions

DuPont Pharma
2655 North Sheridan Way
Suite 180
Mississauga, Ontario
L5K 2P8 Canada

Eastman Chemical Company
Arkansas Eastman Division
P.O. Box 2357
Batesville, AR 72503-2357
Internet site with job bank:
www.eastman.com
fine organic chemicals, process research and development, pharmaceutical intermediates, custom synthesis; solicits scannable résumés by surface mail and offers tips for preparing them
employs interns and co-operative students

Eastman Chemical Company
Carolina Eastman Division
Attn: Employment
P.O. Box 1782
Columbia, SC 29202

Eastman Chemical Company
Corporate Headquarters
Attn: Employment
P.O. Box 1975
Kingsport, TN 37662-5215

Eastman Chemical Company
Texas Eastman Division
Attn: Employment
P.O. Box 7444
Longview, TX 75607

Eisai Research Institute
4 Corporate Drive
Andover, MA 01810-2441
Fax: 978-681-8731
E-mail: hr_eri@eisai.com
Internet site: www.eisai.co.jp

two letters of recommendation and a research summary should accompany a résumé

Emisphere Technologies, Inc.
765 Old Saw Mill River Road
Tarrytown, NY 10591
Fax: 914-347-2498
E-mail: info@emisphere.com
Internet site with job bank:
www.emisphere.com
solicits résumés by surface mail and fax; offers stock options among other benefits

Enanta Pharmaceuticals
750 Main St.
Cambridge, MA 02139
Fax: 617-621-9574

Endocyte, Inc.
1205 Kent Ave
West Lafayette, IN 47906
Fax: 765-463-9271
E-mail: info@endocyte.com
Internet site with job bank:
www.endocyte.com
hires synthetic organic chemists to develop therapeutic and diagnostic agents
solicits résumés by surface or e-mail or fax

EntreMed, Inc.
9600 Medical Center Drive, Suite 104
Rockville, MD 20850
Fax: 301-315-2437
E-mail: susanc@entremed.com
Internet site with job bank:
www.entremed.com
solicits applications by fax or surface or e-mail

Enzon, Inc.
20 Kingsbridge Road
Piscataway, NJ 08854-3969
Internet site with job bank:
www.enzon.com
solicits résumés by surface mail

EnzyMed, Inc.
2501 Crosspark Road
Suite C-150
Iowa City, IA 52242
Fax: 319-626-5410
E-mail: pmichels@enzymed.com
Internet site: www.enzymed.com
invites applications by surface mail
occasionally offers summer jobs to undergraduate chemistry students

Epix Medical, Inc.[37]
71 Rogers Street
Cambridge, MA 02142-1118
Fax: 617-250-6031
E-mail: info@epixmed.com
Internet site with job bank:
www.epixmed.com
invites cover letters and résumés by fax or surface mail; discourages e-mail submissions
offers stock options

Eukarion, Inc.
6F Alfred Cir.
Bedford, MA 01730
Fax: 781-275-0752
Internet site: www.eukarion.com

Evanston Northwestern Healthcare Research Institute
Northwestern University Medical School
1033 University Place, Suite 100
Evanston, IL 60201-3156
E-mail: g-krafft@nwu.edu

FAR Research Inc.
P.O. Box 2278
Morganton, NC 28680
custom manufacturing: fine chemicals and pharmaceutical intermediates, cGMP

FAR Research Inc.
2210 Wilhelmina Court N.E.
Palm Bay, FL 32905

---

[37]Formerly Metasyn, Inc.

FibroPharma, Inc.[38]
260 Littlefield Avenue
South San Francisco, CA 94080
Fax: 650-635-1514

Fish and Neave
1251 Avenue of the Americas
New York, NY 10020
Internet site: www.fishneave.com
employs newly graduated, Ph.D. organic and biochemists in patent law; trains them as patent agents; offers full law school tuition and a starting salary of $63,000.
offers summer internships to undergraduate students

Food and Drug Administration
Center for Drug Evaluation and Research
7520 Standish Place
Metro Park North 1, Rm. 225, HFD-64
Rockville, MD 20855
Fax: 301-827-2860
Job hotline: 301-827-4287
Internet site with job bank: www.fda.gov
posts salaries on Web site and places them in advertisements; employs chemists as regulatory review scientists as civil servants at grades GS12/13, paying $47 to 73K.

Food and Drug Administration
Center for Drug Evaluation and Research
5600 Fishers Lane, Rm. 6B-17; HFD-505
Rockville, MD 20857
Fax: 301-443-4880 and 301-827-3671

Food and Drug Administration
Center for Biologics Evaluation & Research
1401 Rockville Pike, HFM-60
Rockville, MD 20852-1448

---

[38]A subsidiary of FibroGen.

Food and Drug Administration
Center for Drug Evaluation and Research
Office of Pharmaceutical Science
Division of New Drug Chemistry III
Division of Antiviral Drug Products
Rockville, MD 20850

Galderma Research, Inc.
10835 Altman Row, Suite 250
San Diego, CA 92121

Gateway Chemical Technology
11810 Borman Drive
St. Louis, MO 63146
Fax: 314-991-2834
E-mail: RJKaufman@aol.com
Internet site: www.ww-ww.com/gateway/index.html
custom synthesis mg to kg, early-stage process research

GelTex Pharmaceuticals, Inc.
303 Bear Hill Road
Waltham, MA 02154
Fax: 617-290-5890
E-mail: awilson@geltex.com
Internet site with job bank: www.geltex.com
makes polymer hydrogels as effective but nonabsorbed drugs
asks for online résumés

Genelabs Technologies, Inc.
505 Penobscot Drive
Redwood City, CA 94063
Fax: 650-363-6080
E-mail: hr@genelabs.com
Internet site with job bank: www.genelabs.com
Web site details benefits including stock options; stock purchase plan; and help selecting, buying, and setting up employees' own computers
offers summer jobs to undergraduate chemistry students

GeneMedicine, Inc.
Kate Stankis, Vice-President for Human Resources

8301 New Trails Drive
The Woodlands, TX 77381-4248
Fax: 281-364-0858
E-mail: recruitmentmanager
@genemedicine.com
Internet site with job bank:
www.genemedicine.com
*solicits résumés by fax or e-mail;
grants stock options and operates
a stock purchase plan; Web site
summarizes benefits*

Genentech, Inc.
P.O. Box 1950
South San Francisco, CA 94083-1950
E-mail: jobs@gene.com
Internet site with job bank:
www.gene.com
*solicits résumés by surface or e-mail,
discouraging faxes; offers
guidelines for scannable printed
and suitable e-mailed résumés
Web site summarizes benefits,
including 6-week sabbatical
leaves to 6-year employees and
a stock purchase plan
offers summer jobs to undergraduate
chemistry students.*

General Intermediates of Canada, Inc.
17303 108th Avenue
Edmonton, Alberta
T5S 1G2 Canada
Fax: 403-483-2971
*custom synthesis, pharmaceutical
raw materials*

GeneSoft, Inc.
Two Corporate Drive
South San Francisco, CA 94080
*start-up biotechnology company
seeking small-molecule drugs that
interfere with gene expression
offers equity participation*

Genesys Pharma, Inc.
83 Scurfield Boulevard
Winnipeg, Manitoba
R3P 1J6 Canada
Fax: 204-488-6446
E-mail: genesys@pangea.ca
Internet site: www.pangea.ca/~genesys

Genetics Institute, Inc.[39]
87 Cambridge Park Drive
Cambridge, MA 02140
Fax (fine mode): 617-876-8847 or 617-498-8089
E-mail: jobs@genetics.com
Internet site: www.genetics.com
*solicits online résumés; asks that
faxed résumés be followed by
scannable ones sent by mail
Web site gives interviewing tips
offers summer internships,
recruiting minority students*

Gen-Probe, Inc.
10210 Genetic Center Drive
San Diego, CA 92121
Fax: 619-410-8001
E-mail: robinv@genprobe.com
Job line: 619-410-8020
Internet site: www.gen-probe.com
*develops, manufactures and
commercializes diagnostic products*

Gensia Sicor, Inc.
Research Department
9360 Towne Centre Drive
San Diego, CA 92121-3030
Fax: 858-622-5545
E-mail: human.resources
@gensiasicor.com
Internet site with job bank:
www.gensia.com
*develops, manufactures, and markets
pharmaceuticals; engages in
contract manufacture
solicits on-line and e-mailed
applications in the ASCII format
offers stock option, stock purchase,
and profit sharing plans; Web site
cites other benefits*

Genzyme Corp.
One Kendall Square
Cambridge, MA 02139
Fax: 617-374-7200, 617-252-7772

[39]A subsidiary of American Home Products.

E-mail: pharmaceuticals@genzyme.com
Internet site with job bank:
   www.genzyme.com
discovery and development research; custom manufacturing of pharmaceutical actives and intermediates
invites job seekers to send résumés and cover letters by surface mail; accepts on-line applications
offers a stock purchase plan; Web site lists other benefits

Geron Corp.
230 Constitution Drive
Menlo Park, CA 94025
Fax: 650-473-7750
E-mail: hr@geron.com
Internet site: www.geron.com
asks for résumés by surface or e-mail
offers stock incentive and stock purchase programs

Gilead Sciences, Inc.
333 Lakeside Drive
Foster City, CA 94404
Fax: 650-578-9264
E-mail: careers@gilead.com
Internet site with job bank:
   www.gilead.com
invites résumés by e-mail or fax; accepts CVs on-line
offers a stock purchase plan and posts other benefits on Web site

Gilson, Inc.
3000 W. Beltline Hwy.
P.O. Box 620027
Middleton, WI 53562-0828
Fax: 608-836-3620
E-mail: hr@gilson.com
Internet site with job bank:
   www.gilson.com
automated HPLC equipment for combinatorial libraries

Glaxo Wellcome, Inc.[40]
P.O. Box 13398
5 Moore Drive

Research Triangle Park, NC 27709
Job line: 919-483-2565
Fax: 919-315-1053
Internet site with job bank:
   www.glaxowellcome.com
invites inquiries about jobs through its Web site and prefers electronic applications
offers summer jobs including bachelor's- and master's-level opportunities

Gliatech, Inc.
23420 Commerce Park Road
Cleveland, OH 44122
Fax: 216-831-4220
Internet site: www.gliatech.com
asks for names of three references and résumé

GLSynthesis, Inc.
222 Maple Avenue
Shrewsbury, MA 01545
Fax: 508-752-4123
E-mail: glsyn@glsynthesis.com
Internet site: www.glsynthesis.com
custom synthesis: small organic compounds, natural products, heterocycles, nucleosides and nucleotides
offers profit sharing among other benefits

Glycomed
860 Atlantic Avenue
Alameda, CA 94501

Greenfield Laboratories[41]
P.O. Box 708
Greenfield, IN 96190
employs organic chemists in radiolabeling

Gryphon Sciences
250 East Grand Avenue, Suite 90
South San Francisco, CA 94080
Fax: 650-952-3055
E-mail: resume@gryphonsci.com
Internet site with job bank:
   www.gryphonsciences.com

---

[40]See also Burroughs Wellcome.

[41]Part of Eli Lilly and Company.

develops technology for rapid synthesis of functional proteins

Guilford Pharmaceuticals Inc.
6611 Tributary Street
Baltimore, MD 21224
Fax: 410-631-8181
E-mail: hr@guilfordpharm.com
Internet site with job bank:
  www.guilfordpharm.com
offers stock options and bonuses

Gyuran Laboratory & Consulting
35 Felan Crescent
Etobicoke, Ontario
M9V 3A2 Canada
Fax: 416-740-1593
E-mail: gyuran@myna.com
Internet site: www.myna.com/~gyuran
organic synthesis, contract research and development, synthesis and identification of impurities, chemical literature searching, preparation of Material Safety Data Sheets

Hatco
1020 King George Post Road
Ford, NJ 08863
Fax: 908-738-9385
E-mail: info@hatcocorporation.com
Internet site:
  www.hatcocorporation.com
process development, custom organic synthesis

Hauser, Inc.
5555 Airport Boulevard
Boulder, CO 80301-2339
Fax: 303-443-5842
Job line: 303-441-5841, extension 3000
Internet site: www.hauser.com
structure determination, extraction, and purification of natural products

Heico Chemicals, Inc.[42]
Route 611, Post Office Box 160
Delaware Water Gap, PA 18327

Fax: 717-421-9012
E-mail: jdoherty@m1.cambrex.com
Internet site:
  www.heicochemicals.com
supplier of pharmaceutical intermediates and bulk actives

Helios Pharmaceuticals
9800 Bluegrass Parkway
Louisville KY 40299
E-mail: ktaylor@m1.cambrex.com
Internet site with job bank:
  www.helios-pharma.com
asks for a detailed résumé and the names of three references

Hoechst Marion Roussel, Inc.
Route 202-206
P.O. Box 6800
Bridgewater, NJ 08807-0800
Fax: 908-231-4970
E-mail: jobs@hmri.com
Internet site: www.hmri.com
solicits scannable résumés; employs researchers in Bridgewater, NJ, Cincinnati, OH, and in Tucson, AZ
Web site details benefits and research organizational structure, giving application guidelines

Hoechst Celanese Corp.
500 Washington Street
Coventry, RI 02816
bulk pharmaceuticals

Hoechst Marion Roussel, Inc.
2110 East Galbraith Road
Cincinnati, OH 45215

Hoffmann–La Roche[43]
Human Resources
Bldg. 76/508
340 Kingsland Street
Building 76, 5th Floor
Nutley, NJ 07110-1199
Internet site: www.roche.com

---

[42] An operating company of Cambrex Corporation.

[43] A member of the Roche Group; *see also* Roche Bioscience.

The Humphrey Chemical Co., Inc.[44]
45 Devine Street
North Haven, Connecticut 06473
Fax: 203-287-9197
E-mail: jdoherty@m1.cambrex.com
Internet site:
    www.humphreychemical.com

Hybridon, Inc.
155 Fortune Boulevard
Milford, MA 01757
Fax: 508-482-7510
E-mail: jgreco@hybridon.com
Internet site with job bank:
    www.hydridon.com
*solicits résumés by surface or
    e-mail*

Ibis Therapeutics[45]
2292 Faraday Avenue
Carlsbad, CA 92008
Fax 760-603-2700
E-mail: hr@isisph.com
Job line: 760-603-3858
Internet site with job bank:
    www.ibisrna.com
*seeks drugs that bind RNA and
    chemists at all degree levels
asks for detailed CV, cover letter, and
    addresses of three references*

ICAgen, Inc.
P.O. Box 14487
Research Triangle Park, NC 27709
Fax: 919-941-0813
E-mail: info@icagen.com
Internet site with job bank:
    www.icagen.com
offers stock options

ICN Pharmaceuticals, Inc.
3300 Hyland Avenue
Costa Mesa, CA 92626
Fax: 714-641-7268
E-mail: hr-jobs@icnpharm.com
Internet site with job bank:
    www.icnpharm.com

*accepts résumés by e-mail, fax, or
    surface mail*

ICOS Corp.
22021 20th Avenue SE
Bothell, WA 98021
Fax: 425-481-0281
E-mail: hr@icos.com
Jobs hotline: 206-485-1900, extension
    2032
Internet site with job bank:
    www.icos.com

IDUN Pharmaceuticals
11085 N. Torrey Pines Road, Suite 300
La Jolla, CA 92037
Fax: 858-623-2765
Internet site: www.idun.com

Ilex Oncology Inc.
11550 IH-10 West, Suite 300
San Antonio, TX 78230
Fax: 210-949-8483
E-mail: jobs@ilexonc.com
Internet site with job bank:
    www.ilexoncology.com
offers stock options

ImClone Systems Inc.
180 Varick Street
New York, NY 10014
Fax: 212-645-2054
E-mail: hr@imclone.com
Internet site with job bank:
    www.imclone.com

Immunex Corp.
51 University Street
Seattle, WA 98101
Fax: 206-623-4572
E-mail: job-info@immunex.com
Internet site with job bank:
    www.immunex.com

ImmunoPharmaceutics, Inc.
11011 Via Frontera
San Diego, CA 92127

Indofine Chemical Company, Inc.
P.O. Box 473
Somerville, NJ 08876

---

[44]An operating company of the Cambrex Corporation.
[45]A division of ISIS Pharmaceuticals.

Fax: 908-359-1179
E-mail: chemical@indofinechemical.com
Internet site: www.indofinechemical.com
fine chemicals, custom synthesis, and contract manufacturing

Inex Pharmaceuticals Corp.
1779 West 75th Avenue
Richmond, British Columbia
V6P 6P2 Canada
Fax: 604-419-3200
E-mail: careers@inexpharm.com
Internet site with job bank: www.inex.com
employs summer, co-operative, and practicum students as well as interns

Inflazyme Pharmaceuticals, Ltd.
5600 Parkwood Way, Suite 425
Richmond, British Columbia
V6V 2M2 Canada
Fax: 604-279-8711
E-mail: info@inflazyme.com
Internet site with job bank: www.inflazyme.com
accepts résumés by fax or electronic or surface mail

Influx, Inc.
2201 West Campbell Park Drive
Chicago, IL 60612
a start-up pharmaceutical discovery company

Ingelheim Research Inc.[46]
2100 Cunard
Laval, Québec
H7S 2G5 Canada

Innovative Scientific Services, Inc.
810 N 2nd Avenue
Highland Park, NJ 08904-1833
Fax: 732-247-4977
E-mail: issi@pipeline.com

[46]See also Bio-Méga/Boehringer Ingelheim Research, Inc.

metabolite identification, purification, and isolation

Innovir Laboratories, Inc.
510 East 73rd Street
New York, NY 10021

Inotek Corp.
relocating from OH to MA, summer of 1999
Fax: 513-221-8079
pharmaceutical research and development

INSMED Pharmaceuticals, Inc.
800 East Leigh Street, Suite 206
Richmond, VA 23219
Fax: 804-828-6894
E-mail: boatright@insmed.com
Internet site with job bank: www.insmed.com
offers stock options

Integro Scientific Staffing
Gatehall Corporate Center—1
One Gatehall Drive
Parsippany, NJ 07054
Telephone: 973-267-6363
Fax: 973-267-2158
E-mail: resume@integrostaffing.com
Internet site: www.integrostaffing.com
a temporary employment agency placing chemists at all degree levels; has eight offices in New Jersey and nationwide affiliations

Intercardia Research Laboratories
8 Cedarbrook Drive
Cranbury, NJ 08512
Fax: 609-655-6930
E-mail: sofia@irl.intercardia.com
Internet site: www.intercardia.com

IntraBiotics Pharmaceuticals, Inc.
1245 Terra Bella Avenue
Mountain View, CA 94043
Fax: 408-991-6401
E-mail: info@intrabiotics.com
Internet site with job bank: http://www.intrabiotics.com
solicits résumés by surface mail; accepts queries by e-mail

IRIX Pharmaceuticals, Inc.
101 Technology Place
Florence, SC 29501
Fax: 843-673-0052
E-mail: kalaritis@msn.com
Internet site with job bank:
 www.irixpharma.com
offers equity ownership

IRORI
Torrey Pines Science Park
11025 N. Torrey Pines Road
La Jolla, CA 92037-1030
Fax: 858-546-3083
E-mail: info@irori.com
Internet site with job bank:
 www.irori.com
combinatorial chemistry; software development; asks for résumés by surface mail or fax
offers summer jobs to undergraduate chemistry students

Isis Pharmaceuticals
2292 Faraday Avenue
Carlsbad, CA 92008
Fax: 760-931-9639
E-mail: hr@isisph.com
Job line: 760-603-3858
Internet site with job bank:
 www.isip.com
invites résumés by surface mail

Isotec, Inc.
3858 Benner Road
Miamisburg OH 45342-4304
Fax: 937-859-4878
Internet site with job bank:
 www.isotec.com
custom synthesis of stable-isotope-labeled compounds; process development

Institute for Diabetes Discovery[47]
23 Business Park Drive
Branford, CT 06405
Fax: 203-315-4002

[47]One of the Institutes for Pharmaceutical Discovery.

E-mail: info@ipd-discovery.com
Internet site: www.ipd-discovery.com

Institute for Drug Development
14960 Omicron
San Antonio, TX 78245

ISP Fine Chemicals
1979 Atlas Street
Columbus, OH 43228
Fax: 704-846-9065
E-mail: info@ispcorp.com
Internet site: www.isp-pharma.com
pharmaceutical intermediates and bulk actives, GMP, fine chemicals, process development, and custom manufacturing

Janssen Research Foundation[48]
P.O. Box 16597
New Brunswick, NJ 08906-6597
(for positions in Spring House, PA)

Johnson & Johnson Medical, Inc.[49]
2500 Arbrook Boulevard
Arlington, TX 76014-3130
Fax: 817-784-5400

R. W. Johnson Pharmaceutical Research Institute[50]
Dept: PR-I
501 George Street, Room JH215
P.O. Box 16597
New Brunswick, NJ 08906-6597
(for positions in Raritan, NJ, and elsewhere)
Internet site with job bank: www.jnj.com
solicits scannable résumés, offering guidelines for preparing them, but does not scan faxed résumés; invites online submissions of résumés
offers internships and cooperative positions to students; Web site furnishes details

[48]Part of the Johnson & Johnson group of companies.
[49]Part of the Johnson & Johnson group of companies.
[50]Part of the Johnson & Johnson group of companies.

R. W. Johnson Pharmaceutical Research Institute[51]
3535 General Atomics Court, Suite 100
San Diego, CA 92121
Fax: 858-450-2008
E-mail: priljhr@prius.jnj.com
(for jobs in La Jolla)

J-Star Research, Inc.
3001 Hadley Road
South Plainfield, NJ
Telephone: 908-791-9100
Internet site: www.jstar-research.com
custom-synthesis house offering parallel synthesis
benefits include life insurance, membership in a health maintenance organization, and a 401k retirement plan

Karo Bio Inc.
P.O. Box 642311
San Francisco, CA 94164-2311
Internet site: www.karobio.se

Kelly Scientific Resources
140 Route 17, Suite 271
Paramus, NJ 07652
Telephone: 201-599-5959 or 914-683-0087
Fax: 201-599-8470
E-mail: diane_long@kellyservices.com
a temporary employment agency

Kettering Medical Center
3535 Southern Boulevard
Kettering, OH 45429
Fax: 937-297-8004
E-mail: millie_sherron@ketthealth.com
Internet site with job bank:
  www.ketthealth.com
employs radiochemists; offer tips for applying

Kinetix Pharmaceuticals, Inc.
200 Boston Avenue, Suite 4700
Medford, MA 02155
Fax: 781-391-5771

E-mail: hr@kinetixpharm.com
Internet site with job bank:
  www.kinetixpharm.com
solicits résumés by mail, fax, or e-mail; offers "a substantial equity position"

Kiva Genetics, Inc.
2735 Garcia Avenue
Mountain View, CA 94043
Fax: 650-934-9375
E-mail: rachel@kivagen.com
Internet site with job bank:
  www.kivagen.com
develops fluorescent dyes to label nucleic acids

Kosan Biosciences, Inc.
1450 Rollins Road
Burlingame, CA 94010
Fax: 650-343-2931
Internet site with job bank:
  www.kosan.com
genetically engineers microorganisms to produce polyketide lead compounds; Web site names directors of chemistry invites applicants to mail a cover letter, a résumé, and a list of references

Lab Support
26651 West Agoura Road
Calabasas, CA 91302
Telephone: 800-998-3332
Internet site: www.labsupport.com
a temporary employment company with 44 offices in 25 states
benefits include healthcare coverage, stock purchase, and 401k retirement plans

La Jolla Pharmaceutical Co.
6455 Nancy Ridge Drive
San Diego, CA 92121
Fax: 858-625-1055
E-mail: humanresources@ljpc.com
Internet site with job bank:
  www.ljpc.com
sometimes offers summer jobs to undergraduate chemistry students

---

[51]Part of the Johnson & Johnson group of companies.

LeukoSite, Inc.
215 First Street
Cambridge, MA 02142
Fax: 617-621-9349
E-mail: personnel@leukosite.com
Internet site with job bank:
    www.leukosite.com
asks for résumés by surface mail, offering more information by e-mail; asks for three references
offers summer jobs to undergraduate chemistry students

Lexigen Pharmaceuticals Corp.
125 Hartwell Avenue
Lexington, MA 02173
E-mail: info@lexigenpharm.com
Internet site: www.lexigen.com

Ligand Pharmaceuticals
10275 Science Centre Drive
San Diego, CA 92121
Fax: 858-550-5649
E-mail: jobs@ligand.com
Internet site with job bank:
    www.ligand.com
offers stock options

Eli Lilly & Co.[52]
Chemistry Recruiting
U.S. Recruiting and Staffing
Lilly Corporate Center
Indianapolis, IN 46285
Fax: 317-276-9422
E-mail: resume@lilly.com
Job line: 1-800-892-9121
Internet site with job bank:
    www.lilly.com
accepts résumés by fax, e-mail, or surface mail; offers guidelines for submissions; keeps scannable résumés 2 years in a database
offers summer jobs to undergraduate chemistry students

LKT Laboratories, Inc.
2233 University Avenue West
St. Paul, MN 55114-1629

Fax: 612-644-8357
E-mail: info@lktlabs.com
Internet site: www.lktlabs.com
custom synthesis and manufacturing; natural product isolations

Luminide Pharmaceutical Corp.
41 Great Valley Parkway
Malvern, PA 19355
asks for copies of major publications in addition to résumés

L. Y. Research Corp.
67-08 168th Street
Flushing, NY 11365

Magainin Pharmaceuticals, Inc.
5110 Campus Drive
Plymouth Meeting, PA 19462
Fax: 610-941-5399
Internet site with job bank:
    www.magainin.com

Magellan Laboratories, Inc.
P.O. Box 13341
Research Triangle Park, NC 27709
Fax: 919-481-1598
E-mail: jobs@magellanlabs.com
Internet site with job bank:
    www.magellanlabs.com
contract research: development and optimization of syntheses; synthesis of reference standards, markers, and degradants
asks for scannable résumé and salary sought
offers summer jobs to undergraduate chemistry students

Manpower Technical
New Jersey: 732-846-7658
E-mail: manpower@mail.idt.net
New York: 914-896-1354
E-mail: manpower@hvi.net
Internet site: www.manpower.com
temporary employment agency

Martek Biosciences Corp.
6480 Dobbin Road
Columbia, MD 21045

---

[52]See also Sphinx Pharmaceuticals, Tippecanoe Laboratories, and Greenfield Laboratories.

Fax: 410-740-2985
E-mail: martek2000@aol.com
Internet site: www.martekbio.com
*pharmaceutical products,
combinatorial libraries, stable
isotope-labeled compounds*

Maryland Research Laboratories
9900 Medical Center Drive
Rockville, MD 20850

Mass Trace
3-G Gill Street
Woburn, MA
Fax: 781-932-6775
E-mail: mstrace@ix.netcom.com
Internet site: www.masstrace.com
*custom synthesis of tracer
compounds labeled with stable
isotopes
occasionally offers summer jobs to
undergraduate chemistry students*

MAXIA Pharmaceuticals, Inc.
10835 Altman Row, Suite 250
San Diego, CA 92121
Fax: 619-824-1967
offers stock options

Maxygen
3410 Central Expressway
Santa Clara, CA 95051
Fax: 408-481-0385
E-mail: jobs@maxygen.com
Internet site with job bank:
www.maxygen.com

Mayo Clinic Jacksonville
Neurochemistry Research
4051 Belfort Road
Jacksonville, FL 32216
Fax: 904-296-4668
E-mail: careers@mayo.edu
Internet site with job bank:
www.mayo.edu

Mayo Clinic Rochester
Cancer Center/Pharmacology
 Department
200 First Street SW
Rochester, MN 55905

McCarter & English
Four Gateway Center
100 Mulberry Street
Newark, NJ 07102
Fax: (973) 624-7070
E-mail: sboege@mccarter.com
Internet site: www.mccarter.com
*New Jersey firm of 200+ attorneys
employing scientific/medical
analysts
Web site names five branches in four
other states and discloses its
lawyers' starting salary*

MDL Informations Systems, Inc.
14600 Catalina Street
San Leandro, CA 94577
Fax: 510-614-3679
E-mail: jobs@mdli.com
Internet site with job bank:
www.mdli.com
*software development; solicits
résumés and cover letters by
e-mail, fax, or surface mail
Web site offers a guide to job hunting*

MediChem Research, Inc.
12305 South New Avenue
Lemont, IL 60439
Fax: 630-257-4634
E-mail: internet@mcr.medichem.com
Internet site with job bank:
www.medichem.com
*custom synthesis, contract discovery
research, process development;
combinatorial chemistry services*

Medinox, Inc.
11555 Sorrento Valley Road, Suite E
San Diego, CA 92121
Fax: 858-793-4823
E-mail: info@medinox.com
Internet site with job bank:
www.medinox.com
*asks for a résumé and cover letter by
surface mail
offers a stock option program*

Merck Frosst Canada Inc.
16711 Trans Canada Highway
Kirkland, Québec

H9H 3L1, Canada
Fax: 514-428-4940
E-mail: hr_montreal@merck.com
Internet site: www.merckfrosst.ca
invites résumés by surface mail or fax, filing them for 6 months

Merck Research Laboratories
One Merck Drive, P.O. Box 100
Whitehouse Station, NJ 08889
Fax: 908-423-2592
E-mail: resume@merck.com
Internet site: www.merck.com
accepts résumés by fax, electronic or surface mail; offers guidelines for e-mailed résumés; acknowledges receipt of submissions

Merck Research Laboratories
Box 2000
Rahway, NJ 07065
CVs should include the names and addresses of three references

Merck Research Laboratories
P.O. Box 4
West Point, PA 19486

Merck Research Laboratories
Interex Research Division
2201 West 21st Street
Lawrence, KS 66049

Message Pharmaceuticals, Inc.
34 Mt. Pleasant Drive
Aston, PA 19014

Metabasis Therapeutics, Inc.
9390 Towne Centre Drive
San Diego, CA 92121-3015
Fax: 619-458-3504
E-mail: admin@mbasis.com

Metabolex, Inc.
3876 Bay Center Place
Hayward, CA 94545
Fax: 510-293-9090
E-mail: wanhalt@metabolex.com
Internet site with job bank:
  www.metabolex.com
offers stock options

MetaProbe, LLC
Pasadena, CA
start-up making contrast agents for diagnostic radiology
send résumés and cover letters to sreidy@rtech.com or to Stephen Reidy; 101 Wilmot, Suite 600; Tucson, AZ 85711-3365

MetaXen, LLC
650 E. Grand Avenue
South San Francisco, CA 94304
Fax: 650-553-8101
E-mail: webmail@metaxen.com
Internet site: www.metaxen.com

Microcide Pharmaceuticals, Inc.
850 Maude Avenue
Mountain View, CA 94043
Fax: 650-428-3566
E-mail: hr@microcide.com
Internet site with job bank:
  www.microcide.com

MitoKor
11494 Sorrento Valley Road
San Diego, CA 92121
Fax: 760-741-3737
E-mail: hr@mitokor.com
Internet site: www.mitokor.com

Millennium Pharmaceuticals, Inc.
238 Main Street
Cambridge, MA 02142
Fax: 800-370-6925
E-mail: millennium@webhire.com
Internet site with job bank:
  www.mlnm.com
on line applications receive the fastest considerations; résumés may be mailed to Millennium Pharmaceuticals, Résumé Processing Center, P.O. Box 789, Burlington, MA 01803
offers a stock purchase plan

Miravant Medical Technologies
7408 Hollister Avenue
Santa Barbara, CA 93117
Fax: 805-562-9975
E-mail: hr@miravant.com

Internet site with job bank:
www.miravant.com
*develops photodynamic drugs;
accepts e-mailed résumés*

Mitotix, Inc.
One Kendall Square, Building 600
Cambridge, MA 02139
Fax: 617-225-0005
E-mail: walton@mitotix.com
Internet site with job bank:
www.mitotix.com
*combinatorial chemistry, solid-phase organic synthesis
sometimes offers summer jobs to undergraduate chemistry students*

Molecular Probes, Inc.
P.O. Box 22010
4849 Pitchford Avenue
Eugene, OR 97402-0469
Fax: 541-344-6504
Internet site with job bank:
www.probes.com
*manufactures fluorescent probes for biomedical research
employs organic chemists to discover new products or find new applications for existing ones*

Molecular Simulations
9685 Scranton Road
San Diego, CA 92121
Fax: 858-458-0136
E-mail: jobs@msi.com
Internet site: www.msi.com
*development of software for rational drug design
offers cash bonuses to employee-authors of job-related articles as well as stock options and profit shares to them and other employees
sometimes offers summer jobs to undergraduate chemistry students*

Molecumetics
2023 120th Avenue NE, Suite 400
Bellevue, WA 98005
Fax: 425-646-8890
E-mail: mcwhisk@molecumetics.com
Internet site with job bank:
www.molecumetics.com
drug discovery using proprietary chemistry

Monsanto Life Sciences Co.[53]
800 N. Lindbergh Boulevard, Mail Zone C3SB
St. Louis, MO 63167
Fax 314-694-2014
E-mail: monsanto
@ssihiringsolutions.com
Internet site with job bank:
www.monsanto.com
*employs research chemists in Pharma Sector Discovery jobs in St. Louis and Chicago; Pharma process chemists in the Growth Sector work in Los Angeles
asks for scannable résumés by surface or electronic mail. Send documents to monsanto@aon-hros.com, with ad code in subject line; or mail to Monsanto—(ad code), P.O. Box 1262, Findlay, OH 45840
offers tuition reimbursement*

Morflex, Inc.
2110 High Point Road
Greensboro, NC 27403
Fax: 910-854-4058
Internet site: www.reillyind.com
cGMP, process development

Mosaic Technologies, Inc.
1106 Commonwealth Avenue
Boston, MA 02215
Internet site: www.mosaictech.com
clinical diagnostic and research products

MYCOsearch[54]
4727 Pine University Drive, Suite 400
Durham, NC 27707
Fax: 919-490-3745

---

[53]Monsanto employs organic chemists as researchers both in its St. Louis headquarters and its Chesterfield site, which houses G. D. Searle, its subsidiary.
[54]A subsidiary of OSI Pharmaceuticals.

isolation and identification of biologically active natural products elaborated by fungi; offers stock options

3M Pharmaceuticals
NAS Reply Service
Riverview Office Tower
8009 34th Avenue South
Minneapolis, MN 55425-1619
E-mail: careers@mmm.com
Internet site with job bank:
www.3M.com
offers a stock purchase plan

Nanogen, Inc.
10398 Pacific Center Court
San Diego, CA 92126
Fax: 858-546-7718
E-mail: hr@nanogen.com
Internet site with job bank:
www.nanogen.com
makes diagnostic products solicits résumés and cover letters by e-mail, fax, or surface mail

NARCHEM Corp.
3800 West 38th Street
Chicago, IL 60632
Fax: 773-376-8932
E-mail: info@narchem.com
Internet site with job bank:
www.narchem.com
contract research and development for the pharmaceutical industry solicits cover letters and résumés by fax or surface or e-mail

National Institutes of Health
National Institute on Drug Abuse
Division of Intramural Research
Psychobiology Section
P.O. Box 5180
Baltimore, MD 21224
Internet sites: www.nih.gov and
http://helix.niuh.gov.8001/oe/

National Institutes of Health
National Cancer Institute
Developmental Therapeutics Program
Division of Cancer Treatment, Diagnosis and Centers
Laboratory of Drug Discovery Research and Development
Building 1052, Room 121
Frederick, MD 21702-1201

National Institutes of Health
National Cancer Institute
Division of Basic Sciences
Laboratory of Medicinal Chemistry
Building 37
Bethesda, MD 20892

National Institutes of Health
National Cancer Institute
Division of Cancer Treatment
Developmental Therapeutics Program
Drug Synthesis and Chemistry Branch
Bethesda, MD 20892

National Institutes of Health
National Institute of Diabetes and Digestive and Kidney Diseases
Laboratory of Bioorganic Chemistry
Molecular Recognition Section
Bethesda, MD 20892-0810

National Institutes of Health
National Institute of Diabetes and Digestive and Kidney Diseases
Laboratory of Medicinal Chemistry
MSC 0815
8 Center Drive
Bethesda, MD 20892-0815

NATPRO, Inc.
3525 Breakwater Avenue
Hayward, CA 94545

NéoKimia, Inc.
3001 Twelfth North Avenue
Fleurimont, Québec
J1H 5N4 Canada
Fax: 819-820-6841
E-mail: pierre.deslongchamps
@courrier.usherb.ca
specializes in combinatorial chemistry and solicits résumés by mail

Net Chemical Corp.
Fax: 215-387-6453
custom synthesis, bulk pharmaceuticals, chiral intermediates, fine chemicals

NeoRx Corp.
410 West Harrison Street
Seattle, WA 98119
Fax: 206-284-7112
Internet site with job bank:
   www.neorx.com
invites résumés and cover letters by surface mail; files résumés for 2 months
offers stock options

Nepera, Inc.[55]
Route 17
Harriman, New York 10925
Fax: 914-783-9713
E-mail: gwillcox-jones@M1.cambrex.com

Neurex Corp.
3760 Haven Avenue
Menlo Park, CA 94025
Fax: 650-614-1061
E-mail: efeeney@neurex.com
Internet site with job bank:
   www.neurex.com
invites résumés by surface mail and fax
offers stock options and a stock purchase plan

Neurochemistry Research
12 Mershon Drive
Princeton, NJ 08450

Neurocrine Biosciences
10555 Science Center Drive
San Diego, CA 92121
Fax: 858-658-7602
E-mail: hr@neurocrine.com
Internet site with job bank:
   www.neurocrine.com

Neurogen Corp.
35 Northeast Industrial Road
Branford, CT 06405

Fax: 203-481-8683
E-mail: cmarks@nrgn.com
Internet site with job bank:
   www.neurogen.com
invites applications by fax or by surface or e-mail
offers stock options

NeXstar Pharmaceuticals, Inc.
2860 Wilderness Place
Boulder, CO 80301
Fax: 303-546-7894
E-mail: hr-colorado@nexstar.com
Colorado job line: 303-546-7780
Internet site with job bank:
   www.nexstar.com
solicits résumés by surface or e-mail
sometimes offers summer jobs to undergraduate chemistry students

NeXstar Pharmaceuticals, Inc.
650 Cliffside Drive
San Dimas, CA 91773
E-mail: hr-california@nexstar.com
California job line: 909-394-2698
solicits résumés by surface or e-mail

Nipa Hardwicke, Inc.
2114 Larry Jeffers Road
Elgin, SC 29045
Fax: 803-438-4497
E-mail: nipa@aol.com
Internet site: www.nipa.com
manufactures pharmaceutical intermediates and develops chemical processes
solicits résumés by e-mail and offers an online résumé form

NitroMed
12 Oak Park Drive
Bedford, MA 01730
E-mail: djanero@nitromed.com
Internet site: www.nitromed.com
drug discovery but no process development

Novartis Pharmaceuticals Corp.[56]
556 Morris Avenue, Building G2
Summit, NJ 07901

---

[55] An operating company of the Cambrex Corporation.

[56] Formed by the merger of Sandoz and CIBA.

Fax: 908-277-5966
E-mail:
    sandy.latzer@pharma.novartis.com
Internet site: www.novartis.com
asks for separate résumés from applicants interested in different positions

Norse Associates
Thousand Oaks, CA
Fax: 805-375-0812
consulting firm that employs a Ph.D. organic chemist

North Co.
(Midwestern U.S.)
Fax: 715-258-4986
E-mail: North@add-inc.com
carbohydrate-based pharmaceutical ingredients

Nortran Pharmaceuticals, Inc.
3650 Westbrook Mall
Vancouver, British Columbia
V6S 2L2 Canada
Fax: 604-822-9578
Internet site: www.nortran.com

Novalon Pharmaceutical Corp.
4222 Emperor Boulevard, Suite 560
Durham, NC 27703-8466
E-mail: jobs@novalon.com
Internet site with job bank:
    www.novalon.com
solicits applications by surface mail or e-mail
offers equity compensation in addition to salary

Novopharm Limited
575 Hood Road
Markham, Ontario
L3R 4E1 Canada
Fax: 905-475-1012
manufactures generic drugs

Novopharm Limited
54 Nably Court
Scarborough, Ontario
M1B 2K9 Canada
Fax: 416-291-0608

NPS Pharmaceuticals, Inc.
420 Chipeta Way
Salt Lake City, Utah 84108
Fax: 801-583-4961
E-mail: jobs@npsp.com
Internet site: www.npsp.com
combinatorial chemistry, solid-phase organic synthesis; solicits résumés by surface or e-mail

NSC Technologies[57]
601 E. Kensington Road
Mt. Prospect, IL 60056-1300
Internet site: www.nsctech.com
chiral intermediates and auxiliaries, process chemistry, contract research

Nycomed Amersham
466 Devon Park Drive
P.O. Box 6630
Wayne, PA 19087-8630
Fax: 610-225-4404
Internet site: www.nycomed-amersham.com
employs synthetic organic chemists to discover diagnostic imaging agents

Nymox Pharmaceutical Corp.
Dorval, Québec
Canada
Fax: 514-636-3146
E-mail: info@nymox.com
Internet site: www.nymox.com
employment opportunities: send e-mail

Oakwood Products, Inc.
1741 Old Dunbar Road
West Columbia, SC 29169
Fax: 803-739-6957
E-mail: rtracey@oakwoodchemical.com
Internet site:
    www.oakwoodchemical.com
custom synthesis to pilot plant scale, process development, fine chemicals

---

[57] A unit of Monsanto Company and a division of the NutraSweet Company.

Oceanix Biosciences Corp.
7170 Standard Drive
Hanover, MD 21076
Internet site with job bank:
 www.oceanix.com
*isolates drugs from marine
 microbes*
*asks applicants for three
 references*

Ohmeda Inc.
The BOC Group
Technical Center
100 Mountain Avenue
Murray Hill, NJ 07974
Internet site with job bank:
 www.ohmeda.com

Ontogen Corp.
6451 El Camino Real
Carlsbad, CA 92009
Fax: 760-930-0200
E-mail: hr@ontogen.com
Internet site with job bank:
 www.ontogen.com
*chemistry-based drug discovery
 and development*
*solicits résumés by e-mail, fax,
 or surface mail*
offers stock options

Optima Chemical Group LLC
200 Willachoochee Highway
Douglas, GA 31533
Fax: 912-383-0534
Internet site: www.optimachem.com
*pharmaceutical intermediates and
 custom manufacturing*

Oread Laboratories, Inc.
1501 Wakarusa Drive
Lawrence, KS 66047-1803
Fax: 785-832-4395
E-mail: employment@oread
Internet site with job bank:
 www.oread.com
*a contract pharmaceutical
 company*
*solicits résumés or curricula vitae
 by surface mail or as e-mail
 attachments*

Organix, Inc.
240 Salem Street
Woburn, MA 01801
Fax: 781-933-6695
Internet site with job bank:
 www.organixinc.com
*custom synthesis and contract
 discovery and development
 research*
*applications should include a
 résumé, a description of
 experience including reaction
 schemes, and the names of three
 references*

Organomed Corp.
70 Frenchtown Road, Suite 126
North Kingstown, RI 02852
Fax: 401-295-7465
synthesis of fine chemicals

Oridigm Corp.
4010 Stone Way North, Suite 220
Seattle, WA 98103-8012
Fax: 206-675-8881
E-mail: general@oridigm.com
Internet site: www.oridigm.com

OriGenix Technologies, Inc.
230 Bernard Belleau
Laval, Québec
H7V 4A9 Canada

Ortec Inc.
505 Gentry Memorial Highway
P.O. Box 1469
Easley, SC 29641
Fax: 864-859-8580
E-mail: info@ortec.net
Internet site: www.cris.com/~ortec
*custom manufacturing including
 cGMP and pharmaceutical
 intermediates*

ORTECH Corp.
2395 Speakman Drive
Mississauga, Ontario
L5K 1B3 Canada
Fax: 905-823-1446
Internet site: www.info.ortech.on.ca
custom synthesis, process chemistry

OSI Pharmaceuticals, Inc.[58]
106 Charles Lindbergh Boulevard
Uniondale, NY 11553-3649
Fax: 516-222-0114
E-mail: employment@OSI.com
Internet site with job bank:
   www.osip.com
*offers stock options and a stock
   purchase plan*
*employs high school graduates as
   interns*

Otsuka America Pharmaceutical, Inc.
2440 Research Boulevard, Suite 250
Rockville, MD 20850
Fax: 301-212-8647
Internet site: www.otsuka.com
Web site summarizes recruiting areas

Palatin Technologies, Inc.
175 May Street, Suite 500
Edison, NJ 08837
E-mail: info@palatin.com
Internet site: www.palatin.com
*peptides as diagnostic and
   pharmaceutical products*

Paracelsian, Inc.
222 Langmuir Laboratories
95 Brown Road, #1005
Cornell Technology Park
Ithaca, New York 14850
Fax: 607-257-2734
E-mail: paracel@clarityconnect.com
Internet site: www.paracelsian.com

Paratek Pharmaceuticals, Inc.
P.O. Box 1525
Boston, MA 02117
Fax: 617-636-6912

Parke-Davis Pharmaceutical Research[59]
2800 Plymouth Road
Ann Arbor, MI 48105
Fax: 313-998-3394
E-mail: parkedavis@isearch.com

[58]Formerly Oncogene Science, Inc.
[59]A division of Warner-Lambert.

Internet sites: www.warner-
   lambert.com and www.parke-
   davis.com
*Web site lists 13 schools where
   Warner-Lambert recruits and
   solicits scannable résumés,
   offering guidelines for preparing
   them; invites applicants to mail, e-
   mail, or fax scannable résumés to
   Warner-Lambert Processing
   Center, P.O. Box 92242, Los
   Angeles CA 90009-2242; Fax: 888-
   223-8312, and e-mail:
   warner@isearch.com*

Parke-Davis
188 Howard Avenue
Holland, MI 49424

PathoGenesis Corp.
201 Elliot Avenue West, Suite 150
Seattle, WA 98119
Job line: 206-505-6958
Fax: 206-270-3343
Internet site: www.pathogenesis.com
offers stock options

PE Applied Biosystems
850 Lincoln Centre Drive
Foster City, CA 94404
Fax: 650-638-5874
E-mail: biocareer@perkin-elmer.com
Internet site with job bank:
   www.pebio.com
*development of automated
   instruments for drug discovery;
   process research; asks for résumés
   by surface mail*

Pennie & Edmonds LLP
1155 Avenue of the Americas
New York, NY 10036
Fax: 212-869-8864
E-mail: pennie@info.pennie.com
Internet site: www.pennie.com
*law firm hiring chemists to become
   patent agents; has other branches
   in Washington, DC, and Palo Alto*

PerSeptive Biosystems, Inc.
500 Old Connecticut Path

Framingham, MA 01701
Fax: 508-383-7833
E-mail: Michelle_Smith@pbio.com
Internet site with job bank:
 www.perseptive.com
automated instruments for organic synthesis
solicits résumés by electronic or surface mail, posting salary ranges on Web site

Pfanstiehl Laboratories, Inc.
1219 Glen Rock Avenue
Waukegan, IL 60085-0439
Fax: 708-623-9173
E-mail: hr@pfanstiehl.com
Internet site: www.pfanstiehl.com
fine chemicals, carbohydrates, custom synthesis, pharmaceutical intermediates, bulk actives, isolation and purification of natural products

Pfizer Pharmaceuticals, Inc.
Central Research Division
Eastern Point Road
Groton, CT 06340
E-mail: resumes@pfizer.com
Internet site with job bank:
 www.pfizer.com
prefers receiving e-mailed résumés, inviting applicants to compose them on-line and accepting résumés by surface mail; asks applicants to describe their research in a cover letter; offers tips for preparing scannable résumés
Web site lists the campuses where Pfizer recruits; and details benefits, which include complete tuition reimbursement
asks applicants to send résumés to Pfizer, Inc., Central Research; c/o Aon Consulting, P.O. Box 25; Findlay, OH 45839; e-mail: Pfizer@aon-hros.com

Pharm-Eco Laboratories, Inc.
128 Spring Street
Lexington, MA 02421
Fax: 781-676-5393
E-mail: crd@pharmeco.com
Internet site with job bank:
 www.pharmeco.com
contract chemical research and development for the pharmaceutical industry

Pharmaceutical Peptides, Inc.
One Hampshire Street, 5th Floor
Cambridge, MA 02139
Fax: 616-494-8496

Pharmaceutical Product Development, Inc.
3151 South 17th Street Extension
Wilmington, NC 28412
Fax: 910-762-5820
Internet site: www.ppdi.com
contract research including combinatorial chemistry

Pharmacia & Upjohn[60]
Recruiting Office
7000 Portage Road
Kalamazoo, MI 49001
Fax: 616-833-9655
E-mail: recruit@am.pnu.com
Internet site: www.pnu.com
does not respond to unsolicited applications
call 616-833-0550 to inquire about summer internships in the United States

Pharmacopeia, Inc.
101 College Road East
Princeton, NJ 08540
Fax: 609-452-3671
E-mail: recpt@pharmacop.com
Internet site with job bank:
 www.pharmacopeia.com
asks for a cover letter and a résumé; offers stock options among other benefits

Pharmacyclics, Inc.
995 East Arques Avenue
Sunnyvale, CA 94086-4521
Fax: 408-774-0340

---
[60]Parent company of Upjohn.

E-mail: hr@pcyc.com
Internet site with job bank:
  www.pcyc.com
solicits résumés by surface and electronic mail; offers stock options

PharmaGenics, Inc.
4 Pearl Court
Allendale, NJ 07401
Fax: 201-818-9044

Pharmasyn Inc.
1840 Industrial Drive
Libertyville, IL 60048
Fax: 847-918-1610
process development: pharmaceutical intermediates

Phasex Corp.
360 Merrimack Street
Lawrence, MA 01843
Fax: 978-794-9580
E-mail: info@phasex4scf.com
Internet site: www.phasex4scf.com
pharmaceutical process development with supercritical fluids

Pherin Corp.
535 Middlefield Road, Suite 240
Menlo Park, CA 94025
E-mail: drb@pherin.com

Phytera, Inc.
Four Biotech Park
377 Plantation Street
Worcester, MA 01605
Fax: 508-792-1339
E-mail: phyterahjc@aol.com

Picower Institute for Medical Research
350 Community Drive
Manhasset, New York 11030
Fax: 516-365-5090
E-mail: hr@picower.edu
Internet site: www.picower.edu
offers summer jobs to undergraduate chemistry students

Pisgah Labs
795 Old Hendersonville Highway
Pisgah Forest, NC 28768
Fax: 704-884-5540
Internet site: www.pisgahlabs.com
custom organic synthesis, pharmaceuticals, chemical development

polyORGANIX Inc.[61]
9 Opportunity Way
Newburyport, MA 01950
Fax: 508-465-2057

PPG Industries, Inc.
Chemicals Technical Center
440 College Park Drive
Monroeville, PA 15146
E-mail: ppg@recruitmentsolutions.com
Internet site with job bank:
  www.ppg.com
pharmaceutical intermediates, multistep custom synthesis
invites scannable résumés by e-mail and gives tips for composing them; scanned résumés kept 6 months in file
recruits summer interns with varying majors

PPG Industries, Inc.
P.O. Box 995
La Porte, TX 7751
process development

PPD Pharmaco, Inc.[62]
8500 Research Way
Middletown, WI 53562
Fax: 608-827-8806
Internet site with job bank:
  www.ppdpharmaco.com
a contract research organization employing chemists in drug metabolism and radiochemistry

Praecis Pharmaceuticals Inc.
One Hampshire Street
Cambridge, MA 02139

---

[61] A part of Borregard Synthesis of Norway.
[62] A subsidiary of Pharmaceutical Product Development, Inc.

Pressure Chemical Co.
3419 Smallman Street
Pittsburgh, PA 15201
Fax: 412-682-5864
E-mail: service@presschem.com
Internet site with job bank:
   www.presschem.com
*pharmaceutical intermediates:
   process development, scale-up, toll
   manufacturing*

Procept, Inc.
840 Memorial Drive
Cambridge, MA 02139
Fax: 617-491-9019
E-mail: dcook@procept.com
Internet site with job bank:
   www.procept.com

The Procter & Gamble Co.
(Ph.D. recruiting)
P.O. Box 538707
Cincinnati, OH 45253-8707
E-mail: doctoral.im@pg.com.
*e-mailed submissions should use
   plain text or Microsoft Word*

(B.S./M.S. applicants)
Procter & Gamble Pharmaceuticals
Recruiting R & D
P.O. Box 191
Norwich, NY 13815
Fax: 617-335-2400
Internet site with job bank:
   www.pg.com
*asks for cover letter, résumé, and list
   of publications*

(B.S./M.S. applicants)
Procter & Gamble Company
U.S. Recruiting R & D
P.O. Box 599
Cincinnati, OH 45201-0599
employs hundreds of student interns

Prolinx, Inc.
22322 20th Avenue SE
Bothell, WA 98021
Fax: 425-487-9578
E-mail: hr@plinx.com
Internet site with job bank:
   www.prolinxinc.com
*develops tools to tag, isolate, and
   manipulate biological molecules
asks for cover letter and résumé;
   requests that e-mailed résumés be
   composed in rich-text files,
   Wordperfect, or Microsoft Word*

Promega Corp.
2800 Woods Hollow Road
P.O. Box 7879
Madison, WI 53707-7879
Job line: 800-356-9526, option 9
E-mail: hr@promega.com
Internet site with job bank:
   www.promega.com
fine chemicals

Pro-Neuron, Inc.
1530 E. Jefferson Street
Rockville, MD 20852

Pro-Neuron, Inc.
16020 Industrial Drive
Gaithersburg, MD 20877

ProScript, Inc.
38 Sidney Street
Cambridge, MA 02139

Protein Delivery, Inc.
2300 Englert Drive, Suite G
Durham, NC 27713
Fax: 919-361-9765

Protein Design Labs, Inc.
34801 Campus Drive
Fremont, CA 94555
Fax: 510-574-1448
E-mail: hr@pdl.com
Internet site with job bank:
   www.pdl.com
*seeks small-molecule drugs through
   synthetic organic and
   combinatorial chemistry; posts
   current salaries on Web site,
   paying entry-level Ph.D.s 65–75K.
   Invites résumés by surface mail or
   fax, offers information by e-mail
   but discourages e-mailed résumés
offers summer jobs to undergraduate
   chemistry students*

Proteinix, Inc.
16020 Industrial Dr.
Gaithersburg, MD 20877
Fax: 301-947-6998

Protogene Laboratories, Inc.
1454 Page Mill Road
Palo Alto, CA 94304
Fax: 650-842-7110
E-mail: mole@protogene.com
Internet site: www.protogene.com
specializes in automated synthesis and collaborates with pharmaceutical companies in DNA and RNA synthesis programs occasionally offers summer jobs to undergraduate chemistry students

Provid Research[63]
1505 Coles Avenue
Mountainside, NJ 07092
offers stock options

Purdue Pharma L. P.
100 Connecticut Avenue
Norwalk, CT 06850-3590
Fax: 203-851-5300
Internet site with job bank: www.purduepharma.com
accepts résumés by fax or surface mail and seeks staff for research and development in Princeton, NJ, as well as Norwalk and Ardsley

Purdue Pharma L. P.
444 Saw Mill River Road
Ardsley, NY 10502
Fax: 914-709-2590

Quality Chemicals, Inc.
700 North Street
Jackson, MS 39202
Fax: 601-949-0264
Internet site: www.qci.com

pharmaceutical chemicals, process development, custom mfgr

Quality Chemicals, Inc.
1515 Nicholas Road
Dayton, OH 45418

Raylo Chemicals, Inc.
8045 Argyll Road
Edmonton, AB
76C 4A9 Canada
Fax: 403-468-4784
process development for pharmaceuticals

RedCell, Inc.
270-B Littleton Avenue
South San Francisco, CA 94080
Fax: 415-827-0123
E-mail: hr@redcell.com
accepts applications only by fax or e-mail

RedCell Canada, Inc.
6100 Royalmount Avenue
Montréal, Québec
H4P 2R2 Canada

Regis Technologies
8210 North Austin Avenue
Morton Grove, IL 60053-0519
Fax: 847-967-1214
Internet site: www.registech.com
custom synthesis of pharmaceuticals and specialty chemicals; process development, cGMP

Reilly Industries, Inc.
300 N. Meridian, # 1500
Indianapolis, IN 46204
Fax: 317-247-6472
Internet site: www.reillyind.com
custom research, pharmaceutical intermediates

Reliable Biopharmaceutical Corporation
St. Louis, MO
Fax: 314-429-0937

---

[63]A division of Praecis Pharmaceuticals, Inc., to be located near Rutgers University in NJ.

Internet site:
www.reliablebiopharm.com
E-mail: info@reliablebiopharm.com
employs process and production chemists

Repligen Corp.
117 Fourth Avenue
Needham, MA 02494
Fax: 781-453-0048
E-mail: hr@repligen.com
Internet site with job bank:
www.repligen.com
offers equity participation

Research Biochemicals International
One Strathmore Road
Natick, MA 01760
Fax: 508-655-1359
E-mail: hr@resbio.com
Internet site with job bank:
www.callrbi.com
custom synthesis, contract research; accepts résumés by surface or electronic mail or by fax

Research Organics
4353 E. 49th Street
Cleveland, OH 44125
Fax: 216-883-1576
E-mail: annieharlan@resorg.com
Internet site: www.resorg.com
a biochemicals firm employing process chemists; makes biochemicals in bulk or on a laboratory scale; offers custom synthesis

Research Selectide Corp.[64]
1580 East Hanley Boulevard
Tucson, AZ 85737-9525
Fax: 520-575-8283

Research Triangle Institute
Chemistry and Life Sciences
P.O. Box 12194
Research Triangle Park, NC 27709

---
[64] A subsidiary of Hoechst Marion Roussel, Inc.

E-mail: jobs@rti.org
Internet site with job bank: www.rti.org

Rhône-Poulenc Rorer Research and Development
Mailstop H37
P.O. Box 5096
500 Arcola Road
Collegeville, PA 19426-0800
Internet sites: www.rpr.rpna.com and www.rhone-poulenc.com (in French)

Rhône-Poulenc
P.O. Box 2831
Building 2/151
Charleston, WV 25330
process development

Ribi ImmunoChem Research, Inc.
553 Old Corvallis Road
Hamilton, MT 59840
Fax: 406-363-6129
Job line: 406-363-6214, ext. 140
Internet site with job bank:
www.ribi.com
asks for cover letter, résumé, and transcripts

RiboGene, Inc.
26118 Research Road
Hayward, CA 94545
Fax: 510-732-7741
E-mail: hr@ribogene.com
Internet site with job bank:
www.ribogene.com
stock options

Ribozyme Pharmaceuticals, Inc.
2950 Wilderness Place
Boulder, CO 80301
Fax: 303-449-6995
E-mail: jobs@rpi.com
Internet site with job bank:
www.rpi.com
accepts solicited résumés by fax or by electronic or surface mail

Ricerca, Inc.
7528 Auburn Road
P.O. Box 1000

Painesville, OH 44077-1000
Fax: 440-354-4415
E-mail: Werbach_M@Ricerca.com.
Internet site with job bank:
   www.ricerca.com
discovery and development chemical research, clinical supply preparations, intermediates, scale-up and process demonstration, radiochemical and metabolite synthesis; contractual services to the pharmaceutical, agricultural, and specialty chemical industries

Rigel, Inc.
240 East Grand Avenue
South San Francisco, CA 94080
Fax: 408-736-1588
E-mail: jobs@rigel.com
Internet site: www.rigel.com
offers stock options

RoboSynthon, Inc.
1105 Grandview
South San Francisco, CA 94080
Fax: 650-244-0795
E-mail: info@robosynthon.com
Internet site: www.robosynthon.com
multireactors for combinatorial chemistry

Roche Bioscience[65]
A2-HR
3401 Hillview Avenue
Palo Alto, CA 94304
Fax: 650-424-8159
E-mail: opportunities@hr.pal.roche.com
Job line (24 hours): 1-800-400-2448
Internet site with job bank:
   www.roche.com/bioscience
accepts on-line résumés, giving tips for composing them
offers summer jobs to undergraduate chemistry students

Roche Carolina Inc.[66]
Bldg. 802

6173 East Old Marion Highway
Florence, SC 29506-9330
chemical development

Roswell Park Cancer Institute[67]
Chemistry Resource Facility
Elm & Carlton Streets
Buffalo, NY 14263
Internet site: http://
   rpci.med.buffalo.edu

RSP Amino Acid Analogues, Inc.
One Innovation Drive
Worcester, MA 01605
Internet site: www.amino-acids.com
amino acids for pharmaceutical lead discovery

Rütgers Organics Corp.[68]
201 Struble Road
State College, PA 16801
Fax: 814-231-9268
E-mail: humanres@ruetgers-organics-corp.com
Internet site: www.ruetgers-organics-corp.com
manufactures pharmaceutical intermediates; develops chemical processes

Sachem
821 East Woodward
Austin, TX 78704
E-mail: info@sachemusa.com
Internet site: www.sachemusa.com
fine chemicals, pharmaceutical manufacturing

SAIC Frederick[69]
P.O. Box B
Frederick, MD 21702-1201
E-mail: hr@mail.ncifcrf.gov
Internet site with job bank:
   www.saic.com
Web site lists campuses where the employee-owned company

[65]A member of the Roche Group; *see also* Hoffmann–La Roche.
[66]A member of the Roche Group.

[67]A Division of the New York State Department of Health.
[68]Formerly the Ruetgers-Nease Corp.
[69]A division of Science Applications International Corporation.

recruits; invites applicants to
submit résumés online; benefits
include a stock purchase plan

Salsbury Chemicals[70]
1205 11th Street
Charles City, IA 50616
Fax: 515-228-4152
E-mail: dbauer@M1.cambrex.com
*process development for
pharmaceutical intermediates and
bulk actives*

Sandoz Pharmaceuticals Corp.[71]
Staffing Department
Building 122/3
East Hanover, NJ 07936-1080
Internet site: www.novartis.com

SangStat Medical Corp.
1505-B Adams Drive
Menlo Park, CA 94025
Fax: 650-328-8892
E-mail: HR@sangstat.com
Internet site: www.sangstat.com
*human resources department
invites contact by phone
(650-328-0300) and other
traditional means*

Sanofi Winthrop, Inc.
9 Great Valley Parkway
Malvern, PA 19355
Fax: 610-889-8850
Internet site: www.elf.fr

SARCO, Inc.[72]
P.O. Box 14608
Research Triangle Park, NC 27709
Fax 919-485-8440
E-mail: getinfo@sarco.ppdi.com
Internet site: www.sarco.com
*contract research in combinatorial
chemistry; Web site names key
contact personnel in chemistry*

Schein Pharmaceutical, Inc.
P.O. Box 23160
620 N. 51st Avenue
Phoenix, AZ 85063-3160
Fax: 602-447-3385
Internet site: www.scheinpharma.com

Schering-Plough Research Institute
2015 Galloping Hill Road
Kenilworth, NJ 07033-0539
E-mail: spri@spcorp.com
Internet sites: www.sp-research.com;
    job bank at www.careermosaic.
    com/cm/spri/
*invites on-line applications at
    www.sp-research.com; send
    scannable résumés and cover
    letters*
Web site summarizes benefits
*offers summer internships,
    accepting on-line applications*

Schweizerhall Development Co.
1106 Perimeter Road
Greenville, SC 29605
Fax: 864-422-9215
E-mail: sdco@logicsouth.com
Internet site: www.schweizerhall.com
*chemical development in a cGMP
environment*

SciClone Pharmaceuticals
901 Mariner's Island Boulevard
San Mateo, CA 94404
Fax: 650-358-3469
E-mail: humanresources@sciclone.
    com
Internet site: www.sciclone.com
solicits e-mailed résumés

Scientific Staffing
Telephone: 800-363-3111
Fax: 800-480-7555
E-mail: info.corporate.scistaff.com
Internet site:
    www.scientificstaffing.com
*a temporary employment agency
    with 19 offices in 14 states; places
    inexperienced chemists at all
    degree levels*

---

[70] An operating company of Cambrex Corporation.
[71] Part of Novartis; see Ciba Pharmaceuticals Division.
[72] A subsidiary of Pharmaceutical Product Development, Inc.

The Scientists Registry
P.O. Box 784
Cranford, NJ 07016
Telephone: 908-272-8900
Fax: 908-272-0272
E-mail:
 DonaldTruss@ScienceRegistry.com
Internet site: www.scienceregistry.com
an employment agency making temporary and permanent placements

Scios Inc.
2450 Bayshore Parkway
Mountain View, CA 94043
Fax: 650-962-5966
E-mail: jobs@sciosinc.com
Internet site with job bank:
 www.sciosinc.com

Scios Nova Inc.
6200 Freeport Centre
Baltimore, MD 21224

Scripps Research Institute
10550 North Torrey Pines Road, TPC-11
La Jolla, CA 92037
Fax: 858-784-8071
Job line: 858-784-WORK
Internet site with job bank:
 www.scripps.edu
employs a Ph.D. organic chemist in patent liaison; details given on Web site
hires baccalaureate chemists as research technicians and assistants, posting hourly wages ($13.02–23.05, as of 1/99 and depending on grade) on Web site.
scans résumés, keeping them active for 6 months; solicits faxed applications, and offers online pointers for preparing scannable résumés

SCRIPTGEN Pharmaceuticals, Inc.
610 Lincoln Street
Waltham, MA 02451-2186
Fax: 781-393-5629
E-mail: hr@scriptgen.com
Internet site with job bank:
 www.scriptgen.com
accepts online applications
Web site summarizes benefits
offers co-operative positions

G. D. Searle[73]
Searle R & D
700 Chesterfield Village Parkway
St. Louis, MO 63198
Fax: 314-737-6419
Internet sites: www.monsanto.com and
 www.searlehealthnet.com
discovery research in medicinal chemistry

G. D. Searle
4901 Searle Parkway
Skokie, IL 60077
Fax: 847-982-4637
process research

Selectide[74]
1580 E. Hanley Blvd.
Tucson, AZ 85737-9525
provides combinatorial technologies to its parent company; solicits résumés suitable for optical scanning

Sepracor, Inc.
111 Locke Drive
Marlborough, MA 01752
Internet site with job bank:
 www.sepracor.com
Web site summarizes benefits, which include a stock purchase plan

Sepracor Canada Limited
Windsor, Nova Scotia
Canada

Sequus Pharmaceuticals, Inc.
960 Hamilton Court
Menlo Park, CA 94025
Fax: 650-323-9106
Internet site: www.sequus.com

---

[73] A subsidiary of Monsanto Corp.
[74] Part of Hoechst Marion Roussel.

Shearwater Polymers, Inc.
2307 Spring Branch Road
Huntsville, AL 35801
Fax: 256-533-4805
E-mail: info@swpolymers.com
Internet site: www.swpolymers.com
*polymers for pharmaceuticals,
custom synthesis, cGMP
manufacturing*

Sibia Neurosciences, Inc.
505 Coast Boulevard, South, Suite 300
La Jolla, CA 92037
Fax: 858-452-9279
E-mail: lalba@sibia.com
Internet site with job bank:
www.sibia.com
*send a CV including three
references' names*

Sigma Chemical Co.[75]
3050 Spruce Street
St. Louis, MO
Fax: 314-286-7863
E-mail: opportunity@sial.com
Internet site: www.sigma.sial.com
*manufactures and distributes
biochemicals and diagnostic
reagents*

Signal Pharmaceuticals, Inc.
5555 Oberlin Drive
San Diego, CA 92121
Fax: 858-558-7513
E-mail: lcain@signalpharm.com
Internet site with job bank:
www.signalpharm.com/signal
*solicits applications by fax, surface
or electronic mail, using the ASCII
format and containing the word
"resume" as the subject
offers equity participation*

Simulations Plus, Inc.
1220 West Avenue J
Lancaster, CA 93534-2902
Fax: 661-723-5524
E-mail: info@simulations-plus.com

[75]See also The Aldrich Chemical Co.

Internet site with job bank:
www.simulations-plus.com
*software development; solicits
résumés and cover letters by fax
or surface or electronic mail*

**Sloan-Kettering Institute for
Cancer Research**
Laboratory of Organic Chemistry
633 Third Ave., 5th Floor
New York, NY 10017
Job line: 212-639-5627
E-mail: browna@mskcc.org
Internet site with job bank:
www.mskcc.org
*invites résumés by surface or e-mail;
asks for the ASCII format*

**Sloan-Kettering Institute for
Cancer Research**
Laboratory for Bioorganic Chemistry
Box 106
1275 York Ave.
New York, NY 10021

Small Molecule Therapeutics
Dr. Allan Ferguson
Director of Chemistry
11 Deer Park Drive, Suite 116
Monmouth Jct., NJ 08852
Tel.: 732-274-2882, ext. 118
Fax: 732-274-0086
E-mail: ferguson@smtherapeutics.
com

**SmithKline Beecham
Pharmaceuticals**
P.O. Box 40047
Philadelphia, PA 19106
Internet site with job bank:
www.sb.com
*solicits on-line résumés, asking for
the names of three references
offers internships and cooperative
positions to undergraduate
chemistry students*

**SmithKline Beecham
Pharmaceuticals**
P.O. Box 2646
Bala Cynwyd, PA 19004

Solvay Pharmaceuticals
901 Sawyer Road
Marietta, GA 30062
Internet site: www.solvay.com

Southern Research Institute
P.O. Box 55305
Birmingham, AL 35255-5305
Fax: 205-581-2200
E-mail: jobs@sri.org
Internet site with job bank: www.sri.org
solicits résumés by surface mail, fax, or e-mail; asks for ASCII text with "resume" as the subject line;. applications kept active for a year

Specialty-Chem Products Corp.[76]
Two Stanton Street
Marinette, WI 54143
Fax: 715-735-5304
custom synthesis; contractual chemical development

Sphinx Pharmaceuticals[77]
4615 University Drive
Durham, NC 27707

Sphinx Pharmaceuticals[78]
840 Memorial Drive
Cambridge, MA 02139

SRI International
Staffing, AC108
333 Ravenswood Avenue
Menlo Park, CA 94025
Fax: 650-859-4700
E-mail: careers@sri.com
Internet site with job bank: www.sri.com
contractual services in organic and medicinal chemistry; pharmaceutical discovery and development, process research; invites applications by surface and e-mail

SST Corp.
635 Brighton Road
Clifton, NJ 07012
Fax: 973-473-4326
E-mail: sstcorp1@aol.com
pharmaceuticals, intermediates, fine chemicals

Steacie Institute for Molecular Sciences
National Research Council of Canada
100 Sussex Drive
Ottawa, Ontario
K1A 0R6 Canada
Fax: 613-990-7669
E-mail: ra.coordinator@nrc.ca
Internet site: http://gold.sao.nrc.ca/sims
offers renewable 2-year appointments to the staff of the National Research Council; open to Canadian citizens and foreign nationals with Ph.D.s; further information about these Research Associateships from the Program Coordinator

Steroids, Ltd.
2201 West Campbell Park Drive
Chicago, IL 60612
custom synthesis, pharmaceutical intermediates, and finished drugs; cGMP conditions

Structural Bioinformatics, Inc.
10929 Technology Place
San Diego, CA 92127
Fax: 858-451-3828
E-mail: cgaran@strubix.com
Internet site with job bank: www.strubix.com

Sugen, Inc.
230 East Grand Ave.
South San Francisco, CA 94080
Fax: 415-369-8984
E-mail: jobs@sugen.com
Internet site with job bank: www.sugen.com
benefits may include stock options; offers a stock purchase plan

---

[76] A Bayer company.
[77] A division of Eli Lilly and Co.
[78] A division of Eli Lilly and Co.

offers summer internships to undergraduates

Sunesis Pharmaceuticals, Inc.
3696 Haven Avenue, Suite C
Redwood City, CA 94063
Fax: 650-556-8824
E-mail: jobs@sunesis-pharma.com
Internet site: www.sunesis-pharma.com

Symphony Pharmaceuticals
76 Great Valley Parkway
Malvern, PA 19355

Symyx Technologies
420 Oakmead Parkway
Sunnyvale, CA 94086
E-mail: jlindner@symyx.com
Internet site with job bank:
  www.symyx.com
combinatorial chemistry

Synaptic Pharmaceutical Corp.
215 College Road
Paramus, NJ 07652

Synpep Corporation
P.O. Box 94568
Dublin, CA 94568
Fax: 925-803-9301
E-mail: peptide@synpep.com
Internet site with job bank:
  www.synpep.com
custom synthesis of peptides and radiolabeled compounds; analytical services including mass spectrometry; fine chemicals invites applications by e-mail and fax

Synphar Laboratories, Inc.
Taiho Alberta Centre
4290-91A Street
Edmonton, Alberta
T6E 5V2 Canada

Syntex Chemicals, Inc.[79]
2075 N. 55th Street

[79]See also Roche Bioscience, which became Syntex's name.

Boulder, CO 80301
process research

Syntex Discovery Research
Institute of Organic Chemistry
2401 Hillview Avenue
Palo Alto, CA 94304

Syntex Research
2100 Syntex Court
Mississauga, Ontario
Canada

SyntheTech
1290 Industrial Way
P.O. Box 646
Albany, OR 97321
Fax: 541-967-9424
E-mail: careers@synthetech.com
Internet site with job bank:
  www.synthetech.com
pharmaceutical intermediates, chiral synthesis, biotransformations, custom synthesis, process development offers stock options

Synthon Corp.
3900 Collins Road
Lansing, MI 48910
Fax: 517-332-5304
E-mail: coyer@mbi2.mbi.org
Internet site: www.mbi.org/Synthon
fine chemicals for the pharmaceutical industry; offers bonus plus stock options

Syntech Labs, Inc.
100 Jersey Avenue, Building D
New Brunswick, NJ 08901
Fax: 908-545-0120
E-mail: syntchlab@aol.com
custom synthesis

Systems Integration Drug Discovery Co.
Building 40
9040 South Rita Road, #2338
Tucson, AZ 85747
Fax: 520-663-0795

E-mail: junghan@siddco.com
Internet site with job bank:
  www.siddco.com
carries out collaborative contract research with established pharmaceutical companies employs synthetic organic, medicinal, combinatorial, and computational chemists; offers stock options; seeks applicants already permitted to work in the United States

Tanabe Research Laboratories USA, Inc.
4540 Towne Centre Court
San Diego, CA 92121
Fax: 858-558-0650
E-mail: lsmith@trlusa.com
Internet site: www.trlusa.com

TCI America
9211 N. Harborgate Street
Portland, OR 97203
Fax: 503-283-1987
Internet site: www.tciamerica.com
custom synthesis and contract research, fine chemicals, cGMP

Telik, Inc.
750 Gateway Boulevard
South San Francisco, CA 94080
Fax: 650-244-9388
E-mail: inquiry@telik.com
Internet site with job bank:
  www.telik.com
a drug discovery company asks job seekers to mail their résumés

TerraGen Diversity, Inc.
Suite 300-2386 East Mall (UBC)
Vancouver, BC
V6T 1Z3 Canada
Fax: 604-221-8881
E-mail: info@terragen.com
Internet site with job bank:
  www.terragen.com
solicits introductory letter and CV by e-mail; a combinatorial biology company employing natural-products chemists

Terrapin Technologies, Inc.
750 Gateway Boulevard
South San Francisco, CA 94080-7020
Fax: 650-244-9388
E-mail: rgab@mx.trpntech.com
Internet site with job bank:
  www.terrapintech.com

Texas Biotechnology Corp.
7000 Fannin, Suite 1920
Houston, TX 77030
Fax: 713-796-8232
E-mail: jobs@tbc.com
Internet site with job bank:
  www.tbc.com
offers summer jobs to undergraduate chemistry students

Thermogen
2201 W. Campbell Park Drive
Chicago, IL 60612
Fax: 312-226-9686
E-mail: ddem@thermogen.com
Telephone: 312-226-6500
Internet site with job bank:
  www.thermogen.com
biocatalysts for chemical synthesis, catalyst screening and development
solicits résumés by fax and invites inquiries by telephone
occasionally offers summer internships to applicants with laboratory experience

Tippecanoe Laboratories[80]
Lafayette, IN 47099
chemical development and process research

Torcan-Delmar
110 Industrial Pkwy North
Aurora, Ontario
L4G 3H4 Canada
Fax: 905-727-7545
Internet site: www.torcanchemical.on.ca
scale-up to manufacturing

---

[80]Part of Eli Lilly and Co.

Toronto Research Chemicals Inc.
2 Brisbane Road
North York, Ontario
M3J 2J8 Canada
Fax: 416-665-4439
E-mail: torresch@interlog.com
Internet site: www.trc-canada.com
small-molecule custom synthesis, fine chemicals, pilot-plant scale contracting

Trega Biosciences, Inc.[81]
9880 Campus Point Drive
San Diego CA 92121
Fax: 858-410-6651
Internet site with job bank:
  www.trega.com
solid-state organic synthesis, molecular diversity, combinatorial chemistry; discourages faxing of résumés; offers stock options among other benefits

Triad Biotechnology
San Diego, CA
Fax: 858-759-0209
E-mail: jobs@triadbiotechnology.com
Internet site with job bank:
  www.triadbiotechnology.com
solicits on-line résumé submissions

Triangle Pharmaceuticals
4 University Place
4611 University Drive
Durham, NC 27707
Internet site:www.tripharm.com

Trimeris, Inc.
4727 University Drive
Durham, NC 27707
Fax: 919-419-1816
E-mail: info@trimeris.com
Internet site with job bank:
  www.trimeris.com
E-mail: careers@tripharm.com (MS Word 97 if possible)
accepts online applications and solicits them by surface and e-mail

Tripos, Inc.
1699 South Hanley Road, Suite 303
St. Louis, MO 63144
Fax: 314-647-9241
E-mail: tgreer@tripos.com
Internet site with job bank:
  www.tripos.com
compound libraries, molecular modeling, software development; employs both medicinal chemists and computer experts
Web site details benefits including tuition reimbursement and a stock purchase plan

Tripos, Inc.
882 South Matlack Street
West Chester, PA 19380
molecular modeling, software development

Trophix Pharmaceuticals, Inc.[82]
40 Cragwood Road, Suite B
South Plainfield, NJ 07080
Fax: 908-561-1909

Tularik Inc.
Two Corporate Drive
South San Francisco, CA 94080
Fax: 650-829-4303
E-mail: resume@tularik.com
Internet site with job bank:
  www.tularik.com
solicits résumés by fax, surface mail, or e-mail; asks for e-mailed résumés or CVs in MS Word; offers equity participation, including stock options
offers summer jobs to undergraduate chemistry students

Tyger Scientific, Inc.
11 Deer Park Drive, Suite 114
Monmouth, NJ 08852
Mailing address: P.O. Box 2358
Princeton, NJ 08543-2358
Fax: 732-329-8988
E-mail: tygersci@superlink.net
Internet site: www.tygersci.com

---

[81]Formerly Houghten Pharmaceuticals, Inc.

[82]A subsidiary of Allelix Pharmaceuticals.

custom syntheses, contract research, process research and optimization

Ucb Research, Inc.
840 Memorial Drive
Cambridge, MA 02139
Fax: 617-547-8481
E-mail: sally.miller@ucb-group.com
Internet site with job bank: www.ucb-group.com

Upjohn Laboratories
Pharmacia & Upjohn, Inc.
Kalamazoo, MI 49001
Internet site: www.pnu.com

U.S. Patent and Trade Office
(for applications by mail)
Office of Human Resources
Box 171
Washington, DC 20231

(for applications in person)
Office of Human Resources
2011 Crystal Drive, CPKI-707
Arlington, VA

(other)
Internet site with job bank: www.uspto.gov
*entry-level chemical patent examiners meet the requirements of an accredited college for a baccalaureate degree and study chemistry for 30 semester hours; salaries depend on grade level (GS-5 to GS-11) and qualifications and range from $21,421 to $50,029
for application materials, call the Office of Personnel Management (202-606-2700) or the Patent and Trademark Office (800-368-3064)
for further information about patenting in organic chemistry and pharmaceuticals, contact Mukund.Shah@uspto.gov (703-308-4716) or JamesHousel@uspto.gov (703-308-4027)*

United States Surgical Corp.
195 McDermott Road
North Haven, CT 06473
Internet site: www.ussurg.com
synthesis and characterization of organic monomers and polymers; wound-healing products

Versicor Inc.
34790 Ardentech Ct.
Fremont, CA 94555
Fax: 510-739-3003
E-mail: info@versicor.com
Internet site: www.versicor.com
combinatorial chemistry

Vertex Pharmaceuticals Inc.
130 Waverly Street
Cambridge, MA 02139-4211
Fax: 617-577-6645
E-mail: recruiting@vpharm.com
Internet site with job bank: www.vpharm.com
*asks for cover letters and scannable résumés by fax, surface or e-mail; prefers on-line submissions
offers summer jobs to undergraduate chemistry students*

Vical, Inc.
9373 Towne Centre Drive, Suite 100
San Diego, CA 92121-3088
Job line: 858-646-1143
Fax: 858-646-1350
E-mail: cgoodall@vical.com
Internet site with job bank: www.vical.com

Vion Pharmaceuticals, Inc.
Four Science Park
New Haven, CT 06511
Fax: 203-498-4211
E-mail: vioninfo@vionpharm.com
Internet site: www.vionpharm.com

Wacker Chemicals, Inc.
535 Connecticut Ave.
Norwalk, CT 06584
Fax: 203-866-9427
custom synthesis

Walter Reed Army Institute of Research
Division of Experimental Therapeutics
Department of Medicinal Chemistry
Building 500, Room 3
Forest Glen, MD
Fax: 301-295-7755
Internet site: wrair-www.army.mil/wrair.htm
employs civilians as well as soldiers; entry-level jobs for chemists at all degree levels correspond to civil service grades of 12–13, currently (1999) paying $47–73K.

WCRI
219 W. Bel Air Avenue
Aberdeen, MD 21001-3256
nonprofit organization working to discover anticancer drugs

Wright Corp.
102 Orange Street
Wilmington, NC 28401
custom manufacturing, pharmaceutical intermediates

Wright Corp.
333 Neils Eddy Road
Riegelwood, NC 28456
process chemistry, kilo scale synthesis, cGMP manufacturing

Wyeth Ayerst Research[83]
P.O. Box 7886
Philadelphia, PA 19101-7886
(for jobs in NJ, NY, PA or Québec)
Fax: 610-989-4854
E-mail: jobs@ramail1.wyeth.com
Internet site with job bank:
www.ahp.com/wyeth.htm
solicits mailed or faxed résumés (in fine mode), storing them in a centralized research data base; invites cover letters and résumés by e-mail in ASCII characters with no formatting or attachments and with "résumé" as the subject; discourages e-mailed inquiries carries out chemical development in Pearl River and Rouses Point, NY, and in Montréal, Québec; employs organic chemists in discovery research in Princeton, NJ, and Pearl River, NY offers stock options

Wyeth Ayerst Research[84]
401 Middletown Road
Pearl River, NY 10965

Wyeth Ayerst Research[85]
P.O. Box 150
Chazy, NY 12921

Wyeth Ayerst Research[86]
CN 8000
Princeton, NJ 08543-8000

Xechem, Inc.
100 Jersey Ave., Building B, Suite 310
New Brunswick, NJ 08901
Fax: 908-247-4090
E-mail: mlb@xechem.com
Internet site: www.xechem.com
isolation of natural products, contract research

YChem International Corp.
937 Cottonwood Drive
Cupertino, CA 95014
Fax: 408-257-7656
custom synthesis, process development, combinatorial chemistry, racemate resolution

Yukong R & D Laboratories[87]
140A New Dutch Lane
Fairfield, NJ 07004
Fax: 201-227-4488
Internet site: www.sk.co.kr

---

[83]Part of American Home Products.
[84]Part of American Home Products.
[85]Part of American Home Products.
[86]Part of American Home Products.
[87]Biopharmaceutical part of Yukong Limited, a Korean energy and chemicals company.

Zeeland Chemicals, Inc.[88]
215 N. Centennial Street
Zeeland, MI 49464
Fax: 616-772-7344
process development: laboratory through pilot plant to manufacturing of pharmaceutical bulk actives and intermediates

Zeneca Pharmaceuticals[89]
1800 Concord Pike
Wilmington, DE 19850-5437
Fax: 302-886-7124
E-mail: hrposting@phwilm.zeneca.com
Internet site with job bank:
www.AstraZeneca-us.com
offers summer jobs to undergraduate chemistry students

Zeneca Pharmaceuticals
P.O. Box 152
Mt. Pleasant, TN 38474-0152
Fax: 931-379-7124
process research takes place at this site

Zonagen, Inc.
2408 Timberloch Place, B-4
The Woodlands, TX 77380
Fax: 281-719-3446
E-mail: tara@zonagen.com
Internet site: www.zonagen.com

ZymeTx, Inc.
800 Research Parkway, Suite 100
Oklahoma City, OK 73104
Fax: 405-271-1944
E-mail: hammant@zymetx.com
Internet site with job bank:
www.zymetx.com
invites résumés by surface mail and solicits CVs and résumés on-line; keeps file active 3 months to 1 year

ZymoGenetics Corp.[90]
1201 Eastlake Avenue E.
Seattle, WA 98102
Fax: 206-442-6608
E-mail: hr@zgi.com and jobs@zgi.com
for job postings, see www.bio.com

---

[88]An operating company of Cambrex Corporation.
[89]Part of AstraZeneca.

[90]A subsidiary of Novo Nordisk A/S.

INDEX

7-Aminocephalosporin, 203
Abbott Laboratories, 45, 92
Academic degree
  bachelor's
    employment qualifications
      need for research tutoring, 217
      research experience, advantage of, 217
      time needed to find a job, 216
      widest job choice, 216
    employment qualifications, 216–218
    framework, 16
  determinant of
    responsibility, 103
    salary, 16
  doctor's
    challenges, rewards, and opportunities, 17
    employment qualifications 219–220
      judgment 219
    lifetime earnings differential, 17
  master's
    employment qualifications
      bonuses
        research experience, 218–219
        technical know-how, 218–219
      employment qualifications, 218–219
    finding employment, 19
    not prerequisite to a doctor's, 19
    stigma, 19
Accomplishments, tracking, 106
Advanced Chem Tech, 94
Advanced Scientific Professionals, 90
Advice, social
  sipping not slurping, 230
Agencies, temporary, 90

Agouron, 251
Albany Molecular Research, 92
Alberty, R. A., 201
American Immigration Center, 222
Amersham Pharmacia Biotech, 93
Amoxicillin, clarithromycin and lansoprazole combined, stomach ulcers, 52
Annual Reports in Medicinal Chemistry, 154
Anormed, 49
Anti-cancer Drug Design, 288
Antiviral Chemistry and Chemotherapy, 288
Argonaut Technologies, 95
Artzneimittel Forschung, 154, 288
Aspirin, 78
AstraZeneca, 92, 251
Atevirdine, autoimmune disease, 65
Atom utilization, 202
Attractions
  benefits
    aid to choosing and hunting suitable jobs, 13
    defining corporate culture, 27
    twenty most common, 14
    variability, determinants and exclusivity, 15
  benefits, 13–28
  benefits, educational
    books, 23
    chemical societies, reimbursed membership, 23
    courses, 21–22
    graduate school
      attending, 18–19
      what to study, 19–21

349

350 Index

Attractions (*continued*)
   libraries, corporate, 23–24
   meetings, scientific, reimbursed
      attendance, 24–25
   tuition reimbursement, 16
   who needs a Ph. D., 16–17
  benefits, educational, 15–25
  benefits, financial, 25, *see also* Chapter 8
  benefits, miscellaneous
   beer garden, 27
   communications, 28
   company store, 28
   relocation assistance, 27–28
   sabbatical leaves, 27
  benefits, miscellaneous, 27–28
  benefits, social, 26
  benefits, temporal
   flexible working hours, 26
   holidays, 26–27
   vacations, 27
  benefits, temporal, 26–27
  salaries
   annual ACS survey, 35–36
   determinants
    academic degree, 29–30
    company size, 33
    employer type, 29
    geography, 32–33
    industrial division, 30
    sex, 31
   determinants, 29–33
   going offers, ascertaining, 35
   median 1998 starting, for experienced
    B. S. chemists, 32
   median 1998 starting, for experienced
    M. S. chemists, 31
   median 1998 starting, for experienced
    Ph. D. chemists, 30
   median 1998 starting, for
    inexperienced chemists of any
    degree, 33
   median 1998 starting, for
    inexperienced chemists, 30
   nonimmigrant wages, learning, 36–37
   online calculator, 34
   starting pay, negotiating, 34
   upper and lower limits, 29
  salaries, 28–37, *see also* Internet posting, 238
Attractions, 13–37
Aventis, 92

Back-up drug
  characteristics, 52
  discovery of cimetidine, second, 69
Bargaining power, 258
Barton, D. H. R., 170
Bayer, 92
Beilstein, 159
*Berichte der Deutschen Chemischen
  Gesellschaft*, 146
Betamethasone, inflammatory disease, 89,
  see also Steroids
Big things from small companies, 235–236
Biological activity
  sample purity, 177
Biology, knowledge of, and chemists'
  progress, 179
*Bioorganic and Medicinal Chemistry
  Letters*, 154, 288
Bioorganic and Medicinal Chemistry,
  154, 288
Biotechnology companies, profit and loss,
  11
Black, J., 68
Boehringer-Ingelheim, 45, 204
Bohdan, 95
Brimblecombe, R., 9
Bristol-Myers Squibb, xviii, 6, 45
Bristol-Myers Squibb, *see also* Stock
  options, 256–257
Brown, H. C., 20
Bulk actives, 94
Burn rate, 251

C/D/N Isotopes, 93
Cambridge Isotope Laboratories, 94
CambridgeSoft, 95
Campbell and Flores, 94
Canadian Chemical News, 288
Captopril, high blood pressure, 7
Career development
  climbing academic ladder, 3
  consultants, volunteer, 230
  internships, 19
  learning drug discovery and development,
   68, 263
  productivity, 3
  promotion and salary, 17
  publications and patents, 19
  publications and promotions, 19
  snaring a senior position, 17
  synthetic chemistry, mastery, 3

Index 351

undergraduate research, 19, *see also* Summer internships, 76
Career Resource Library, 230
Career transitions
  chemists' versatility, aid, 67
  law school, 85
  managerial achievements, 246
  part-time study, 16
  prerequisites, 3
  reevaluation, 245
  transferability of chemists' skills, *see* Drug discovery and development
  tuition reimbursement, 16
Carlson, M., 249
Catalytica, Inc., 94, 192
Cavalla, D., 5
Chemical Abstracts Service, 98, 285
Chemical Abstracts, 159, 243, 262
Chemical and Engineering News, 221, 243, 288
Chemical Design, 95
Chemical development
  information resources, 184
  scope, 184–185
  ten-liter reactor, 183
  timing, 185–186
Chemical development, 182–186
Chemistry and Industry (London), 184
Chemistry, manufacturing and control, 88, 209
Chemists
  manufacturing, generic drug houses, 48
  process, generic drug houses, 48, 49
Chemists' jobs
  abstracting, *see* Chemical Abstracts Service, 98
  analytical chemistry
    pervasiveness, 81
    techniques, 82
    temporariness, 82
  baccalaureate holders, 73
  cheminformatics, 83–84
  chemometrics, *see* cheminformatics and Jobs in the drug industry
  clinical research organizations, 48, 55, 62
  coordinating patents, 77
  custom-synthesis firms, 60
  drug discovery companies, 49
  drug discovery or development, most, 41
  editing, *see* Merck Index, 98
  entry-level research, 77, 78
  fine-chemical supply houses, 60
  Food and Drug Administration, 96
  generic drug companies, 48
  government agencies, 73
  information science, 85–86
  molecular modeling, computational 79
  natural products, isolating, 79
  patent law firms, 73
  performance evaluations, 104
  pharmaceutical companies, non-fully-integrated, 48
  protein crystallography, 79–80
  registries, 77, 83
  regulatory affairs, 77
  research institutions, nonprofit, 73
  scientific writing, 98
  start-up pharmaceutical companies, 49
  technology transfer, 97
  temporary assignments, *see* Jobs in the drug industry
  U. S. Patent and Trade Office, 96–97
Chemists' work
  autonomy, 149
  communication, 150
  process research, 194
  restrictions, 149
  summary, 73
  supervision, 150
  target compounds, 149
Chemists, manufacturing, 286
ChemSyn, 94
ChemWeb, 239
Chiral intermediates, 94
Choice of employer, irreversibility, 14
Christ, C., 203
CHUL Research Center, 285
Churchill, W., 143
CIBA-Geigy, 283
Cimetidine
  against stomach ulcers, 51
Cimetidine, 68, 69, *see also* Drug discovery
Clark, G., 14
Clinical development
  drug formulation, 56
  effect on patent lifetime, 70
  numbers of human patients, 56–57
  placebos, 56
  progress and success, 56–57
  regulatory approval, 57
  stages, 56–57

Clinical trials
    far-flung treatment centers, 187
    phases I and II, 187
    phases I to IV, 56–57
Cole, G., 276
Coleridge, S. T., 23
Collaboration between discovery and development chemists, 186
Combinatorial chemistry, 43–44, 49
Communicating results, plans, problems, and persona
    publishing, 112–114
    reporting results, 111–112
    the importance of being visible, 114
Communicating results, plans, problems, and persona, 111–115
Computer literacy
    indecisiveness in obtaining job offers, 217–218
Consolidation and fragmentation in the pharmaceutical industry
    economic forces, 70
    outsourcing science, 70
    start-up companies, 71
Consolidation and fragmentation in the pharmaceutical industry, 70–71
Contract manufacture, 94
Corey, E. J., 20
Cornforth, J. W., 160
Cortisone, arthritis, 9
Cox, B., 41
Cram, D. J., 20
Criteria for evaluating companies
    benefits beyond salary and other compensation
        accrual of cash balances, 271
        accruals of tax-deferred, inflation-adjusted cash-balance and 401k pensions, 272
        features of selected pension plans, 265
        group insurance, 264
        growth of tax-deferred and tax-paid savings, 275
        retirement
            defined-benefit pension plans, 266–273
            inflation-ravaged pension benefits, 269
        retirement, 264–277
    benefits beyond salary and other compensation, 263–277
    career prospects

choosing your boss, 262–263
hypothetical distribution of highest attained ranks and total numbers of filled positions on a scientific ladder, 260
lateral moves, 261
patenting and publishing, 261–262
scientific and managerial ladders, 258–259
career prospects, 258–263
corporate status and future considerations
    how many drugs for sale, 248
    drugs under development, 249
    headlines, 250
    in-licensing, 249–250
    one-drug companies, 249
    patent expirations of best-selling drugs, 248–249
    survival and risk, 250–251
corporate status and future, 248–252
equivalent salaries in selected U. S. and Canadian cities home to pharmaceutical companies and related organizations, 253
forms of compensation other than salary
    cost-of-living adjustments, 256
    hiring bonuses, 256
    jobs for spouses, 257–258
    stock options, 256–257
    stock ownership plans, 257
    year-end bonuses, 256
forms of compensation other than salary, 255–258
geography
    but can you afford to live there, 252–255
    geography, 252–255
    median prices of homes in selected states during the first quarter of 1999, 255
Criteria for evaluating companies, 248–277
Crum-Brown, A., 2
Cuatrecasas, P., 8, 41
Current good manufacturing practices, 209
Current Medicinal Chemistry, 288
Current Pharmaceutical Design, 288
Cusumano, J. A., 201
Cyclosporin A, psoriasis, 50

Daniels, F., 201
Defense research, government investment, 7
Degussa, 94

Demon, arithmetic, 59, 182, *see also* Woodward, R. B., 205
Development research
  adoption of the research synthesis or a new one, 185
  applied, 184
  basics, 156, 194
  bulk supplies, 186–187
  campaign prerequisites, 185
  clarity of goals, 185
  concentration of efforts, 185
  contrasts to university and chemical discovery research, 181
  cost of single experiments, 182
  defensive, 189
  duration of a project, 185
  goals, 104, 194
  regulation, 182
  scale, 182
  success and criteria, 8
  teamwork, 105
  urgency, 181
  work assignments, 67
Dexamethasone, inflammatory disease, 89, *see also* Steroids
Dihyrdoavermectin, river blindness, 45
Discovery research
  basics, 156
  common goals of discovery chemists, 151–152
  decisions, 149
  inventorship, test, 57
  lead compounds
    defined, 152–155
    medicinal chemists, 155
    origins, 154–155
    prerequisites, 153–154
  lead compounds, 152–155
  learning medicinal chemistry, 20
  making target compounds
    devising syntheses
      adaptability, 174
      brevity, 173
      convergent and linear syntheses of structure ABCDEF, 175
      criteria, 172–173
      irrelevant criteria, 174
      linearity versus convergency, 174
      robustness, 173–174
      simplicity, 173
      versatility, 173

  devising syntheses, 172–175
  elucidating structures
    four principles, 169–170
    X-ray analysis, 170
  elucidating structures, 168–172
  running reactions
    characterizing and identifying samples
      characterization, 165–166
      identification, 166–167
      relations between sample identifications and structure assignments, 166
      tips and traps, 167–168
    characterizing and identifying samples, 165–168
    determining purity and purifying
      methods of choice, 164–165
      rules of thumb, 162–163
    determining purity and purifying, 161–165
    isolating products, 161
    persistent efforts and satisfactory experiments, 160–161
    resources, 157–160
    selections from the literature surveying organic synthesis, 159
    starting materials, 157
    typical search query, 160
  running reactions, 156–161
  suitable target compounds, 156
  making target compounds, 155–175
  pragmatic principles, 157
  progression of events, 151
  quantifying success, 151
  resynthesis, 58–60
  satisfactions and successes, 177–178
  stages and timing, 59
  submitting compounds for biological testing
    company code names, 176
    registration forms
      content, 176
      what to do about impurities, 177
    registration forms, 175–177
  submitting compounds for biological testing, 175–177
  success and criteria, 8
  work assignments, 67
Discovery research, 151–179

Doyle, A. C., 164
Dozeman, G. J., 208
Drews, J., 248
Drug action
    duration
        interspecies variation, 55
        need for, 54
    margin of safety, 54
    mechanisms, advantage of different, 7, 11, 52
    selectivity of, 54
Drug design, 141
Drug discovery and development
    basis, 2
    chemists' contributions, 5
    chemists' insight and creativity, 8–9
    cost of one new medicine, 6
    custom synthesis houses, 60
    dose-response experiments, 54, 64, 65
    drug metabolism, radiochemisty and radioimmunology, 50
    elements, summarized, 8
    failure, 43, 54
    governments, 5
    influence, academic research, 6
    innovation, 8, 11
    kilo laboratories, 61
    metabolites
        pharmacological activity, 61
        structure elucidation, 61
        toxicity, 61
    number of drugs, 6
    oral pharmacological activity, 52, 53
    organization of chemical research
        function and matrix organization, 69
        functional organization within a FIPCO, 69
        goals, inexperienced researchers' contributions, 66
        skill transferability, 67
        the project team as an organizing principle, 67–69
        work assignments, 66
    organization of chemical research, 66–69
    outsourcing of research, 70–71
    patentable structural novelty, 52
    pharmaceutical companies, 5
    political will, 6
    progress from hands-on efforts, 216
    project changes, abrupt, 108
    project lifetime, 2
    renewable enthusiasm, 44
    serendipity, 119
    stratagems
        medical needs, 50–51
        research goals, 52–53
        therapeutic advantage, 52
        therapeutic approaches, 51
    stratagems, 50–53
    synthetic chemists' work, 4
    tactics
        biological discovery groups, 53–54
        chemical development groups
            clinical supplies, 63–64
            drug specifications, 65
            hazards, 63
            manufacturing synthesis, 63
        chemical development groups, 62–66
        chemical discovery groups
            degradants, 62
            metabolites, 61
            substrates for radioimmunoassays, 61–62
            supplies, 58
        chemical discovery groups, 57–62
    time to launch, 6
    universities, 5
Drug discovery and development, 50–66
Drug discovery companies
    customers, 49
Drug Discovery Online, 239
Drug discovery, chemical
    cimetidine project
        lifetime, 68
        number of compounds, 68
    empiricism, 155
Drug metabolism studies
    contract research organizations, 55
    structural elucidation, 55
    toxicological species, 55
Drug News and Perspectives, 154, 288
Drug sales, global markets, 11
Drugs
    best-selling, income, 6
    trace impurities
        concentration and number, 66
        determined by manufacturing synthesis, 65
        regulation, 65
Drugs in Development, 154
DuPont, 44

Index

Eastern Analytical Symposium, 241
Eastman Chemical Company, 92
Efficiency, volume, 199, 208
Elements of a suitable manufacturing synthesis
  average yields per step influencing the overall yield, 206
  costing, 204–205
  efficiency, brevity, and the arithmetic fiend, 205–207
  number of steps influencing the overall yield, 207
  regulation, 209–210
  robustness, 208
  safety, 200–201
  soundness, 201–202
  sourcing and cost of raw materials, 203–204
  specifications, 199–200
  throughput, 207–208
Elements of a suitable manufacturing synthesis, 199–210
Eli Lilly, 9
Ellin, A., 257
Emerging Pharmaceuticals, 154, 288
Employment
  delays in finding, 76
  identifying, 74
  industrial, advantages, 4, 10
  small company or large, 235–236
  start-up firms, 75
  summer internships, 76
  top-ten pharmaceutical companies, 75
  undergraduate research, 76
Employment applications, timing, 241, *see also* Finding job openings, 234–243
Enthusiasm, voicing, 76
Enticements
  altruism defined, 12
  altruism in chemical research vs. medical practice, 13
  altruism, 12–13
  financial
    benefits, retirements and salaries, 4
  intellectual preoccupation, 2
  relief of suffering by making drugs, 2
Enticements, 1–37
Environmental load factor, 202
Enzyme inhibition, therapy, 51
Etchells, J. C., 200
European Journal of Medicinal Chemistry, 154, 288

European Journal of Organic Chemistry, 217
Evidence
  negative, 168, 169
  positive, 169
  positive, from nuclear magnetic resonance spectra, 171
Experiments, satisfactory, 106

F1 Student visas, 222
Faraday, M., 109
Finders' fees, 243
Finding job openings
  corporate Web sites
    application guidelines, 238–239
    finding Web addresses, 238
    job alerts by e-mail, 240
    job banks, 239
    job lines, 240
  networking
    informational interviews, 242
    names for networks, 242
    network members, 241
  networking, 240–243
  corporate Web sites, 238–239
  other job banks, 239
  print advertisement for chemists
    what to look for, 237–238
  print advertisements for chemists, 236–238
Finding job openings, 234–243
Fine-chemicals suppliers, 157
Fish and Neave, 94
Fish, R. A., 247
Flexibility, 108
Food and Drug Administration, 96, 209, 285
Foreign languages, reading knowledge of selected, necessity, 158
Fourneau, 7, *see also Maurois, A.*, 7
Fraser, T. R., 2

Ganellin, R., 20, 68
Generic drug companies
  patents, 48
Gilson, 95
Glaxo, 6, 52, 195, 202, 203, 206
Goals
  features, 107
  success, 108
God, talking to, 170
Good laboratory practices, 209

Good manufacturing practices, 284
Graduate school, part-time
　considerations for choosing, 20, 21
　time to degree, 18
Green cards, 223
Greenfield Laboratories, 94
Groban, R. S., 222
Group insurance, 263
Grubb, P. W., 133

H1B visas
　delays, 225
　prerequisites to employment, 223
　renewals, 223
　sponsorship, 223
　terms, 223
　time and employment contingencies, 222
Halberstam, M. J., 8
Hammond, G. S., 139
Headhunters, 241
Hoechst Marion Roussell, 203
Hoechst, 45, 202
Hoekstra, M. S., 205
Hoffman-La Roche, 248, *see also* Roche
Holmes, S., 163–164
Horton, B. J., 106

Immigration and Naturalization Act of 1990, 224
Immigration and Naturalization Service, 222
Impressions, interviewers', 103
Impurities
　professionalism in reporting, 177
　profile, change of, in an approved drug, 188
　profile, in an approved drug, 188, *see also* Patents, 189
Indices of the North American pharmaceutical industry
　"...a tide...leads on to fortune", 287
　typical entry in the name index, 286
　contents of each entry, 285
　excluded information, 286
　kinds of indexed organizations
　　selected government and private institutes, 285
　kinds of indexed organizations, 284
　sources, 287–288
Indinavir, ritonavir and saquyinavir, human immunodeficiency virus, 9
Inflation, effect on salaries, 33–34
Insulin, diabetes, 50
Integrilin, angina, 250

Integro Scientific Staffing, 90
Interferons, hepatitis C and cancers, 47
Internships, summer, *see* Employment
Interviews
　mock, 230
　team, 233
Inventorship, test of
　attribution of drug action to structure, 57
　attribution of therapy to mechanism, 131
Investigational New Drug Applications, 96
Investigational New Drug, 209
IRORI, 95

J-Star Research, 235
Jacobs, M., 213
Job banks and Internet addresses, 232
Job exchanges, 231
Job offers
　analysis, method of, 247
　contingent, 214
　full-time employment, of inexperienced chemists, 246
　large or small companies, 248
　more than one, 245
Job satisfactions
　determinants, 246
Jobs
　changes, arduousness, 14
　offers, comparing and contrasting, 29
　stability and small-molecule drugs, 10
Jobs in the drug industry
　abundance, for chemists, 10, *see also* Chemists' jobs
　development research
　　bioorganic catalysis, 89–90
　　natural products, 89
　　safety, 90
　　temporary jobs, 90–91
　development research, 88–91
　discovery research
　　drug metabolism, 80–81
　　medicinal chemistry, 78
　　natural products, 78–79
　　radiochemistry, 81
　long-term, 10
　mergers and outsourcing, 70
　over- and underqualifications, 77
　prerequisites and experience, 76–77
　research classifications, 77–78
　satellite companies, government agencies and nonprofit institutes
　　automated synthesis, 95

chemical development and
manufacturing, 94
fine chemicals, 95
Food and Drug Administration, 96
miscellaneous posts, 98
National Institutes of Health, 97–98
satellite companies, functions, 93
scientific advising, 94
service firms, 92–93
software development and molecular
modeling, 95
United States Patent and Trade Office,
96
Walter Reed Army Institute of
Research, 97
satellite companies, government agencies,
and nonprofit institutes, 91–98
service groups
analytical chemistry, 81–82
chemical information, 85–86
cheminformatics, 83–84
chromatography and separation
science, 87
compound registration, 82–83
patent agents and attorneys, 85
patent coordinators, 84–85
profiling and identification, 86
regulatory affairs, 87–88
synthetic services, 86
suitability, for chemists, 12
Jobs, summer, *see* Employment
Journal of Heterocyclic Chemistry, 184
Journal of Medicinal Chemistry, 154, 288
*Journal of Pharmacology and
Experimental Therapeutics*, 154
Justus Liebigs Annalen der Chemie, 217

Kehrer, D., 266
Kelly Scientific Resources, 90
Ketting Medical Center, 94
Kilo laboratory, work, 63
Kimmel, B. B., 222
Kinds of drugs, companies, and therapeutic
areas
biotechnology and biopharmaceutical
companies, 48
clinical research organizations, 48
diagnostic companies, 47
drug categories, 46–47
drug discovery firms and start-up
pharmaceutical companies, 49
fully integrated companies, 47

generic drug companies, 48
nucleus of the pharmaceutical industry,
47–49
therapeutic areas, 50
Kinds of drugs, companies, and therapeutic
areas, 46–50
Kirk, K., 97
Kirkup, M. P., 235
Klaus, S., 17
Knight, H. J., 118
Koster, W. H., xvii–xviii
Kubie, L. S., 108

Lab Support, 90
Labor Condition Applications, 223
Laird, T., 182, 184, 201, 204
Layoffs, 10, 246, 250, 251
Lead compounds
definition, 53, 152
drug discovery companies, 49
faults, 152
finding, 2, *see also* Discovery research
from old samples, 175
indispensability, 9, 154
molecular size, 152
Lead compounds, 152–155
Leaves of absence, arranging, 18
Lee, S., 184, 200
Lester, T., 201, 202
Libraries, chemical, 23, 145, 158
Likelihood of success, 235–236, *see also*
Profile of the pharmaceutical business,
43–44, 246
Liu, C., 194
Lubiner, A. M., 222
Lybrook Associates, 241

Machin, P. J., 69
Manpower Technical, 90
Markush structures, 125–126
Markush structure and protected
compound, 125
Marshall, D. R., 195
Marvell, A., 181
Mass spectrometry, invalid conclusions, 34
Mass Trace, 93
Maurois, A., 7
Maynard, J. T., 123
MDL Information Systems, 95
MediChem Research, 92
Medicinal Chemistry Research, 288
Medicinal Chemistry Reviews, 154

Meetings, internal
  consultants, 109
  duration, 109
  interdisciplinary, 109
  obligatory, 109
  participants, 109
  reporting research results, 109
  research plans, 109
Merck Index, 98
Merck, 44, 45
Metamorphosis, 178–179
Microcide Pharmaceuticals, 49
Mitscher, L. A., 20
Molecular modeling, *see* Jobs in the drug industry and Chemists' jobs
Molecular modifications
  biological activity, 147
  clinical improvement, 8
  effect, on physical properties, 79
  lead compounds, 153
  toxic metabolites, 61
Molecular Simulations, 95
Monsanto Life Sciences, 44, 194
Montelukast, asthma, 44
Morphine, pain, 10
Multitasking, 106, 149, 164, 168
Myco Pharmaceuticals, 49

National Institutes of Health, 6, 285
Nature, 154, 288
Needle and Rosenberg, 94
NéoKimia, 93
Networking
  American Chemical Society members, 228
  patent inventors' names and addresses, 125
  salary information, 37
  scientific conventions, 24
  workshop, 230
Networking, 251, 258, 259
New chemical entities, 5
New Drug Approvals
  Chemistry, manufacturing and control in, 96
  reviewers of, 96
New Drug Approvals, 88, 209
Newman, B., 225
North American Free Trade Agreement, 222
Novartis, 283
NSC Technologies, 94
Nuclear magnetic spectra
  recording written notes, 172
  versatility, 170
Nugent, T. C., 192

Occam's razor, shaving with, 171
Olah, G., 20
One day in the work of a medicinal chemist
  getting started, 149–151
  organization, or who chooses what, 149
  preparations for a discovery research project, 150
One day in the work of a medicinal chemist, 139–149
Operators, chemical, 286
Organic Process Research and Development, 184, 288
Organic stereochemistry, 218
Organization, of chemical development
  assignments, 189–191
  developmental project teams, 191–192
  discipline, 189
  matrix structure in composition and staffing of developmental project teams, 191
Orphan drugs, 51
Orphan Medical, 8
Outsourcing
  defined, 70
  drug development, 74
  drug discovery, 74
  effects on employment, 74

Page, I. H., 7
Parallel synthesis, 49
Parke-Davis, 8, 205, 208
Pasteur, L., 143
Patent applications
  examiners, 96
  prosecutions, 96
Patent Gazette, 288
Patenting your work
  inventorship
    controversy, 133–134
    definition, 133
    glory, 134
    rewards, 134
    test of, 131, 57
  inventorship, 133–136
  laboratory notebooks within industry, 134–136
  organizational structure of patents
    background and summary, 126
    bibliography, 124–125
    claims, 130–133
    detailed description, 126–130
    Markush structures, 125–126

organizational structure of patents, 124–133
patentability
  novelty and nonobviousness, 120–121
  reduction to practice, 122–123
  remarkable claims and paper patents, 123–124
  unpatentable ideas, 120
  utility, 121–122
patentability, 119–124
patents defined
  costs, 118
  disclosure, 117
  term, 117
patents defined, 116–118
speed readers' guide to composition-of-matter patents, 131
types of patents
  fortress of claims, 118–119
  patentable utility inventions, 119
types of patents, 118–119
what chemists should understand about patents, 115
Patenting your work, 115–136
Patents
  abandonment, 135
  advantages, 117
  applications, prolonged prosecution, 103
  applications, responsibility for, 115
  assignments of rights in, 125, 134
  best modes
    disclosure, 119
    identification, 119
    superiority, 120
    types, 119
  best modes, 122, 127, 132
  business strategy, 122, 123
  claims
    compositions of matter, 117
    devices, 118
    fingerprint, 123
    formulations, 117–118
    generic, 132
    narrowing, 132
    processes
      chemical reactions, 118
      methods of treatment, 118
    processes, 117, 118
    product by process, 123
    repetitiousness, 132
    specific, 132

structure elucidation, 123
competitors' research interests, 125
compounds of the invention, 125
conjecture, 128
constructive reduction to practice, 123
controversies, 133
copies of, obtaining, 117
copyright analogy, 120
definitions, 127
design, 118
diligence, 135
disclosures, 126, 133
employment credentials, 116
evidentiary standards, 121
examiners, 119, 120, 135
examples
  paper, 123
  preparative, 129
  working, 129
extrinsic utility, 120
fees, 118
filing dates, 117
first to file, 135
first to invent, 135
foreign, 118
Gallery of Obscure, 117
generic drug houses, 115
generic structures, 125
homologs, 121
identifiers
  Arabic numerals, 129
  chemical names, systematic, 129
  Roman numerals, 129
  structural drawings, 129–130
infringements, determination from impurity profile, 189
inoperability, 126
interferences, 135
interferences, infrequency of, 123–124
isomers, 121, 126, 127
issue dates, 117
Markush structures, 124, 127
medical utility, 128
monopolies from, 115
myths
  inventorship, 115
  prestige, compensation and inventorship, 133
notebook keeping, 135
obviousness
  shoals of, 121
  paper, 123

Patents (*continued*)
  pharmaceutical
    types
      chemical processes, 130
      formulations, 130
      individual compounds, 130
      medical devices, 130
      methods of use, 130
    physical forms, 121
    plants, 118
    prior art, 120
    process
      breadth, 191
      marketed drugs, 116
    reading
      care in, 129
      last items first, 130
      one phrase at a time, 130
      standard wording, 128, 132
    reading competitors', 126
    reading, 124
    reproducibility, 122, 128
    right to exclude others, 116
    salts, 121, 127
    scientific publishing, effect on, 113
    selection inventions, 121
    solvates, 127
    specification, 127
    specifications versus claims, 127
    structure of, 124
    summary of the invention, 126
    teachings, 127
    terms, effective, 117
    terms, statutory, 117
    timing of patent grants and of chemical, pharmaceutical and clinical development, 188
    titles, 124
    unauthorized practice, 116
    utility, 118
    validity, 116, 124
    value to medicinal chemists, 116
PE Applied Biosystems, 95
Penicillins, bacterial infections, 12, 78
Pennie and Edmonds, 94
Pension Benefit Guarantee Corporation, 266, 273
Pensions
  401k plans
    online calculator, 275
  401k plans, 274–276
  benefits accrual, 268
  cash-balance, 270–273
  defined benefit, 266–273
  defined contribution, 273–277
  eligibility, 267
  entry date, 266
  inflation, ravages, 269–270
  Internal-Revenue-Service qualified, 265
  maximum benefits, calculation, 267
  minimum service requirement, 266
  money-purchase plans, 277
  principle, 264
  profit-sharing plans, 276
  service, length of, counted, 267
  sizes, 268
  vesting, 267
Pensions, 264–277
Permanent residency, *see* Why read about visas, 221
Perrault, W. R., 182
PerSeptive Biosystems, 95, 238
Persistence, 108
Personnel recruiters, 286, *see also* Personnel representatives
Personnel representatives, 263, 264
Peters, M., 123
Pfizer, 42, 45, 58, 92
Pharmaceutical companies
  businesses, 41
  chemists' indispensability, 7, 8
  complexity, 7
  effective drugs, 8
  fully integrated, 41
Pharmaceutical industry
  chemists' skills, 8
  component institutions, diversity, 7
  consolidation and fragmentation, 70–71
  elastic definition, 287
  mergers and outsourcing, 70
  microbiological fermentation, 90
  nucleus, North American, 48
  opportunities, chemists', 7
Pharmaceutical Manufacturers Association of Canada, 288
Pharmaceutical Research and Manufacturers Association, 9, 288
Pharmacia and Upjohn, 65, 182
Pharmacopeia, 93
Pharmaprojects, 154
Pharmascope, 154
Phase I-to-IV trials, *see* Clinical trials

Phenacetin, influenza, 2
Phytera, 49
Pilot plant, work, 63
Pittsburgh Conference, 241
PPD Pharmaco, 94
Problems
  circumventing, 108, 219
  worthy, 219
Process research and development
  areas of right circular cylinders and a
    sphere of equal volume, 197
  changed techniques and conditions,
    196–197
  distinctions between process research
    and process development, 193
  equipment, 195–196
  process research versus process
    development, 192–194
  reforms, 198–199
Process research and development, 192–199
Procter & Gamble, 45
Productivity
  services, 106
  target compounds, 104
Profile of the pharmaceutical businesses
  benefits of patents, 42–43
  competition
    by therapeutic area and
      pharmacological mechanism, 44
    from FIPCOS and start-ups, 44
  competition, 44–45
  complexity
    coordination, 45
    headcount, 45
    regulatory filings, 45
    variety of staff, 45
  complexity, 45–46
  likelihood of success
    chemist-years per marketed drug, 43
    fraction of new compounds becoming
      drugs, 43
    reasons to halt development, 44
  likelihood of success, 43–44
  time to market, 42
  what drug discovery and development
    cost, 42
Profile of the pharmaceutical businesses,
  42–46
Profiling services, 238
Proof of concept
  finding, 53

  high drug doses in establishing, 58
Protein Design, 238
Proton nuclear magnetic resonance spectra
  coalescence, 172
  nonequivalence, 172
  Overhauser effect, 171
Publications
  advantages, 112–113, *see also* Career
    development
  approvals, 113
  delays, 113, 221
  subjects, 114
  time course, 113
Purity
  determining, 162
  enantiomeric excess, 163
  intermediate compounds, 161
  measurement techniques
    microcombustion analyses,
      applicability, 165
  measurement techniques, 164–165
  measurements, qualitative or quantitative,
    163
  multiple tests of, 163
  target compounds, 161, 172
  wasted efforts, 162
Purposes of chemical development
  bulk supplies
    clinical study, 186
    toxicity studies, 186
  bulk supplies, 186–187
  manufacturing synthesis, 187–189
Purposes of chemical development, 186–189

Ranitidine, stomach ulcers, 51
REACCS, 159
Reaction mechanisms, 194, 218, 220
Reactions, chemical
  basics, of running, 156–157
  no general, 160
  runaway, 196, 198
Receptor agonism, therapy, 51
Receptor antagonism, therapy, 51
Record keeping, 166
Recruiting firms, 241
Repic, O., 184
Reports
  benefits, 112
  good oral, 112
Research management, effectiveness,
  249–250

Research, cost
  drug discovery and development, 6, 9
  semiconductor and computer industries, 9
Résumés, reviews of, 230
Retirement ages, 247
Ricerca, Inc., 94
Ringrose, P. S., 73
Robinson, G., 184, 200
Robinson, R., 148, 160
RoboSynthon, 95
Roche, 6, *see also* Hoffman-La Roche, 248
Rohm and Haas, 259
Role of the American Chemical Society in job searches
  Career Resource Center, 230–231
  Career Services Department, 229–234
  Chemical and Engineering News, 228–229
  National and Regional Employment Clearing Houses, 231–234
  selected publications of the ACS Department of Career Services, 229
Role of the American Chemical Society in job searches, 228–234
Roosevelt, F. D., 247
Roseman, E., 259
Roth, G. P., 204

Salaries
  bonuses or raises, 259
  postings
    corporate Web sites, 238
  starting
    compensation specialists, 13
    underlying calculations, 13
Salaries, *see also* Benefits
  Ph. D. versus B. S. chemists', 17
Salicylic acid, 78
Salvemini, D., 12
Sample characterization
  infrared spectra, fingerprint region, 167
  mass spectrometry, 168
Sample identification
  danger of indirect, 167
  mass spectrometry, 168
  typical spectrometric peaks, 170
  wishful thinking, 167
Santayana, G., xix
SARCO, 93
Satisfactions and successes, 177–178, 210
Schering-Plough, 45
Science, 154, 221, 288

Scientific ladder, staying on, consequences, 15
Scientific Staffing, 90
Screening
  high throughput, 155
  millions of compounds, 155
  source of drugs, 79
SCRIP Magazine, 249, 288
SCRIP, 154, 288
Searle, G. D., 194
Securities and Exchange Commission, 251
Serendipity
  clinical development or pharmacological discovery research, 185
Serendipity, 104
Sheldon, J. A., 201
Sheldon, R. A., 182, 202
Siegel, J. S., 257, 274
Silva, A., 222
Simulations Plus, 95
SmithKline Beecham, 270
SmithKline, 52, 68
Social Security Act, 247
Society for Chemical Industry, 239
Spilker, B., 8, 41
Standard and Poor's, 273
Start-up firms
  California, 75
  Massachusetts, 75
Steroids, asthma, 78
Stinson, S. C., 94–95
Stock options
  Bristol-Myers Squibb, 257
  strike price, 257
Stock options, *see also* Appendix B
  Wyeth-Ayerst, 257
Structures
  drawing, 111
  molecular activity map, 147–148
Sulfonamides, bacterial infections, 2
Summer internships
  essential to résumés, 217
  interview determinants, 217
Sutton, W., 4
Syntheses, chemical
  automated, *see* Jobs in the drug industry
  changes to manufacturing, 188
  changing criteria for suitable, 199
  conciseness, 104
  cost per kilo, 105
  ease, 55, 104, 172
  efficiency, in discovery research, 172

effluents, control of, 210
emissions, control of, 210
environmental friendliness, 105
expertise and employment, 20
integration of a manufacturing, 195
knowledge of, no help in choosing clinical candidates, 178
labeled drugs, 81
logistics, 105
manufacturing, scale, 63, 105
natural products department, 79
number of steps versus average yield per step, 206
protecting groups, the albatross of, 206
social importance, 9
steps defined, 205
sterilization as an operational step, 207
successes and failures, 161
superiority of manufacturing to research, 195
telescoping, 204
throughput, 105
tolerable inefficiency, 187
unoptimized, use, 64
Syntheses, chemical, *see also* Discovery research 3
Synthetic Organic Chemical Manufacturers Association, 288
Synthon, 94

Takeda, 134
Target compounds
  characteristics, 156
  defined, 156
  failures, 152
Temporary jobs
  compensation for uneven workload, 45
Testa, B., 5
Tetracyclines, infections, 78
Thayer, A. M., 258
The Scientists Registry, 90
The Scientist, 288
The Wall Street Journal, 250
Todd, A. R., 160
Toll synthesis, 94
Topliss, J. G., 20
Trade secrets
  obstacles to publishing, 221
Tripos, 95
Trovafloxacin
  bacterial infections, 45
  clinical trials, 42

U. S. Patent and Trade Office, 262, 285
Unemployment rate, chemists', 10

Vaccines, smallpox, polio and measles, 47
ValueLine, 252
Vanguard Medical Group, 9
Victories, Pyrrhic, 107, 220
Visa forms
  form 129, 224
  form 539, 223
  form ETA-750A, 225
  form ETA-750B, 225
  form ETA-9035, 223
  form I-20, 227
  form I-140, 226
Vitamins B_{12}, C and D, folate anemia, scurvy and rickets, 50

Wages, nonimmigrant
  ascertaining
    ACS salary survey, 35
    corporate bulletin boards, 36
    corporate Web sites, 238
    networking, 37
    U. S. Department of Labor, 37
Waiver of labor certification, 227
Want ads
  abbreviations, 76
  Chemical and Engineering News, 228, 236
  filled positions, 225, 237–238
  incomplete information, 237
  instructions for responding, 237
  job banks, 232, 297–348
  job codes, 228, 237
  kinds, 228
  link to pharmaceutical-industry feature articles, 229
  scarcity, for older experienced chemists, 15
  Science, 236
  sources of wonderment, 221
Warner-Lambert, 251
Weinstock, L. M., 195
Wernick, A., 222
Werth, B., 23
What you need to know and show
  common prerequisites
    complete educations, 214
    English commmunication, 214–215
    finished theses, 214–215
  common prerequisites, 214–216

What you need to know and show
    (*continued*)
  personal qualities
    crucial traits, 215
    nonintellectual abilities, 216
  personal qualities, 215–216
  qualifications
    B. S. chemists, 216–218
    M. S. chemists, 218–219
    Ph. D. chemists, 219–220
    pros and cons of postdoctoral research, 220–221
  qualifications, 216–220
What you need to know and show, 214–221
What you should know about the job you seek
  duties of discovery and development chemists, 102–103
  the basics
    aids to productivity
      embracing services, 106–107
      handling paper, 107
      seeking tempi, 106
    aids to productivity, 106–107
    budgeting, 108–109
    goal setting, 107–108
    internal meetings, 109–111
    performance and productivity
      development research, 104–105
      discovery research, 104
    performance and productivity, 104–105
    who participates in crucial meetings?, 110
  the basics, 103–111
What you should know about the job you seek, 101–103

Why read about visas
  Work visas in the United States
    do you need a lawyer?, 227–228
    employment-based visas in the United States
      selected divisions and subdivisions, 223
    H1B Visas, 224
    how to obtain INS forms, 227
    immigrant visas, 225–226
    information sources, 222
    Labor Condition Application, 223–224
    national interest waivers, 226–227
    nonimmigrant visas, 222–223
    selected routes to permanent residency with preference categories, 226
  Work visas in the United States, 222–228
Why read about visas?, 221–228
Wilde, O. F. O'F. W., 88
Williams, T. I., 160
Williams, T., 245
Wilson, E. M., 188, 203, 206
Wishful thinking, 167
Woods, S., 91
Woodward, R. B., 101, 153, 160, 181, 205–206, 220, 283
Working visas, *see* Why read about visas, 221
Workshops, job hunters', 230
World Patent Alert, 154
Wyeth-Ayerst, 58, 257, *see also* Stock options, 256–257

Yahoo!, 288
Yahoo!Finance, 251

Zeneca, 44
Zufirlukast, asthma, 44